VOLUME EIGHTY

ADVANCES IN BOTANICAL RESEARCH

Glucosinolates

ADVANCES IN BOTANICAL RESEARCH

Series Editors

Jean-Pierre Jacquot
Professor, Membre de L'Institut Universitaire de France, Unité Mixte de Recherche INRA, UHP 1136 "Interaction Arbres Microorganismes", Université de Lorraine, Faculté des Sciences, Vandoeuvre, France

Pierre Gadal
Honorary Professor, Université Paris-Sud XI, Institut Biologie des Plantes, Orsay, France

VOLUME EIGHTY

ADVANCES IN BOTANICAL RESEARCH

Glucosinolates

Volume Editor

STANISLAV KOPRIVA

Botanical Institute,
Cluster of Excellence on Plant Sciences (CEPLAS),
University of Cologne, Germany

ELSEVIER

AMSTERDAM • BOSTON • HEIDELBERG • LONDON
NEW YORK • OXFORD • PARIS • SAN DIEGO
SAN FRANCISCO • SINGAPORE • SYDNEY • TOKYO

Academic Press is an imprint of Elsevier

Academic Press is an imprint of Elsevier
125 London Wall, London EC2Y 5AS, United Kingdom
The Boulevard, Langford Lane, Kidlington, Oxford OX5 1GB, United Kingdom
50 Hampshire Street, 5th Floor, Cambridge, MA 02139, United States
525 B Street, Suite 1800, San Diego, CA 92101-4495, United States

First edition 2016

Notices
Knowledge and best practice in this field are constantly changing. As new research and experience broaden our understanding, changes in research methods, professional practices, or medical treatment may become necessary.

Practitioners and researchers must always rely on their own experience and knowledge in evaluating and using any information, methods, compounds, or experiments described herein. In using such information or methods they should be mindful of their own safety and the safety of others, including parties for whom they have a professional responsibility.

To the fullest extent of the law, neither the Publisher nor the authors, contributors, or editors, assume any liability for any injury and/or damage to persons or property as a matter of products liability, negligence or otherwise, or from any use or operation of any methods, products, instructions, or ideas contained in the material herein.

ISBN: 978-0-08-100327-5
ISSN: 0065-2296

Publisher: Zoe Kruze
Acquisition Editor: Kirsten Shankland
Editorial Project Manager: Hannah Colford
Production Project Manager: Magesh Kumar Mahalingam
Designer: Greg Harris

Typeset by TNQ Books and Journals

CONTENTS

CONTRIBUTORS

P. Bednarek
Institute of Bioorganic Chemistry, Polish Academy of Sciences, Poznań, Poland

M. Burow
University of Copenhagen, Frederiksberg, Denmark

N.I. Cacho
Universidad Nacional Autónoma de México, Mexico City, Mexico

H. Frerigmann
University of Cologne, Cologne, Germany

J. Gershenzon
Max Planck Institute for Chemical Ecology, Jena, Germany

T. Gigolashvili
University of Cologne, Cologne, Germany

B.A. Halkier
University of Copenhagen, Frederiksberg, Denmark

S. Haneklaus
Julius Kuehn-Institut (JKI), Braunschweig, Germany

A.-M. Herfurth
Technische Universität Braunschweig, Braunschweig, Germany

V. Jeschke
Max Planck Institute for Chemical Ecology, Jena, Germany

D.J. Kliebenstein
University of California, Davis, Davis, CA, United States; University of Copenhagen, Frederiksberg, Denmark

S. Kopriva
University of Cologne, Cologne, Germany

E. Kurzbach
Technische Universität Braunschweig, Braunschweig, Germany

M. Pastorczyk
Institute of Bioorganic Chemistry, Polish Academy of Sciences, Poznań, Poland

E. Schnug
Julius Kuehn-Institut (JKI), Braunschweig, Germany

E.J. Stauber
Technische Universität Braunschweig, Braunschweig, Germany

M.H. Traka
Institute of Food Research, Norwich, United Kingdom

D.G. Vassão
Max Planck Institute for Chemical Ecology, Jena, Germany

U. Wittstock
Technische Universität Braunschweig, Braunschweig, Germany

PREFACE

The present volume (#80) of the *Advances in Botanical Research*, '*Glucosinolates*', is dedicated to a fascinating group of natural products. Glucosinolates have a large range of functions in plants and beyond. Apart from being important components of plant defence systems against various pathogens and pests, they are responsible for the flavour of cruciferous vegetables and one of the mechanisms underlying the 5-a-day importance of vegetables in our diet. It is therefore surprising that unlike other groups of secondary metabolites, terpenes, flavonoids, etc., glucosinolates are restricted to a single taxonomic order, the Capparales. However, as the model plant *Arabidopsis thaliana* is also classified in this order and forms glucosinolates, and presumably because the number of glucosinolate structures is far lower than the variety of terpenes or flavonoids, glucosinolates have become probably the most studied class of secondary metabolites. This book brings together world's leading experts in different aspects of glucosinolate biology to provide the readers with the current state of the art in glucosinolate research.

The book is divided into 10 chapters. In the first, introductory chapter Barbara Halkier gives a short overview of the newest developments in glucosinolate research, the progress in understanding the synthesis pathway, recently discovered glucosinolate transporters and attempts of glucosinolate synthetic biology. Second chapter by Meike Burow describes the environmental factors that are responsible for the variation in glucosinolate accumulation in plants. She focuses on the mechanisms that control the dynamic responses of glucosinolate patterns to complex environmental signals. The next chapter from Daniel Kliebenstein and Ivalú Cacho explains how the large variation in glucosinolate composition in *Arabidopsis* can be caused by a small number of genetic loci. The genetic and molecular information is compared and discussed in the frame of ecological and evolutionary observations. Henning Frerigmann in the fourth chapter then focuses on the transcriptional regulation of the pathway. He describes the transcription factors involved in this control and their interplay and interdependence in regulation of the complex pathway. The fifth chapter by Stanislav Kopriva and Tamara Gigolashvili shows how glucosinolate synthesis is embedded in the general plant metabolism, particularly the connections to sulphate assimilation. This chapter also highlights the function of various transporters involved in transport of glucosinolates and intermediates in their synthesis.

The sixth chapter is the first one focussing more on glucosinolate function. Ute Wittstock et al. address the breakdown of glucosinolates into the active volatile isothiocyanates and other compounds that form the molecular basis of the 'mustard oil bomb'. The seventh chapter by Marta Pastorczyk and Paweł Bednarek describes the recent progress in deciphering the role of glucosinolates in plant immunity and defence against various pathogens. The eighth chapter from Verena Jeschke et al. takes the reader beyond the plant kingdom. It shows different strategies among insects to metabolize and detoxify glucosinolates and fascinating examples of evolution and adaptation to plant defence compounds. The ninth chapter by Maria Traka is also concerned with nonplant processes. It describes the health benefits of glucosinolates and their degradation products, including the molecular mechanisms. In the 10th and last chapter Ewald Schnug and Silvia Haneklaus take the readers on a tour of a different, more applied aspect of glucosinolate metabolism. Their overview explains, for example, the role glucosinolates played in development of modern oilseed rape varieties.

We hope that this book will find broad audience as an overview of an interesting and important research area that touches our everyday life. We believe that it will be of interest to graduate students and researchers in a wide range of disciplines including plant nutrition, plant physiology, plant biochemistry and agriculture and that the readers will find the book to be an informative, interesting and enjoyable reading.

CHAPTER ONE

General Introduction to Glucosinolates

B.A. Halkier
University of Copenhagen, Frederiksberg, Denmark
E-mail: bah@plen.ku.dk

Contents

Abstract

As the major specialized metabolites of the model plant *Arabidopsis thaliana*, glucosinolates have become model specialized metabolites with all the advantages this brings to glucosinolate research, e.g. in the form of extensive 'omics tools, natural variation and mutant collections. Here will be presented a general introduction to glucosinolates ranging from the evolution of glucosinolates to the many roles glucosinolates have for humans as well as an overview of the current knowledge on the orchestration of the glucosinolate biosynthetic pathway. The latter includes an introduction to the genes in the biosynthetic pathway, localization of the enzymes at the cellular level and the storage of the glucosinolates in the S-cells. In addition, the status on transport of glucosinolates is presented with focus on the lessons learnt from the first identified glucosinolate transporters, namely the importers localized to the plasma membrane. Examples of how the knowledge gained from basic research has been translated into applied glucosinolate research through pathway and transport engineering will be presented.

Advances in Botanical Research, Volume 80
ISSN 0065-2296
http://dx.doi.org/10.1016/bs.abr.2016.07.001

1. INTRODUCTION

Glucosinolates are a fascinating class of specialized metabolites characteristic of the Brassicales order (Agerbirk & Olsen, 2012). They are part of a two-component, bioactivatable defence system, the mustard bomb, that comes into play when thioglucosidases hydrolyse the glucosidic linkage and bioactive hydrolysis products, typically isothiocyanates and nitriles, are formed (Halkier & Gershenzon, 2006). Many of us are familiar with the characteristic pungent flavour of wasabi, rich in allyl isothiocyanate, and mustard, rich in benzyl isothiocyanate. The presence of glucosinolates in cabbage vegetables has received a lot of attention that is associated with health-promoting properties related with intake of cruciferous vegetables, particularly broccoli (Cartea & Velasco, 2008; Traka & Mithen, 2009). Inversely, glucosinolates in *Brassica* oilseeds are antinutritional factors with growth inhibitory effects (Mailer, McFadden, Ayton, & Redden, 2008). Additionally, the hydrolysis products of glucosinolates, especially the isothiocyanates, have status as natural biopesticides when leached into the soil in rotational agriculture or when glucosinolate-producing plants are mulged into the soil in a process called biofumigation (Gimsing & Kirkegaard, 2009).

Glucosinolates are organic anions with an amino acid-derived side chain and oxime moiety with an *S*-linked thioglucose and an *O*-linked sulphate group. Glucosinolates have evolved from the cyanogenic glucosides that are present all over the plant kingdom from ferns to gymnosperms and angiosperms, whereas the more recent glucosinolates are limited to the Brassicales order and the *Drypetes* genus (Johnson, Griffiths, Peter, & Lawes, 2009). Whereas only Val, Ile, Phe and Tyr are precursor amino acids for cyanogenic glucosides (Møller, 2010), glucosinolates have evolved to be derived from many more protein amino acids Ala, Val, Leu, Ile, Met, Tyr, Phe and Trp, as well as chain-elongated derivatives of methionine (up to nine additional methylene groups) and homophenylalanine (Agerbirk & Olsen, 2012; Fahey, Zalcmann, & Talalay, 2001). Curiously, cyanogenic glucosides and glucosinolates do not coexist in plants, except for one species, *Carica papaya*, that produces both a phenylalanine-derived cyanogenic glucosides and glucosinolates (Bennett, Kiddle, & Wallsgrove, 1997).

The first glucosinolate structures to be elucidated were the structure of sinigrin and sinalbin in 1956 (Ettlinger & Lundeen, 1956), and the term glucosinolate appeared first time in 1961 (Ettlinger & Kjaer, 1968). By mid-2014, around 136 different glucosinolate structures were identified (Agerbirk & Olsen, 2015). The high number of glucosinolates is due to side chain modification, e.g., *S*-oxygenation, alkenylation, hydroxylation, as well as substitutions with acyl conjugation with, e.g., benzoic acid, in *Arabidopsis thaliana* on the sugar moieties. There are excellent reviews of all identified structures until 2001 (Fahey et al., 2001) and until 2011 (Agerbirk & Olsen, 2012).

It is fortunate for glucosinolate research that the glucosinolate-producing *A. thaliana* was selected as the first model plant to have its genome sequenced, a choice that was based on its small genome, short life cycle and genetics (Meyerowitz, 2001). In 2000, when the genome sequence was published (Kaul et al., 2000), the molecular glucosinolate research era started. This research has benefitted tremendously from the extensive 'omics databases, bioinformatics tools, natural variation and mutant collections that are now available for *A. thaliana*. In this introductory chapter, I will focus on *A. thaliana* and briefly describe the different developments in glucosinolate research that lead to the elucidation of the biosynthetic pathway, which naturally leads to the questions about where these compounds are synthesized and stored in the plant, and the transport processes that move them from site of synthesis to site of storage. In addition, examples of applied research linked to the many activities of glucosinolates (or rather their hydrolysis products) will be introduced.

2. BIOSYNTHESIS

Prior to the availability of genomic sequences, the knowledge on the biosynthesis of glucosinolates (Fig. 1) was limited (Halkier & Du, 1997). Historically, glucosinolate biosynthesis was elucidated by first identifying intermediates by in vivo feeding of candidate intermediates to plants as a means to investigate whether a given compound could serve as precursor for glucosinolate end products (Underhill, Wetter, & Chisholm, 1973). Biochemical approaches included characterization of the biosynthetic enzymes in plant extracts (Du, Lykkesfeldt, Olsen, & Halkier, 1995). Early genetic analyses were based on glucosinolate analysis of recombinant populations of *Brassica napus* and *A. thaliana* developed from parental lines which varied in their glucosinolate content. This identified *GSL-ELONG*, *GSL-OHP*, *GSL-ALK* and *GSL-OH* (Magrath et al., 1994). These genes

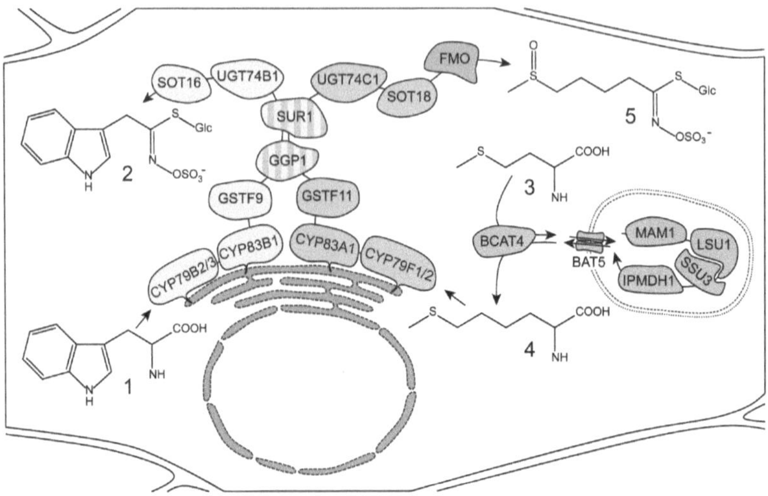

Figure 1 The biosynthetic pathway of the aliphatic glucoraphanin and indolyl-3-glucosinolate. The designated genes are from *Arabidopsis thaliana* Col-0. Homologous enzymes are catalysing the reactions in each pathway, except for two enzymes, GGP1 and SUR1, that are shared between the pathway. Structures: 1: tryptophan; 2: indolyl-3-glucosinolate; 3: methionine; 4: dihomomethionine; 5: glucoraphanin. *BAT5*, bile acid transporter 5; *BCAT4*, branched chain amino acid transferase 4; *FMO*, flavin monooxygenase; *GGP1*, gamma glutamyl peptidase; *GST*, glutathione-*S*-transferase; *IPMDH*, isopropylmalate dehydrogenase; *IPMI-LSU1*, isopropylmalate isomerase large subunit1; *IPMI-SSU*, isopropylmalate isomerase small subunit; *MAM1/3*, methylthioalkylmalate synthase1/3; *SOT*, sulphotransferase; *SUR1*, *C-S* lyase; *UGT74*, UDP-glucosyl transferase. The *blue/green pathway* is for the aliphatic and the *yellow* is for indolyl glucosinolate.

were later mapped by exploiting the natural variation in *A. thaliana*, and *GSL-ELONG* was identified as one of the methylthioalkylmalate synthase (MAM) enzymes (Kroymann et al., 2001), *GSL-ALK* was identified as *AOP2* and *GSL-OH* as *AOP3* (Kliebenstein, Lambrix, Reichelt, Gershenzon, & Mitchell-Olds, 2001). Several glucosinolate biosynthetic genes came out of screens aimed at identifying genes in the biosynthetic pathway of the plant hormone auxin. These include the oxime-metabolizing CYP83B1 (SUPERROOT2, allelic to RUNT1) and the *C-S* lyase (SUPERROOT1 allelic to ROOTY/ABERRANT LATERAL ROOT FORMATION/HOOKLESS3) which both gave the characteristic high-auxin phenotype with multiple roots (for review Nafisi et al., 2006). The phenotype was an effect of blocking the indole glucosinolate pathway downstream of indole-acetaldoxime, which resulted in overflow of indole-acetaldoxime to IAA. Similarly, the mutation in the *CYP79F1* gene (allelic to BUSHY/SUPERSHOOT gene) gave a strong bushy phenotype due to increased levels of cytokinin (Tantikanjana, Mikkelsen, Hussain, Halkier, & Sundaresan, 2004).

The sequencing of *A. thaliana* genome (Kaul et al., 2000) opened up for reverse genetic approaches to identify genes in the biosynthetic pathway. This was particularly true for the CYP79 enzymes that catalyse the conversion of amino acids to oximes, the first committed step in the synthesis of both cyanogenic glucosides and glucosinolates (Wittstock & Halkier, 2002). Here, it became a question of which amino acid a given CYP79 has as substrate. Today, only the CYP79Cs have not yet been assigned a function, and at the same time we have the homophenylalanine precursor for phenylethyl glucosinolate that has not yet been assigned a CYP79. Following the CYP79-catalysed oxime formation, the cyanogenic and glucosinolate pathways diverge to produce distinct glucosidic compounds (Rodman, Soltis, Soltis, Sytsma, & Karol, 1998). The oxime-metabolizing enzymes in the glucosinolate pathway – the CYP83s that are related to the CYP71s in the cyanogenic pathway – produce the reactive nitrile oxide, rather than the hydroxyl nitrile in the cyanogenic pathway (Hansen et al., 2001). Accordingly, the enzymes in the remaining part of the glucosinolate pathway leading to the glucosinolate structure are hypothesized to be recruited from the general detoxification pathway, starting with the glutathione-S-transferase conjugation of glutathione, followed by cleavage with gamma-glutamate peptidase, cleavage of the cysteine side chain with *C-S* lyase and then glucosylation with a UDP-glucosyl transferase and sulphation by sulphotransferases (Hansen et al., 2001). With an increasing number of genes identified and the development of coexpression databases, it became increasingly easy to identify candidate genes of biosynthetic pathway based on the 'guilt-by-association' principle from coexpression databases to be validated biochemically and *in planta*. This has resulted in an inventory of biosynthetic genes in *A. thaliana* (Sønderby, Geu-Flores, & Halkier, 2010). Identification of the genes in the pathway provided the molecular tools necessary for being able to engineer the pathway into a nonnative host organism (see later discussion).

3. LOCALIZATION OF THE BIOSYNTHESIS

Besides knowing which genes are involved in the glucosinolate pathway, it is essential from a biological perspective to know where the glucosinolates are being synthesized to understand how spatial glucosinolate distribution in the plant is obtained. Promoter-GUS data with many of the biosynthetic genes in the core structure pathway indicate an association of glucosinolate biosynthesis with the vasculature (Chen et al., 2003; Douglas Grubb et al., 2004; Mikkelsen, Hansen, Wittstock, & Halkier, 2000; Reintanz et al., 2001; Tantikanjana et al., 2004). Recently, the localization of the biosynthetic proteins at the cellular and whole plant level was investigated using several fluorophore-tagged enzymes (Nintemann and Halkier, unpubl. res.). Using the CYP83A1 and CYP83B1 enzymes in the biosynthesis of the aliphatic and indole glucosinolates as markers, these pathways were localized to distinct cells in separate as well as overlapping vascular tissues, with the indole pathway being only in the phloem whereas the aliphatic pathway is both in the phloem and the xylem. Interestingly, localization of glucosinolate biosynthetic enzymes in parenchymatic cells of the vasculature may assign new functions in defence chemistry to these cell types in what was hitherto thought of as undifferentiated ground tissue.

4. STORAGE

In *A. thaliana*, glucosinolates have been reported to be stored to high levels (>130 mM) in specific laticifer-like sulphur-rich cells called S-cells (Koroleva, Gibson, Cramer, & Stain, 2010). These cells are localized in the phloem cap along the vasculature and along the leaf margins (Koroleva et al., 2010). Measurement of glucosinolate content in the leaf apoplast revealed that less than 0.12% of the total glucosinolate content was in the apoplast, suggesting that the majority is in the symplasm (Madsen, Olsen, Nour-Eldin, & Halkier, 2014). It is currently not known whether glucosinolates primarily are stored in the vacuoles of the S-cells or also in other cells. To fully understand how the plant orchestrates its chemical defences, we need to understand what happens to the glucosinolate once it is being synthesized, i.e., understand its route from site of synthesis to site of storage and potential (re)mobilization upon attack.

5. PATHWAY ENGINEERING

The health-promoting properties of glucosinolates have spurred an interest in increasing the intake of these compounds, e.g. by increasing the levels of glucosinolates in the vegetables. Historically, broccoli that is rich in glucoraphanin has received a lot of attention, which is at least partially due to one particular rat study (Fahey, Zhang, & Talalay, 1997). In this study, rats were chemically induced to get breast cancer and also given different levels of extract from broccoli seedlings. The number of tumours per rat was reduced upon intake of broccoli extract, and even more noticeable was that the reduction was proportional to the intake of the extract.

In the superbroccoli Beneforté, high levels of glucoraphanin was obtained through classical breeding by crossing to wild relatives (Mithen, 2012; Sivapalan, Melchini, Traka, Saha, & Mithen, 2015). Subsequent transcriptomic analysis showed that superbroccoli was upregulated in the transcription factor MYB28 that is a master regulator of aliphatic glucosinolates (Traka et al., 2013). This suggests that GM technology overexpressing MYB28 as a transgene could be a way to enrich for aliphatic glucosinolates not only in superbroccoli, but possibly also in other crucifer vegetables.

Alternative means to obtain high amounts of health-promoting glucosinolates is through engineering of the biosynthetic pathway into a fermentable microbial host organism suitable for large-scale production. This will provide a cost-efficient, stable source and enable intake of well-defined dose of these compounds. The feasibility of engineering glucosinolate production in a non-cruciferous host was demonstrated by *Agrobacterium*-mediated transient expression of both glucoraphanin, benzyl and indole glucosinolate pathway in *Nicotiana benthamiana* (Geu-Flores et al., 2009; Mikkelsen, Olsen, & Halkier, 2010; Pfalz et al., 2011). In *Saccharomyces cerevisiae*, stable integration of genes in the indole glucosinolate pathway into the yeast genome provided proof of concept for indole glucosinolate production in yeast (Mikkelsen et al., 2012). The glucoraphanin biosynthetic pathway poses additional challenges as it consists of 13 biosynthetic enzymes, with the chloroplast-localized chain elongation machinery providing the chain-elongated methionine-derivative dihomomethionine that subsequently is converted to the glucoraphanin by the core structure pathway and the side chain modifying flavin monooxygenase (Fig. 1) (Mikkelsen et al., 2010). Using *Escherichia coli* as host organism, five genes (*BCAT4*, *MAM1*, *IPMI-LSU1*, *IPMI-SSU3* and *IPMDH*) catalysing the chain elongation of methionine were successfully engineered to produce dihomomethionine (Mirza, Crocoll, Olsen, & Halkier, 2015). It remains to be shown if the compartmentalized glucoraphanin pathway can uncompartmentalize in the cytosol of a microbial organism.

6. TRANSPORT PROCESSES

With the identification of the biosynthetic genes and their localization, it became increasingly important to understand the transport processes responsible for translocating the glucosinolates within the plants. Early studies showed that there is no de novo biosynthesis in the seeds and that glucosinolates in seeds were maternally derived (Magrath & Mithen, 1993). Evidence for long-distance transport was provided in a study showing that radio-labelled precursor amino acid tyrosine in vivo-fed to *A. thaliana* was converted into *p*-hydroxybenzyl glucosinolate in young rosettes. Upon bolting, the radio-labelled glucosinolates were transported into the seeds (Chen et al., 2000). Biochemical characterization of glucosinolate transport was obtained from experiments in *B. napus*, where uptake of glucosinolates into embryos and protoplasts was shown to be dependent on a proton gradient, but not ATP hydrolysis, suggesting the involvement of H^+/glucosinolate symporters (Chen & Halkier, 2000; Gijzen, McGregor, & Séguin-Swartz, 1989).

6.1 Glucosinolate Transporter-Mediated Transport

To date, one set of *A. thaliana* **gl**ucosinolate **tr**ansporters (GTRs), GTR1 and GTR2, have been identified by employing a functional genomics approach (Nour-Eldin et al., 2012). The method is based on expression of an *A. thaliana* cDNA transporter library in *Xenopus laevis* oocytes followed by screening for uptake activity of a given compound, here glucosinolates. The GTRs belonging to the nitrate peptide family are importers localized at the plasma membrane. Identification of GTR1 and GTR2 provided an important first molecular tool to investigate the transport processes involved in shaping glucosinolate distribution patterns. Many lessons (and surprises) were learnt (Fig. 2) (Jørgensen, Nour-Eldin, & Halkier, 2015). The first discovery showed unequivocally that GTR1 and GTR2 were essential for transport of glucosinolates — both aliphatic and indole glucosinolates — to seeds (Nour-Eldin et al., 2012). Furthermore, the glucosinolates overaccumulating in rosette leaves of *gtr1gtr2* were first believed to accumulate because the phloem loading route was blocked in the absence of GTR1 and GTR2. However, the glucosinolates that accumulated were not the typical leaf glucosinolates, but the long-chained aliphatic glucosinolates that normally dominate in roots. This suggested that roots were sink tissue for long-chained, aliphatic glucosinolates produced in the rosette. However,

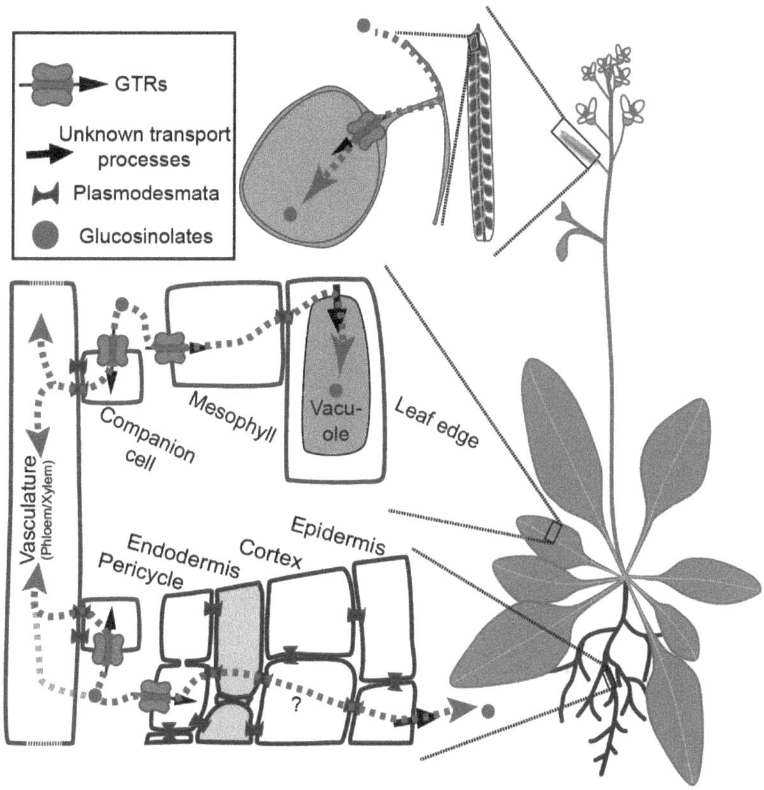

Figure 2 Learnt lessons in glucosinolate transport processes using the glucosinolate transporters (GTRs) as molecular tool. In seeds, import is mediated by GTRs. In leaves, glucosinolates are either phloem loaded by GTRs for long-distance transport to other organs or moved to the leaf edge via plasmodesmata along a concentration gradient imposed by vacuolar import processes. In roots, GTR-mediated import is essential for retention of glucosinolates in the roots. Additionally, GTRs may also play a role in import into the symplasmic domain for exudation to the rhizosphere (route with question mark). In the *gtr1gtr2* mutant the glucosinolates are moved from root to shoot via xylem (lighter stippled route). *Blue*: GTRs; *orange*: plasmodesmata; *red dots*: glucosinolates; *black arrows*: not yet identified transporters.

micrografting experiments between rosette and roots using all combinations of three genotypes (wild type, biosynthetic *null* mutant and *gtr1gtr2*) showed that roots were the source for long-chained, aliphatic glucosinolates overaccumulating in rosette of *gtr1gtr2* (Andersen et al., 2013). This revealed a critical role for GTR1 and GTR2 in import of glucosinolates into storage cells in roots by removing the compounds from the apoplast. Importantly, it also showed that long-chained, aliphatic glucosinolates are exported from the biosynthetic cells in the root per default and unless imported by GTRs into storage cells they are moved to aerial parts via the xylem. The latter is evidenced by higher glucosinolate level in root xylem of *gtr1gtr2* (Madsen et al., 2014). Alternatively, it cannot be excluded that the root-synthesized glucosinolates that overaccumulate in rosette *gtr1gtr2* were meant to be exudated to the rhizosphere (Fig. 2). In contrast, short-chained, aliphatic glucosinolates were shown to be mainly produced in the rosette and transported to the roots (Andersen et al., 2013). Thus, not only do indole glucosinolates behave differently from aliphatic glucosinolates, but also short-chained and long-chained aliphatic glucosinolates have different source—sink tissue relationship and are to be viewed as different classes.

In the arms race between plants and insects during coevolution, *A. thaliana* developed a strategy to defend its leaves against herbivores by accumulating glucosinolates to higher concentrations along the margin of the leaves. However, as there is no synthesis at the leaf margin, this raises the question how glucosinolates are translocated to the margin. When the distinct intra-leaf distribution of glucosinolates were investigated in the leaf of *gtr1gr2,* the level of glucosinolates at the leaf edge unexpectedly (at least intuitively) was increased in the *gtr1gtr2* mutant rather than decreased, suggesting that the GTRs were not directly involved in moving glucosinolates from site of synthesis at the vasculature to the leaf edge (Madsen et al., 2014). The increased levels of glucosinolates in *gtr1gtr2* leaf edges originated from the root, where the GTRs in the wild type function to import glucosinolates from apoplast into storage cells. Based on in vivo feeding studies of exogenous glucosinolates, a model was proposed for glucosinolate leaf distribution including transport through plasmodesmata.

Taken together, crosstalk between glucosinolate synthesis, storage and transport creates a dynamic source—sink relationship between organs. Plants monitor defence compound levels in storage sites (often tissues of high fitness value) and can activate local biosynthesis if supply from transport should become inhibited. We are only beginning to unravel the intricate dynamics of glucosinolate transport using *A. thaliana* as model plant. The GTRs are plasma-localized glucosinolate importers, and additional transport processes are responsible for export over the plasma membrane as well as in and out of the vacuolar storage compartment. The molecular tools, 'omics databases and grafting techniques available for *A. thaliana* creates a foundation for studying transport of specialized metabolites using glucosinolates as model compounds.

The protein-rich meal from *Brassica* oilseeds has currently reduced value due to the presence of antinutritional factors such as glucosinolates. Previous attempts to remove antinutritional glucosinolates in *Brassica* include breeding towards reduced glucosinolate biosynthesis. This approach was not successful as the reduced levels of glucosinolates caused the 'disarmed' sprouts to be eaten by generalist insects. However, a Polish spring-rape variety – the Bronowski cultivar – evolved in the late 1960s with reduced glucosinolates in the seeds but not reduced glucosinolates in the vegetative part of the plant, a phenotype that resembles the phenotype of the *gtr1gtr2* mutant in *A. thaliana*. Due to the complex genome structure of *B. napus*, the molecular identification of the mutation has remained elusive, but it is tempting to suggest that it is within a glucosinolate transporter.

7. CONCLUSIONS

In this book, the reader will be exposed to experts review on many aspects of glucosinolate research. This includes the role of glucosinolates in plant–environment interactions, the exploitation of natural variation in glucosinolate research, insects' way of detoxifying glucosinolates, the health benefits of glucosinolates, transcriptional regulation of glucosinolates, glucosinolates in plant immunity, glucosinolate breakdown, as well as glucosinolates in agriculture.

The view on glucosinolates (and specialized metabolites in general) as defence compounds has changed over the years. In addition to being a source of S and N for the germinating seedling, these compounds are being an integrated part of the plant's balance of growth and development. Although genomes and transcriptomes are becoming available for many other plant species, research on *A. thaliana* and glucosinolates as model metabolites with all its tools available will have enormous advantages also in the future.

REFERENCES

Agerbirk, N., & Olsen, C. E. (2012). Glucosinolate structures in evolution. *Phytochemistry, 77*, 16–45.

Agerbirk, N., & Olsen, C. E. (2015). Glucosinolate hydrolysis products in the crucifer *Barbarea vulgaris* include a thiazolidine-2-one from a specific phenolic isomer as well as oxazolidine-2-thiones. *Phytochemistry, 115*, 143–151.

Andersen, T. G., Nour-Eldin, H. H., Fuller, V. L., Olsen, C. E., Burow, M., & Halkier, B. A. (2013). Integration of biosynthesis and long-distance transport establish organ-specific glucosinolate profiles in vegetative *Arabidopsis*. *The Plant Cell, 25*(8), 3133–3145.

Bennett, R. N., Kiddle, G., & Wallsgrove, R. M. (1997). Biosynthesis of benzylglucosinolate, cyanogenic glucosides and phenylpropanoids in *Carica papaya*. *Phytochemistry, 45*(1), 59–66.

Cartea, M. E., & Velasco, P. (2008). Glucosinolates in *Brassica* foods: bioavailability in food and significance for human health. *Phytochemistry Reviews,* 7(2), 213–229.

Chen, S., Glawischnig, E., Jørgensen, K., Naur, P., Jørgensen, B., Olsen, C. E., … Halkier, B. A. (2003). CYP79F1 and CYP79F2 have distinct functions in the biosynthesis of aliphatic glucosinolates in *Arabidopsis*. *The Plant Journal, 33*(5), 923–937.

Chen, S., & Halkier, B. A. (2000). Characterization of glucosinolate uptake by leaf protoplasts of *Brassica napus*. *Journal of Biological Chemistry, 275*(30), 22955–22960.

Douglas Grubb, C., Zipp, B. J., Ludwig-Müller, J., Masuno, M. N., Molinski, T. F., & Abel, S. (2004). Arabidopsis glucosyltransferase UGT74B1 functions in glucosinolate biosynthesis and auxin homeostasis. *The Plant Journal, 40*(6), 893–908.

Du, L., Lykkesfeldt, J., Olsen, C. E., & Halkier, B. A. (1995). Involvement of cytochrome P450 in oxime production in glucosinolate biosynthesis as demonstrated by an in vitro microsomal enzyme system isolated from jasmonic acid-induced seedlings of *Sinapis alba* L. *Proceedings of the National Academy of Sciences of the United States of America, 92*(26), 12505–12509.

Ettlinger, M., & Kjaer, A. (1968). Sulfur compounds in plants. *Recent Advances in Phytochemistry, 1*(59), 144.

Ettlinger, M. G., & Lundeen, A. J. (1956). The structures of sinigrin and sinalbin; an enzymatic rearrangement. *Journal of the American Chemical Society, 78*(16), 4172–4173.

Fahey, J. W., Zalcmann, A. T., & Talalay, P. (2001). The chemical diversity and distribution of glucosinolates and isothiocyanates among plants. *Phytochemistry, 56*(1), 5–51.

Fahey, J. W., Zhang, Y., & Talalay, P. (1997). Broccoli sprouts: an exceptionally rich source of inducers of enzymes that protect against chemical carcinogens. *Proceedings of the National Academy of Sciences, 94*(19), 10367–10372.

Geu-Flores, F., Nielsen, M. T., Nafisi, M., Møldrup, M. E., Olsen, C. E., Motawia, M. S., & Halkier, B. A. (2009). Glucosinolate engineering identifies a γ-glutamyl peptidase. *Nature Chemical Biology, 5*(8), 575–577.

Gijzen, M., McGregor, I., & Séguin-Swartz, G. (1989). Glucosinolate uptake by developing rapeseed embryos. *Plant Physiology, 89*(1), 260–263.

Gimsing, A. L., & Kirkegaard, J. A. (2009). Glucosinolates and biofumigation: fate of glucosinolates and their hydrolysis products in soil. *Phytochemistry Reviews, 8*(1), 299–310.

Halkier, B. A., & Du, L. (1997). The biosynthesis of glucosinolates. *Trends in Plant Science, 2*(11), 425–431.

Halkier, B. A., & Gershenzon, J. (2006). Biology and biochemistry of glucosinolates. *Annual Review of Plant Biology, 57*, 303–333.

Hansen, C. H., Du, L., Naur, P., Olsen, C. E., Axelsen, K. B., Hick, A. J., … Halkier, B. A. (2001). CYP83B1 is the oxime-metabolizing enzyme in the glucosinolate pathway in *Arabidopsis*. *Journal of Biological Chemistry, 276*(27), 24790–24796.

Johnson, S. D., Griffiths, M. E., Peter, C. I., & Lawes, M. J. (2009). Pollinators, "mustard oil" volatiles, and fruit production in flowers of the dioecious tree *Drypetes natalensis* (Putranjivaceae). *American Journal of Botany, 96*(11), 2080–2086.

Jørgensen, M. E., Nour-Eldin, H. H., & Halkier, B. A. (2015). Transport of defense compounds from source to sink: lessons learned from glucosinolates. *Trends in Plant Science, 20*(8), 508–514.

Kaul, S., Koo, H. L., Jenkins, J., Rizzo, M., Rooney, T., Tallon, L. J., … Lin, X. (2000). Analysis of the genome sequence of the flowering plant *Arabidopsis thaliana*. *Nature, 408*(6814), 796–815.

Kliebenstein, D. J., Lambrix, V. M., Reichelt, M., Gershenzon, J., & Mitchell-Olds, T. (2001). Gene duplication in the diversification of secondary metabolism: tandem

2-oxoglutarate–dependent dioxygenases control glucosinolate biosynthesis in *Arabidopsis*. *The Plant Cell, 13*(3), 681–693.
Koroleva, O. A., Gibson, T. M., Cramer, R., & Stain, C. (2010). Glucosinolate-accumulating S-cells in Arabidopsis leaves and flower stalks undergo programmed cell death at early stages of differentiation. *The Plant Journal, 64*(3), 456–469.
Kroymann, J., Textor, S., Tokuhisa, J. G., Falk, K. L., Bartram, S., Gershenzon, J., & Mitchell-Olds, T. (2001). A gene controlling variation in Arabidopsis glucosinolate composition is part of the methionine chain elongation pathway. *Plant Physiology, 127*(3), 1077–1088.
Madsen, S. R., Olsen, C. E., Nour-Eldin, H. H., & Halkier, B. A. (2014). Elucidating the role of transport processes in leaf glucosinolate distribution. *Plant Physiology, 166*(3), 1450–1462.
Magrath, R., Bano, F., Morgner, M., Parkin, I., Sharpe, A., Lister, C., … Mithen, A. (1994). Genetics of aliphatic glucosinolates. I. Side chain elongation in *Brassica napus* and *Arabidopsis thaliana*. *Heredity, 72*, 290–299.
Magrath, R., & Mithen, R. (1993). Maternal effects on the expression of individual aliphatic glucosinolates in seeds and seedlings of *Brassica napus*. *Plant Breeding, 111*(3), 249–252.
Mailer, R. J., McFadden, A., Ayton, J., & Redden, B. (2008). Anti-nutritional components, fibre, sinapine and glucosinolate content, in Australian canola (*Brassica napus* L.) meal. *Journal of the American Oil Chemists' Society, 85*(10), 937–944.
Meyerowitz, E. M. (2001). Prehistory and history of Arabidopsis research. *Plant Physiology, 125*(1), 15–19.
Mikkelsen, M. D., Buron, L. D., Salomonsen, B., Olsen, C. E., Hansen, B. G., Mortensen, U. H., & Halkier, B. A. (2012). Microbial production of indolylglucosinolate through engineering of a multi-gene pathway in a versatile yeast expression platform. *Metabolic Engineering, 14*(2), 104–111.
Mikkelsen, M. D., Hansen, C. H., Wittstock, U., & Halkier, B. A. (2000). Cytochrome P450 CYP79B2 from Arabidopsis catalyzes the conversion of tryptophan to indole-3-acetaldoxime, a precursor of indole glucosinolates and indole-3-acetic acid. *Journal of Biological Chemistry, 275*(43), 33712–33717.
Mikkelsen, M. D., Olsen, C. E., & Halkier, B. A. (2010). Production of the cancer-preventive glucoraphanin in tobacco. *Molecular Plant, 3*(4), 751–759.
Mirza, N., Crocoll, C., Olsen, C. E., & Halkier, B. A. (2015). Engineering of methionine chain elongation part of glucoraphanin pathway in *E. coli*. *Metabolic Engineering, 35*.
Mithen, R. (2012). Development and commercialisation of 'Beneforte' broccoli and potential health benefits. In *Paper presented at the VI International Symposium on Brassicas and XVIII Crucifer Genetics Workshop 1005*.
Møller, B. L. (2010). Functional diversifications of cyanogenic glucosides. *Current Opinion in Plant Biology, 13*(3), 337–346.
Nafisi, M., Sønderby, I. E., Hansen, B. G., Geu-Flores, F., Nour-Eldin, H. H., Nørholm, M. H. H., … Halkier, B. A. (2006). Cytochromes P450 in the biosynthesis of glucosinolates and indole alkaloids. *Phytochemistry Reviews, 5*(2), 331–346.
Nour-Eldin, H. H., Andersen, T. G., Burow, M., Madsen, S. R., Jørgensen, M. E., Olsen, C. E., … Halkier, B. A. (2012). NRT/PTR transporters are essential for translocation of glucosinolate defence compounds to seeds. *Nature, 488*(7412), 531–534.
Pfalz, M., Mikkelsen, M. D., Bednarek, P., Olsen, C. E., Halkier, B. A., & Kroymann, J. (2011). Metabolic engineering in *Nicotiana benthamiana* reveals key enzyme functions in Arabidopsis indole glucosinolate modification. *The Plant Cell, 23*(2), 716–729.
Reintanz, B., Lehnen, M., Reichelt, M., Gershenzon, J., Kowalczyk, M., Sandberg, G., … Palme, K. (2001). Bus, a bushy Arabidopsis CYP79F1 knockout mutant with abolished synthesis of short-chain aliphatic glucosinolates. *The Plant Cell, 13*(2), 351–367.

Rodman, J., Soltis, P., Soltis, D., Sytsma, K., & Karol, K. (1998). Parallel evolution of glucosinolate biosynthesis inferred from congruent nuclear and plastid gene phylogenies. *American Journal of Botany, 85*(7), 997.

Sivapalan, T., Melchini, A., Traka, M., Saha, S., & Mithen, R. (2015). Investigating the bioavailability of phytochemicals and minerals from broccoli soups. *Proceedings of the Nutrition Society, 74*(OCE3), E191.

Sønderby, I. E., Geu-Flores, F., & Halkier, B. A. (2010). Biosynthesis of glucosinolates – gene discovery and beyond. *Trends in Plant Science, 15*(5), 283–290.

Tantikanjana, T., Mikkelsen, M. D., Hussain, M., Halkier, B. A., & Sundaresan, V. (2004). Functional analysis of the tandem-duplicated P450 genes SPS/BUS/CYP79F1 and CYP79F2 in glucosinolate biosynthesis and plant development by Ds transposition-generated double mutants. *Plant Physiology, 135*(2), 840–848.

Traka, M., & Mithen, R. (2009). Glucosinolates, isothiocyanates and human health. *Phytochemistry Reviews, 8*(1), 269–282.

Traka, M. H., Saha, S., Huseby, S., Kopriva, S., Walley, P. G., Barker, G. C., … Constant, H. (2013). Genetic regulation of glucoraphanin accumulation in Beneforté® broccoli. *New Phytologist, 198*(4), 1085–1095.

Underhill, E. W., Wetter, L. R., & Chisholm, M. D. (1973). Biosynthesis of glucosinolates. *Biochemical Society Symposium, 38*, 303–326.

Wittstock, U., & Halkier, B. A. (2002). Glucosinolate research in the Arabidopsis era. *Trends in Plant Science,* 7(6), 263–270.

CHAPTER TWO

Complex Environments Interact With Plant Development to Shape Glucosinolate Profiles

M. Burow
University of Copenhagen, Frederiksberg, Denmark
E-mail: mbu@plen.ku.dk

Contents

Abstract

Facing complex environments, plants answer with equally complex metabolic outputs to optimize the use of their resources despite ecological and metabolic trade-offs. These metabolic outputs include dynamic changes in levels and composition of the constitutive glucosinolate defence compounds in response to biotic and abiotic environmental changes. Glucosinolate responses to insects, pathogens, drought, heat and irradiation involve distinct quantitative fluctuations of individual, few or all glucosinolate structures in a given tissue. Changes in glucosinolate profiles do not seem to be strongly determined by specific environmental factors. Instead, they are driven by the plant's genotype, physiology and development, which determine the metabolic constraints for glucosinolate biosynthesis. Fine-tuning of the glucosinolate defence compounds relies on regulatory feedback loops linking the pathway to networks controlling primary metabolism, hormone signalling, growth and the onset of flowering. Recent findings on the dynamics of glucosinolate profiles have brought us closer to understanding how plants optimize the use of resources for defensive metabolites.

Advances in Botanical Research, Volume 80
ISSN 0065-2296
http://dx.doi.org/10.1016/bs.abr.2016.06.001

1. INTRODUCTION

In their natural and agricultural habitats, plants are exposed to dynamically changing environmental conditions, and they must complete their life cycle in these unpredictable environments. As a consequence, plants have evolved the ability to optimize their metabolism across diverse combinations of biotic and abiotic factors (Burow, Halkier, & Kliebenstein, 2010; Paul-Victor, Züst, Rees, Kliebenstein, & Turnbull, 2010; Woods, Hastings, Turley, Heard, & Agrawal, 2011). To cope with threats by pathogens and insect herbivores, plants allocate resources to produce bioactive metabolites, either as constitutive or inducible chemical defences. The production of these defence compounds can be costly due to the resources required for their biosynthesis and storage and/or due to ecological trade-offs (Frerigmann & Gigolashvili, 2014b; Gershenzon, Fontana, Burow, Wittstock, & Degenhardt, 2012; Heil, 2002; Strauss, Rudgers, Lau, & Irwin, 2002; Wittstock & Gershenzon, 2002). Yet, unless plant growth is limited by an imbalance in the availability of resources, metabolic costs of chemical defences may not be critical (Kliebenstein, 2016). The use of resources for constitutive and inducible defence compounds must therefore be regulated depending on the interaction of biotic and abiotic environmental factors and be coordinated with plant development (Fig. 1).

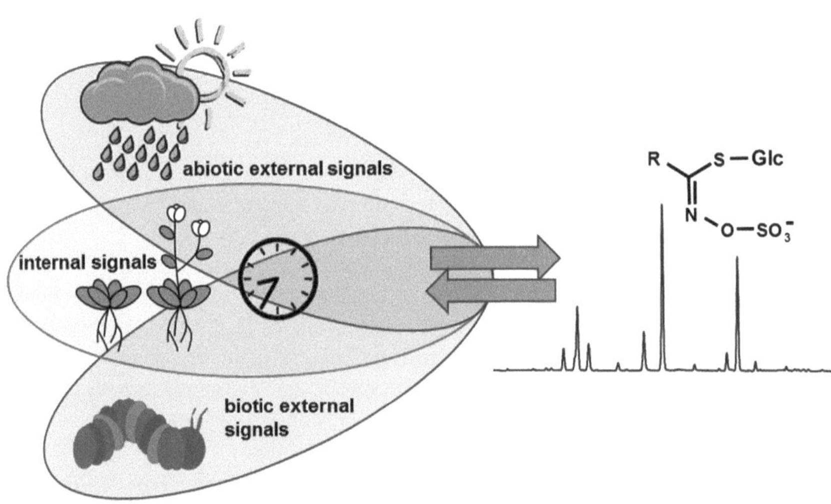

Figure 1 ***Interactions of external and internal signals shape glucosinolate responses.*** Plants integrate abiotic and biotic environmental factors with internal signals to optimize glucosinolate defences. Signal integration involves the circadian clock. The chemical structure depicts the core structure common to all glucosinolate (R, amino acid–derived variable side chain).

The glucosinolate defence compounds can generally be produced from six to nine amino acids (Agerbirk & Olsen, 2012). Apart from the amino acid precursor, formation of each glucosinolate structure requires two sulphur donors, glutathione and 3′-phosphoadenosine-5′-phosphosulfate (Kopriva et al., 2012; Sønderby, Geu-Flores, & Halkier, 2010), entailing the need to tightly regulate the sulphur influx from general metabolism and potentially remobilization of sulphur through glucosinolate catabolism (Falk, Tokuhisa, & Gershenzon, 2007; Zhang et al., 2011). Similarly, glucosinolate accumulation is linked to general nitrogen metabolism (He, Liang, Li, Wang, & Yu, 2014; Marino et al., 2016), another reason why glucosinolate are considered metabolically costly (Züst, Joseph, Shimizu, Kliebenstein, & Turnbull, 2011). In *Arabidopsis thaliana*, the biosynthesis of glucosinolates primarily from methionine and tryptophan (Sønderby et al., 2010) was estimated to require 15% or more of the total photosynthetic energy in leaf tissue (Bekaert, Edger, Hudson, Pires, & Conant, 2012).

The metabolism of cruciferous plants does, however, not only accommodate constitutive accumulation of glucosinolates in all tissues at all developmental stages (Brown, Tokuhisa, Reichelt, & Gershenzon, 2003; van Dam, Tytgat, & Kirkegaard, 2009; Smallegange et al., 2007). In addition, higher levels can be induced when the plant is under attack (Abdalsamee & Muller, 2012; Brader, Tas, & Palva, 2001; Kliebenstein, Rowe, & Denby, 2005; Mathur et al., 2012; Mewis, Appel, Hom, Raina, & Schultz, 2005; Riach, Perera, Florance, Penfield, & Hill, 2015; Schweizer et al., 2013; Skirycz et al., 2006), indicating that additional resources are allocated when higher glucosinolate levels are needed as defence against insect and microbial attackers. Yet, challenging environmental conditions such as drought, cold or limited nutrient availability that introduce additional metabolic trade-offs can interfere with plants' capacity to increase glucosinolate levels. Complex environments therefore require equally complex regulatory networks that allow cruciferous plants to integrate diverse external signals and adjust their metabolism accordingly.

2. ALL CLASSES OF GLUCOSINOLATES CAN CHANGE IN RESPONSE TO PATHOGENS AND HERBIVORES

Glucosinolates have been most intensely studied for their roles as chemical defences against insects and pathogens (Bednarek et al., 2009; Chew, 1988; Hopkins, van Dam, & van Loon, 2009; Sarwar, Kirkegaard, Wong, & Desmarchelier, 1998; Stotz et al., 2011). In most cases, these biological activities arise from the products of glucosinolate breakdown mediated by myrosinases (thioglucoside glucohydrolases) and specifier proteins when the plant is attacked (Wittstock & Burow, 2010; Wittstock, Kliebenstein, Lambrix, Reichelt, & Gershenzon, 2003). As preformed defences, intact glucosinolates are stored, transported and turned over (Andersen et al., 2013; Barth & Jander, 2006; Brown et al., 2003; Koroleva et al., 2000; Nour-Eldin et al., 2012; Piotrowski, 2008; Zhang et al., 2011) as needed to optimize the plant's defence and metabolism.

Several studies have shown that differences in constitutive amounts and variation in glucosinolate structure can influence plant–pathogen interactions (Stotz et al., 2011; Tierens et al., 2001). In terms of the plants' glucosinolate response to pathogen infection, increased levels of all or specific indolic glucosinolate have been reported most frequently (Bednarek et al., 2011; Brader et al., 2001; Kliebenstein et al., 2005; Stotz et al., 2011), while no clear pattern appears to arise from aliphatic glucosinolates. In *A. thaliana* (Columbia-0), aliphatic glucosinolates were found to be unchanged 48 h after infection with *Erwinia carotovora* (Brader et al., 2001); specific glucosinolates increased or decreased 48 h after infection with *Sclerotinia sclerotiorum* (Stotz et al., 2011), or all aliphatic glucosinolates were reported to increase in response to *Botrytis cinerea* after 72 h (Kliebenstein et al., 2005). These different changes in glucosinolate profiles depend on the genotype, the pathogen and plant host and reflect the complexity of their interaction (Bednarek et al., 2011; Kliebenstein et al., 2005). It should also be noted that glucosinolates are of course not the only defence compounds that protect plants from pathogens. In *A. thaliana*, camalexin has been demonstrated to be induced in leaf tissue directly surrounding *Botrytis*-induced regions (Kliebenstein et al., 2005; Schuhegger et al., 2006), while glucosinolates, flavonol and sinapate show opposite spatial patterns (Kliebenstein et al., 2005). Interestingly, pathogen-induced changes in glucosinolate profiles were more pronounced in systemic leaves than in infected leaves (Stotz et al., 2011), further underlining the importance to assess spatial aspects of plant responses when studying biotic interactions.

Similar to pathogens, insects have been reported to induce all classes of glucosinolates (with most studies on either aphids and/or lepidopteran larvae; Table 1). Yet, increases in indolic glucosinolates have been found more frequently, possibly because this class of glucosinolates often shows a higher fold-change (Textor & Gershenzon, 2008). Aphid feeding frequently leads to an increase in all or individual indolic glucosinolates (Cao, Liu, Zhang, & Liu, 2016; Kim & Jander, 2007), while chewing insects typically

Table 1 Changes in Glucosinolate Levels in Response to Insect Feeding

Ref.	Plant species	Insect species	AG	IG	Duration of insect treatment (days)
1	*Arabidopsis thaliana*	*Myzus persicae*	↓	↓↑	3
2	*Arabidopsis thaliana*	*Myzus persicae*	–	–	3
3	*Brassica oleracea*	*Myzus persicae*	–	↑	3
4	*Brassica pekinensis*	*Myzus persicae*	–	↑	5
5	*Arabidopsis thaliana*	*Myzus persicae*, *Brevicoryne brassicae*	↑	↑	7
6	*Arabidopsis thaliana*	*Myzus persicae*	n.a.	–	7
7	*Arabidopsis thaliana*	*Spodoptera littoralis*	–	↑	1
8	*Arabidopsis thaliana*	*Spodoptera exigua*, *Pieris rapae*	↑	–	2
9	*Arabidopsis thaliana*	*Spodoptera littoralis*	↑	↑	2
10	*Arabidopsis thaliana*	*Spodoptera littoralis*	↑	↑	2
11	*Brassica oleracea*, *Cleome spinosa*, *Lunaria annua*	*Pieris rapae*	–	↑	2-3
12	*Arabidopsis thaliana*	*Spodoptera exigua*, *Pieris rapae*	↑	↑	3
13	*Arabidopsis thaliana*	*Spodoptera exigua*, *Pieris rapae*	↑	↑	5
14	*Brassica juncea*	*Plutella xylostella*	↑	–	20

AG, aliphatic glucosinolates; *IG*, indolic glucosinolates; *n.a.*, not available. References are as follows: 1, Kim and Jander (2007); 2, Madsen, Kunert, Reichelt, Gershenzon, and Halkier (2015); 3, Kim and Jander (2007); 4, Cao et al. (2016); 5, Mewis et al. (2006); 6, Hillwig et al. (2016); 7, Skirycz et al. (2006); 8, Mewis et al. (2005); 9, Schweizer et al. (2013); 10, Glauser et al. (2012); 11, Riach et al. (2015); 12, Mewis et al. (2006); 13, Kos et al. (2012); 14, Mathur et al. (2012).

induce distinct changes in glucosinolates from all classes (Beekwilder et al., 2008; Glauser, Schweizer, Turlings, & Reymond, 2012; Kos et al., 2012; Mathur et al., 2012; Mewis et al., 2006; Schweizer et al., 2013). Different feeding modes and consequently different signalling pathways in the host plant may thus play a certain role for the type of glucosinolate response to insect feeding or at least the timing of these responses, but specific changes in glucosinolate profiles must be driven by additional factors.

Both the constitutive and inductive transcriptional control of glucosinolate biosynthesis have been linked to jasmonate signalling, often based on increased expression glucosinolate genes and increased glucosinolate levels after application of exogenous jasmonates (Brader et al., 2001; Frerigmann & Gigolashvili, 2014a; Hirai et al., 2007; Jost et al., 2005; Kliebenstein, Figuth, & Mitchell-Olds, 2002; Mewis et al., 2005;

Mikkelsen et al., 2003). In addition, triple knockouts of three basic helix–loop–helix transcription factors (MYC2, MYC3, MYC4) described to have additive functions in jasmonate-mediated defence responses are devoid of both aliphatic and indolic glucosinolates (Frerigmann, Berger, & Gigolashvili, 2014; Schweizer et al., 2013). Jasmonate signalling is therefore considered as the most critical signalling pathway at play when plants increase glucosinolate levels upon herbivory. Whereas insects are allowed to feed continuously over a certain time period, jasmonates are often applied by spraying. The changes observed after this one-time treatment include increased levels of indolic glucosinolates, while aliphatic glucosinolates are unchanged (Jost et al., 2005; Kliebenstein et al., 2002; Mewis et al., 2005; Mikkelsen et al., 2003). Prolonged jasmonate treatment (over 2 days) has been reported to affect the levels of both classes of glucosinolates (Frerigmann & Gigolashvili, 2014a) as also observed after herbivore feeding. Intriguingly, the glucosinolate biosynthetic gene *AOP2* has recently shown to be part of a positive feedback loop linking glucosinolates and jasmonate signalling (Burow et al., 2015). This regulatory link could function to amplify jasmonate-related signalling when the external trigger persists, for example, when an insect herbivore keeps feeding and causes more substantial damage. Future studies will be needed to decipher the regulatory networks that mediate early versus late glucosinolate responses comprising changes in biosynthesis, transport and breakdown.

3. CHANGES IN ABIOTIC ENVIRONMENTAL FACTORS MODULATE GLUCOSINOLATE RESPONSES

The overall observation that plants respond to insects and pathogens with distinct, but not predictable changes in glucosinolate profiles suggests that these responses are shaped by the plant's physiology and abiotic environment. Even without a biotic attacker, abiotic factors can alter accumulation of individual or all glucosinolates as it has, for example, been reported for different temperature regimes (Rosa & Rodrigues, 1998; Steindal, Rødven, Hansen, & Mølmann, 2015), increased UV-B irradiance (Mewis, Schreiner, et al., 2012), drought (Gill, Haugen, Matzner, Barakat, & Siemens, 2016; Khan, Ulrichs, & Mewis, 2010; Mewis, Khan, Glawischnig, Schreiner, & Ulrichs, 2012; Tariq, Rossiter, Wright, & Staley, 2013; Tong et al., 2014), salt stress (Guo, Yuan, & Wang, 2013; Khan et al., 2014; Qasim, Ashraf, Ashraf, Rehman, & Rha, 2003; Yuan, Wang, Guo, & Wang, 2010), oxidative stress (Lehmann et al., 2009) and high ozone levels

(Khaling et al., 2015). This responsiveness of the regulatory networks controlling the accumulation of glucosinolates is not at all unexpected as any environmental change perceived by the plant will lead to physiological changes and alter metabolic trade-offs. Like metabolic consequences of plant—biotic interactions, responses to changes in abiotic factors are characterized by overlapping signalling pathways, i.e., no specific genes or physiological changes can be matched with specific abiotic factors (Knight & Knight, 2001).

Typically, unfavourable environmental conditions cause imbalances in plant metabolism associated with fluctuations in cellular redox and altered phytohormone signalling (Spoel & van Ooijen, 2014). These metabolic changes can affect precursor availability for glucosinolate biosynthesis or flux through the pathway due to cross talk with other pathways. In *A. thaliana*, the cytochrome P450 monooxygenase CYP83A1 is involved in the biosynthesis of both aliphatic glucosinolates and phenylpropanoids (Hemm, Ruegger, & Chapple, 2003). Increased UV-B irradiation triggers accumulation of the phenylpropanoid sinapoylmalate as UV-light protectant (Hemm et al., 2003; Landry, Chapple, & Last, 1995) and levels of aliphatic glucosinolates (Mewis, Schreiner, et al., 2012). By contrast, phenylpropanoid accumulation can be limited by the indolic glucosinolate precursor indole-3-acetaldoxime (IAOx), the substrate of CYP83B1 due to cross talk through the Mediator complex (Kim, Dolan, Anderson, & Chapple, 2015).

In contrast to laboratory studies using controlled growth conditions, natural environments plants face combinations of abiotic and biotic factors, which moreover change continuously. When plants are subjected to changes in two or more environmental factors, the treatments often interact, i.e., their consequences cannot be predicted from the impact of each environmental factor tested in isolation (Suzuki, Rivero, Shulaev, Blumwald, & Mittler, 2014), underlining the ability of plants to integrate external cues and respond to complex environments. Studies that systematically addressed changes in glucosinolate profiles in response to combinations of environmental factors are relatively scarce, but some investigated combinatorial effects of drought stress and insect feeding (Table 2). Changes in glucosinolate levels have been reported for aliphatic and indolic glucosinolates depending on the plant genotype (Haugen et al., 2008), the insect species (Khan et al., 2010; Khan, Ulrichs, & Mewis, 2011) and overall indicated nonadditive effects of water availability and insect feeding on glucosinolates. Thus, drought stress can modulate glucosinolate responses to insect attack.

Table 2 Combinatorial Effects of Water Availability and Insect Feeding on Glucosinolates

Plant Species	Insect Species	Drought Treatment	Effect on GLS Levels	References
Arabidopsis thaliana (Col-0)	*Mamestra brassicae*, *Myzus persicae*	Low, continuous, high drought	AG and IG changed under drought stress (especially in aphid-infested plants)	Pineda et al. (2016)
A. thaliana (Col-0)	*M. brassicae*, *M. persicae*	Well-watered, drought, water-logged	AG and IG induction in response to aphid feeding varied with water availability	Mewis, Khan, et al. (2012)
Brassica oleracea var. *italica*	*Brevicoryne brassicae*, *M. persicae*	Well-watered, drought, water-logged	IG induction by *M. persicae* modulated water availability	Khan et al. (2011)
Brassica oleracea var. *gemmifera*	*Delia radicum*	Standard water regime, medium and high drought stress	Mostly IG induction affected by *D. radicum*	Tariq et al. (2013)
Boechera stricta	*Spodoptera exigua*	Continuous drought treatment	Drought impacted herbivore-induced changes in GLS ratio response (branched chain to methionine-derived straight chain)	Gill et al. (2016)
B. stricta, *Boechera holboellii*	*Plutella xylostella*	Standard water regime vs drought	Inducibility of GLS by drought was dependent on plant genotype	Haugen et al. (2008)

AG, aliphatic glucosinolates; *GLS*, glucosinolate; *IG*, indolic glucosinolates.

When water availability is limited, plants need to adjust stomatal conductance to balance photosynthetic carbon uptake (Pinheiro & Chaves, 2011). Yet, in addition to changes in photosynthetic rate, drought stress has also been associated with alterations in sulphur metabolism, partially for glutathione-dependent detoxification of reactive oxygen species (Chan, Wirtz,

Phua, Estavillo, & Pogson, 2013). These drought-induced alterations in general sulphur metabolism might predict changes in glucosinolate levels as (1) glucosinolates can account for up to 30% of total sulphur in plant tissues (Falk et al., 2007), (2) sulphur fertilization typically leads to increased glucosinolates (Falk et al., 2007) and (3) glucosinolates can be utilized as sulphur source under conditions of severe sulphur deficiency (Zhang et al., 2011). Yet, plant species and tissues vary in their glucosinolate response to both drought and sulphur deficiency, illustrating the need for further investigation of the underlying regulatory mechanisms.

4. PLANTS INTEGRATE EXTERNAL AND INTERNAL SIGNALS TO OPTIMIZE THEIR METABOLISM

Steady-state levels of glucosinolates in cruciferous plants vary between plant species, tissues, developmental stages and environmental conditions (Andersen & Halkier, 2014; Andersen et al., 2013; Brown et al., 2003; van Dam et al., 2009; Fahey, Zalcmann, & Talalay, 2001; Petersen, Chen, Hansen, Olsen, & Halkier, 2002), and the same degree of variability is associated with absolute or relative changes in glucosinolates in response to single or multiple biotic and abiotic factors. To ensure efficient use of resources, this dynamic regulation of glucosinolate accumulation requires tissue-specific regulatory networks that integrate environmental signals with internal signals reflecting the plant's metabolism. In turn, glucosinolates must provide feedback information to primary metabolism indicating that glucosinolates fulfil biological functions beyond their role as precursors of chemical defences. In agreement with this model, glucosinolates or their breakdown products have been reported to be involved in ABA-mediated stomatal aperture (Zhao, Zhang, Stanley, & Assmann, 2008), regulation of water transport under sulphur deficiency (Martínez-Ballesta et al., 2014), feedback regulation of JA signalling (Burow et al., 2015), biomass responses (Francisco et al., 2016a, 2016b; Maharjan et al., 2014) and the onset of flowering (Atwell et al., 2010; Jensen, Jepsen, Halkier, Kliebenstein, & Burow, 2015; Kerwin et al., 2011) illustrating cross talk between glucosinolates and regulatory networks controlling plant physiology and development.

Environmental and internal signals are to a large extent integrated by circadian clocks regulating biological processes with a day's period (Harmer, 2009; Hsu & Harmer, 2014). The circadian clock provides input to redox- and hormone-mediated responses to stressful environmental conditions, and these clock outputs feedback regulate the clock (Seo & Mas, 2015; Spoel &

van Ooijen, 2014). This feedback regulation seems vital to anticipate and counteract abiotic and biotic factors that fluctuate throughout the day, including irradiation, temperature, insects and pathogens, as plants that are impaired in circadian clock control or are clock-shifted compared to insect herbivores show reduced resistance (Goodspeed, Chehab, Min-Venditti, Braam, & Covington, 2012; Wang et al., 2011). Basal levels of glucosinolates fluctuate only moderately during the day (Rosa & Rodrigues, 1998), but whether the inducibility of glucosinolates is clock-controlled remains to be tested. In *A. thaliana*, feedback regulation from the glucosinolate pathways to the circadian clock involves the glucosinolate biosynthetic gene *AOP2* (Kerwin et al., 2011) establishing a regulatory link between glucosinolates and the circadian clock.

5. CONCLUDING REMARKS

Although glucosinolates are constitutive defences in cruciferous plants, glucosinolate profiles represent a complex phenotype that changes dynamically in response to virtually any change in the abiotic or biotic environment and to combinations of environmental factors. This likely reflects the ecological trade-offs associated with high glucosinolate levels as much as the metabolic costs, depending on the glucosinolate composition. Due to their dynamic nature, the regulatory networks shaping glucosinolate profiles in different plant genotypes and tissues and controlling glucosinolate responses under diverse condition provide an exceptional model to study the complexity of signal integration in plants.

ACKNOWLEDGEMENT

Financial support was provided by the Danish National Research Foundation (DNRF99).

REFERENCES

Abdalsamee, M. K., & Muller, C. (2012). Effects of indole glucosinolates on performance and sequestration by the sawfly *Athalia rosae* and consequences of feeding on the plant defense system. *Journal of Chemical Ecology, 38*(11), 1366–1375.

Agerbirk, N., & Olsen, C. E. (2012). Glucosinolate structures in evolution. *Phytochemistry, 77*, 16–45.

Andersen, T. G., & Halkier, B. A. (2014). Upon bolting the GTR1 and GTR2 transporters mediate transport of glucosinolates to the inflorescence rather than roots. *Plant Signaling & Behavior, 9*(1), e27740.

Andersen, T. G., Nour-Eldin, H. H., Fuller, V. L., Olsen, C. E., Burow, M., & Halkier, B. A. (2013). Integration of biosynthesis and long-distance transport establish organ-specific glucosinolate profiles in vegetative *Arabidopsis*. *Plant Cell, 25*(8), 3133–3145.

Atwell, S., Huang, Y. S., Vilhjalmsson, B. J., Willems, G., Horton, M., Li, Y., ... Nordborg, M. (2010). Genome-wide association study of 107 phenotypes in *Arabidopsis thaliana* inbred lines. *Nature, 465*(1476–4687), 627–631.

Barth, C., & Jander, G. (2006). *Arabidopsis* myrosinases TGG1 and TGG2 have redundant function in glucosinolate breakdown and insect defense. *Plant Journal, 46*(4), 549–562.

Bednarek, P., Piślewska-Bednarek, M., Svatoš, A., Schneider, B., Doubský, J., Mansurova, M., ... Schulze-Lefert, P. (2009). A glucosinolate metabolism pathway in living plant cells mediates broad-spectrum antifungal defense. *Science, 323*(5910), 101–106.

Bednarek, P., Pislewska-Bednarek, M., Ver Loren van Themaat, E., Maddula, R. K., Svatos, A., & Schulze-Lefert, P. (2011). Conservation and clade-specific diversification of pathogen-inducible tryptophan and indole glucosinolate metabolism in *Arabidopsis thaliana* relatives. *New Phytologist, 192*(3), 713–726.

Beekwilder, J., van Leeuwen, W., van Dam, N. M., Bertossi, M., Grandi, V., Mizzi, L., ... Bovy, A. (2008). The impact of the absence of aliphatic glucosinolates on insect herbivory in *Arabidopsis*. *PLoS One, 3*(4).

Bekaert, M., Edger, P. P., Hudson, C. M., Pires, J. C., & Conant, G. C. (2012). Metabolic and evolutionary costs of herbivory defense: systems biology of glucosinolate synthesis. *New Phytologist, 196*(2), 596–605.

Brader, G., Tas, É., & Palva, E. T. (2001). Jasmonate-dependent induction of indole glucosinolates in *Arabidopsis* by culture filtrates of the nonspecific pathogen *Erwinia carotovora*. *Plant Physiology, 126*(2), 849–860.

Brown, P. D., Tokuhisa, J. G., Reichelt, M., & Gershenzon, J. (2003). Variation of glucosinolate accumulation among different organs and developmental stages of *Arabidopsis thaliana*. *Phytochemistry, 62*(3), 471–481.

Burow, M., Atwell, S., Francisco, M., Kerwin, R. E., Halkier, B. A., & Kliebenstein, D. J. (2015). The glucosinolate biosynthetic gene *AOP2* mediates feed-back regulation of jasmonic acid signaling in *Arabidopsis*. *Molecular Plant, 8*, 1201–1212.

Burow, M., Halkier, B. A., & Kliebenstein, D. J. (2010). Regulatory networks of glucosinolates shape *Arabidopsis thaliana* fitness. *Current Opinion in Plant Biology, 13*(3), 348–353.

Cao, H. H., Liu, H. R., Zhang, Z. F., & Liu, T. X. (2016). The green peach aphid *Myzus persicae* perform better on pre-infested Chinese cabbage *Brassica pekinensis* by enhancing host plant nutritional quality. *Scientific Reports, 6*, 21954.

Chan, K. X., Wirtz, M., Phua, S. Y., Estavillo, G. M., & Pogson, B. J. (2013). Balancing metabolites in drought: the sulfur assimilation conundrum. *Trends in Plant Science, 18*(1), 18–29.

Chew, F. S. (1988). Biological effects of glucosinolates. In H. G. Cutler (Ed.), *Biologically active natural products* (pp. 155–181). Washington, DC: American Chemical Society.

van Dam, N. M., Tytgat, T. O. G., & Kirkegaard, J. A. (2009). Root and shoot glucosinolates: a comparison of their diversity, function and interactions in natural and managed ecosystems. *Phytochemistry Reviews, 8*(1), 171–186.

Fahey, J. W., Zalcmann, A. T., & Talalay, P. (2001). The chemical diversity and distribution of glucosinolates and isothiocyanates among plants. *Phytochemistry, 56*(1), 5–51.

Falk, K. L., Tokuhisa, J. G., & Gershenzon, J. (2007). The effect of sulfur nutrition on plant glucosinolate content: physiology and molecular mechanisms. *Plant Biology (Stuttgart), 9*(5), 573–581.

Francisco, M., Joseph, B., Caligagan, H., Li, B., Corwin, J. A., Lin, C., Kerwin, R. E., Burow, M., & Kliebenstein, D. J. (2016a). The defense metabolite, Allyl glucosinolate, modulates *Arabidopsis thaliana* biomass dependent upon the endogenous glucosinolate pathway. *Frontiers in Plant Science, 7*.

Francisco, M., Joseph, B., Caligagan, H., Li, B., Corwin, J. A., Lin, C., Kerwin, R. E., Burow, M., & Kliebenstein, D. J. (2016b). Genome wide association mapping in *Arabidopsis thaliana* identifies novel genes involved in linking allyl glucosinolate to altered biomass and defense. *Frontiers in Plant Science, 7*.

Frerigmann, H., Berger, B., & Gigolashvili, T. (2014). bHLH05 is an interaction partner of MYB51 and a novel regulator of glucosinolate biosynthesis in *Arabidopsis. Plant Physiology*.

Frerigmann, H., & Gigolashvili, T. (2014a). MYB34, MYB51, and MYB122 distinctly regulate indolic glucosinolate biosynthesis in *Arabidopsis thaliana. Molecular Plant, 7*(5), 814–828.

Frerigmann, H., & Gigolashvili, T. (2014b). Update on the role of R2R3-MYBs in the regulation of glucosinolates upon sulfur deficiency. *Frontiers in Plant Science, 5*, 626.

Gershenzon, J., Fontana, A., Burow, M., Wittstock, U., & Degenhardt, J. (2012). Mixtures of plant secondary metabolites: metabolic origins and ecological benefits. In G. R. Iason, M. Dicke, & S. E. Hartley (Eds.), *Ecology of plant secondary metabolites: From genes to global processes* (pp. 56–77). New York: Cambridge University Press. the Pitt Building, Trumpington St, Cambridge Cb2 1rp, Cambs, UK.

Gill, G., Haugen, R., Matzner, S., Barakat, A., & Siemens, D. (2016). Effect of drought on herbivore-induced plant gene expression: population comparison for range limit inferences. *Plants, 5*(1), 13.

Glauser, G., Schweizer, F., Turlings, T. C., & Reymond, P. (2012). Rapid profiling of intact glucosinolates in *Arabidopsis* leaves by UHPLC-QTOFMS using a charged surface hybrid column. *Phytochem Analysis, 23*(5), 520–528.

Goodspeed, D., Chehab, E. W., Min-Venditti, A., Braam, J., & Covington, M. F. (2012). *Arabidopsis* synchronizes jasmonate-mediated defense with insect circadian behavior. *Proceedings of the National Academy of Sciences of the United States of America, 109*(12), 4674–4677.

Guo, R.-F., Yuan, G.-F., & Wang, Q.-M. (2013). Effect of NaCl treatments on glucosinolate metabolism in broccoli sprouts. *Journal of Zhejiang University Science B, 14*(2), 124–131.

Harmer, S. L. (2009). The circadian system in higher plants. *Annual Review of Plant Biology, 60*(1), 357–377.

Haugen, R., Steffes, L., Wolf, J., Brown, P., Matzner, S., & Siemens, D. H. (2008). Evolution of drought tolerance and defense: dependence of tradeoffs on mechanism, environment and defense switching. *Oikos, 117*(2), 231–244.

He, H., Liang, G., Li, Y., Wang, F., & Yu, D. (2014). Two young MicroRNAs originating from target duplication mediate nitrogen starvation adaptation via regulation of glucosinolate synthesis in *Arabidopsis thaliana. Plant Physiology, 164*(2), 853–865.

Heil, M. (2002). Ecological costs of induced resistance. *Current Opinion in Plant Biology, 5*(4), 345–350.

Hemm, M. R., Ruegger, M. O., & Chapple, C. (2003). The *Arabidopsis* ref2 mutant is defective in the gene encoding CYP83A1 and shows both phenylpropanoid and glucosinolate phenotypes. *Plant Cell, 15*(1), 179–194.

Hillwig, M. S., Chiozza, M., Casteel, C. L., Ting Lau, S., Hohenstein, J., Hernandez, E., … Macintosh, G. C. (2016). Abscisic acid deficiency increases defence responses against *Myzus persicae* in *Arabidopsis. Molecular Plant Pathology, 17*(2), 225–235.

Hirai, M. Y., Sugiyama, K., Sawada, Y., Tohge, T., Obayashi, T., Suzuki, A., … Saito, K. (2007). Omics-based identification of *Arabidopsis* Myb transcription factors regulating aliphatic glucosinolate biosynthesis. *Proceedings of the National Academy of Sciences of the United States of America, 104*(15), 6478–6483.

Hopkins, R. J., van Dam, N. M., & van Loon, J. J. (2009). Role of glucosinolates in insect-plant relationships and multitrophic interactions. *Annual Reviews in Entomology, 54*, 57–83.

Hsu, P. Y., & Harmer, S. L. (2014). Wheels within wheels: the plant circadian system. *Trends in Plant Science, 19*(4), 240–249.

Jensen, L. M., Jepsen, H. K. S., Halkier, B. A., Kliebenstein, D. J., & Burow, M. (2015). Natural variation in cross-talk between glucosinolates and onset of flowering in *Arabidopsis. Frontiers in Plant Science, 6.* Article 697.

Jost, R., Altschmied, L., Bloem, E., Bogs, J., Gershenzon, J., Hähnel, U., … Hell, R. (2005). Expression profiling of metabolic genes in response to methyl jasmonate reveals regulation of genes of primary and secondary sulfur-related pathways in *Arabidopsis thaliana. Photosynthesis Research, 86*(3), 491–508.

Kerwin, R. E., Jimenez-Gomez, J. M., Fulop, D., Harmer, S. L., Maloof, J. N., & Kliebenstein, D. J. (2011). Network quantitative trait loci mapping of circadian clock outputs identifies metabolic pathway-to-clock linkages in *Arabidopsis. Plant Cell, 23*(2), 471–485.

Khaling, E., Papazian, S., Poelman, E. H., Holopainen, J. K., Albrectsen, B. R., & Blande, J. D. (2015). Ozone affects growth and development of *Pieris brassicae* on the wild host plant *Brassica nigra. Environmental Pollution, 199*, 119–129.

Khan, M. A. M., Ulrichs, C., & Mewis, I. (2010). Influence of water stress on the glucosinolate profile of *Brassica oleracea* var. *italica* and the performance of *Brevicoryne brassicae* and *Myzus persicae. Entomologia Experimentalis et Applicata, 137*(3), 229–236.

Khan, M. A. M., Ulrichs, C., & Mewis, I. (2011). Water stress alters aphid-induced glucosinolate response in *Brassica oleracea* var. *italica* differently. *Chemoecology, 21*(4), 235–242.

Khan, N. A., Khan, M. I. R., Asgher, M., Fatma, M., Masood, A., & Syeed, S. (2014). Salinity tolerance in plants: revisiting the role of sulfur metabolites. *Journal of Plant Biochemistry & Physiology, 02*(01).

Kim, J. H., & Jander, G. (2007). *Myzus persicae* (green peach aphid) feeding on *Arabidopsis* induces the formation of a deterrent indole glucosinolate. *Plant Journal, 49*(6), 1008–1019.

Kim, J. I., Dolan, W. L., Anderson, N. A., & Chapple, C. (2015). Indole glucosinolate biosynthesis limits phenylpropanoid accumulation in *Arabidopsis thaliana. Plant Cell, 27*(5), 1529–1546.

Kliebenstein, D. J. (2016). False idolatry of the mythical growth versus immunity tradeoff in molecular systems plant pathology. *Physiological and Molecular Plant Pathology.*

Kliebenstein, D. J., Figuth, A., & Mitchell-Olds, T. (2002). Genetic architecture of plastic methyl jasmonate responses in *Arabidopsis thaliana. Genetics, 161*(4), 1685–1696.

Kliebenstein, D. J., Rowe, H. C., & Denby, K. J. (2005). Secondary metabolites influence *Arabidopsis/Botrytis* interactions: variation in host production and pathogen sensitivity. *Plant Journal, 44*(1), 25–36.

Knight, H., & Knight, M. R. (2001). Abiotic stress signalling pathways: specificity and cross-talk. *Trends in Plant Science, 6*(6), 262–267.

Kopriva, S., Mugford, S. G., Baraniecka, P., Lee, B. R., Matthewman, C. A., & Koprivova, A. (2012). Control of sulfur partitioning between primary and secondary metabolism in *Arabidopsis. Frontiers in Plant Science, 3*, 163.

Koroleva, O. A., Davies, A., Deeken, R., Thorpe, M. R., Tomos, A. D., & Hedrich. (2000). Identification of a new glucosinolate-rich cell type in *Arabidopsis* flower stalk. *Plant Physiology, 124*(2), 599–608.

Kos, M., Houshyani, B., Wietsma, R., Kabouw, P., Vet, L. E., van Loon, J. J., & Dicke, M. (2012). Effects of glucosinolates on a generalist and specialist leaf-chewing herbivore and an associated parasitoid. *Phytochemistry, 77*, 162–170.

Landry, L. G., Chapple, C., & Last, R. L. (1995). *Arabidopsis* mutants lacking phenolic sunscreens exhibit enhanced Ultraviolet-B injury and oxidative damage. *Plant Physiology, 109*(4), 1159–1166.

Lehmann, M., Schwarzlander, M., Obata, T., Sirikantaramas, S., Burow, M., Olsen, C. E., … Laxa, M. (2009). The metabolic response of *Arabidopsis* roots to oxidative stress is distinct from that of heterotrophic cells in culture and highlights a complex relationship between the levels of transcripts, metabolites, and flux. *Molecular Plant, 2*(3), 390–406.

Madsen, S. R., Kunert, G., Reichelt, M., Gershenzon, J., & Halkier, B. A. (2015). Feeding on leaves of the glucosinolate transporter mutant gtr1gtr2 reduces fitness of *Myzus persicae*. *Journal of Chemical Ecology, 41*(11), 975–984.

Maharjan, P. M., Dilkes, B. P., Fujioka, S., Pencik, A., Ljung, K., Burow, M., … Choe, S. (2014). Arabidopsis gulliver1/SUPERROOT2-7 identifies a metabolic basis for auxin and brassinosteroid synergy. *Plant Journal, 80*(5), 797–808.

Marino, D., Ariz, I., Lasa, B., Santamaria, E., Fernandez-Irigoyen, J., Gonzalez-Murua, C., & Aparicio Tejo, P. M. (2016). Quantitative proteomics reveals the importance of nitrogen source to control glucosinolate metabolism in *Arabidopsis thaliana* and *Brassica oleracea*. *Journal of Experimental Botany*.

Martínez-Ballesta, M. D. C., Muries, B., Moreno, D.Á., Dominguez-Perles, R., García-Viguera, C., & Carvajal, M. (2014). Involvement of a glucosinolate (sinigrin) in the regulation of water transport in *Brassica oleracea* grown under salt stress. *Physiologia Plantarum, 150*(2), 145–160.

Mathur, V., Tytgat, T. O. G., de Graaf, R. M., Kalia, V., Sankara Reddy, A., Vet, L. E. M., & van Dam, N. M. (2012). Dealing with double trouble: consequences of single and double herbivory in *Brassica juncea*. *Chemoecology, 23*(2), 71–82.

Mewis, I., Appel, H. M., Hom, A., Raina, R., & Schultz, J. C. (2005). Major signaling pathways modulate *Arabidopsis* glucosinolate accumulation and response to both phloem-feeding and chewing insects. *Plant Physiology, 138*(2), 1149–1162.

Mewis, I., Khan, M. A. M., Glawischnig, E., Schreiner, M., & Ulrichs, C. (2012). Water stress and aphid feeding differentially influence metabolite composition in *Arabidopsis thaliana* (L.). *PLoS One,* 7(11), e48661.

Mewis, I., Schreiner, M., Nguyen, C. N., Krumbein, A., Ulrichs, C., Lohse, M., & Zrenner, R. (2012). UV-B irradiation changes specifically the secondary metabolite profile in broccoli sprouts: induced signaling overlaps with defense response to biotic stressors. *Plant & Cell Physiology, 53*(9), 1546–1560.

Mewis, I., Tokuhisa, J. G., Schultz, J. C., Appel, H. M., Ulrichs, C., & Gershenzon, J. (2006). Gene expression and glucosinolate accumulation in *Arabidopsis thaliana* in response to generalist and specialist herbivores of different feeding guilds and the role of defense signaling pathways. *Phytochemistry, 67*(22), 2450–2462.

Mikkelsen, M. D., Petersen, B. L., Glawischnig, E., Jensen, A. B., Andreasson, E., & Halkier, B. A. (2003). Modulation of CYP79 genes and glucosinolate profiles in *Arabidopsis* by defense signaling pathways. *Plant Physiology, 131*(1), 298–308.

Nour-Eldin, H. H., Andersen, T. G., Burow, M., Madsen, S. R., Jorgensen, M. E., Olsen, C. E., … Halkier, B. A. (2012). NRT/PTR transporters are essential for translocation of glucosinolate defence compounds to seeds. *Nature, 488*(7412), 531–534.

Paul-Victor, C., Züst, T., Rees, M., Kliebenstein, D. J., & Turnbull, L. A. (2010). A new method for measuring relative growth rate can uncover the costs of defensive compounds in *Arabidopsis thaliana*. *New Phytologist, 187*(4), 1102–1111.

Petersen, B., Chen, S., Hansen, C., Olsen, C., & Halkier, B. (2002). Composition and content of glucosinolates in developing *Arabidopsis thaliana*. *Planta, 214*(4), 562–571.

Pineda, A., Pangesti, N., Soler, R., Dam, N. M. V., Loon, J. J. A. V., & Dicke, M. (2016). Negative impact of drought stress on a generalist leaf chewer and a phloem feeder is

associated with, but not explained by an increase in herbivore-induced indole glucosinolates. *Environmental and Experimental Botany, 123*, 88–97.
Pinheiro, C., & Chaves, M. M. (2011). Photosynthesis and drought: can we make metabolic connections from available data? *Journal of Experimental Botany, 62*(3), 869–882.
Piotrowski, M. (2008). Primary or secondary? Versatile nitrilases in plant metabolism. *Phytochemistry, 69*(15), 2655–2667.
Qasim, M., Ashraf, M., Ashraf, M. Y., Rehman, S.-U., & Rha, E. S. (2003). Salt-induced changes in two canola cultivars differing in salt tolerance. *Biologia Plantarum, 46*(4), 629–632.
Riach, A. C., Perera, M. V., Florance, H. V., Penfield, S. D., & Hill, J. K. (2015). Analysis of plant leaf metabolites reveals no common response to insect herbivory by *Pieris rapae* in three related host-plant species. *Journal of Experimental Botany, 66*(9), 2547–2556.
Rosa, E. A. S., & Rodrigues, P. M. F. (1998). The effect of light and temperature on glucosinolate concentration in the leaves and roots of cabbage seedlings. *Journal of the Science of Food and Agriculture, 78*(2), 208–212.
Sarwar, M., Kirkegaard, J. A., Wong, P. T. W., & Desmarchelier, J. M. (1998). Biofumigation potential of brassicas. *Plant and Soil, 201*(1), 103–112.
Schuhegger, R., Nafisi, M., Mansourova, M., Petersen, B. L., Olsen, C. E., Svatoš, A., … Glawischnig, E. (2006). CYP71B15 (PAD3) catalyzes the final step in camalexin biosynthesis. *Plant Physiology, 141*(4), 1248–1254.
Schweizer, F., Fernandez-Calvo, P., Zander, M., Diez-Diaz, M., Fonseca, S., Glauser, G., … Reymond, P. (2013). *Arabidopsis* basic helix-loop-helix transcription factors MYC2, MYC3, and MYC4 regulate glucosinolate biosynthesis, insect performance, and feeding behavior. *Plant Cell, 25*(8), 3117–3132.
Seo, P. J., & Mas, P. (2015). STRESSing the role of the plant circadian clock. *Trends in Plant Science, 20*(4), 230–237.
Skirycz, A., Reichelt, M., Burow, M., Birkemeyer, C., Rolcik, J., Kopka, J., … Witt, I. (2006). DOF transcription factor AtDof1.1 (OBP2) is part of a regulatory network controlling glucosinolate biosynthesis in *Arabidopsis*. *The Plant Journal: for Cell and Molecular Biology, 47*(1), 10–24.
Smallegange, R. C., van Loon, J. J., Blatt, S. E., Harvey, J. A., Agerbirk, N., & Dicke, M. (2007). Flower vs. leaf feeding by *Pieris brassicae*: glucosinolate-rich flower tissues are preferred and sustain higher growth rate. *Journal of Chemical Ecology, 33*(10), 1831–1844.
Sønderby, I. E., Geu-Flores, F., & Halkier, B. A. (2010). Biosynthesis of glucosinolates–gene discovery and beyond. *Trends in Plant Science, 15*(5), 283–290.
Spoel, S. H., & van Ooijen, G. (2014). Circadian redox signaling in plant immunity and abiotic stress. *Antioxidants & Redox Signaling, 20*(18), 3024–3039.
Steindal, A. L. H., Rødven, R., Hansen, E., & Mølmann, J. (2015). Effects of photoperiod, growth temperature and cold acclimatisation on glucosinolates, sugars and fatty acids in kale. *Food Chemistry, 174*, 44–51.
Stotz, H. U., Sawada, Y., Shimada, Y., Hirai, M. Y., Sasaki, E., Krischke, M., … Kamiya, Y. (2011). Role of camalexin, indole glucosinolates, and side chain modification of glucosinolate-derived isothiocyanates in defense of *Arabidopsis* against *Sclerotinia sclerotiorum*. *Plant Journal, 67*(1), 81–93.
Strauss, S. Y., Rudgers, J. A., Lau, J. A., & Irwin, R. E. (2002). Direct and ecological costs of resistance to herbivory. *Trends in Ecology & Evolution, 17*(6), 278–285.
Suzuki, N., Rivero, R. M., Shulaev, V., Blumwald, E., & Mittler, R. (2014). Abiotic and biotic stress combinations. *New Phytologist, 203*(1), 32–43.
Tariq, M., Rossiter, J. T., Wright, D. J., & Staley, J. T. (2013). Drought alters interactions between root and foliar herbivores. *Oecologia, 172*(4), 1095–1104.

Textor, S., & Gershenzon, J. (2008). Herbivore induction of the glucosinolate–myrosinase defense system: major trends, biochemical bases and ecological significance. *Phytochemistry Reviews, 8*(1), 149–170.

Tierens, K. F., Thomma, B. P., Brouwer, M., Schmidt, J., Kistner, K., Porzel, A., … Broekaert, W. F. (2001). Study of the role of antimicrobial glucosinolate-derived isothiocyanates in resistance of *Arabidopsis* to microbial pathogens. *Plant Physiology, 125*(4), 1688–1699.

Tong, Y., Gabriel-Neumann, E., Ngwene, B., Krumbein, A., George, E., Platz, S., … Schreiner, M. (2014). Topsoil drying combined with increased sulfur supply leads to enhanced aliphatic glucosinolates in *Brassica juncea* leaves and roots. *Food Chemistry, 152*, 190–196.

Wang, W., Barnaby, J. Y., Tada, Y., Li, H., Tor, M., Caldelari, D., … Dong, X. (2011). Timing of plant immune responses by a central circadian regulator. *Nature, 470*(7332), 110–114.

Wittstock, U., & Burow, M. (2010). Glucosinolate breakdown in *Arabidopsis*: mechanism, regulation and biological significance. Arabidopsis *Book*, e0134.

Wittstock, U., & Gershenzon, J. (2002). Constitutive plant toxins and their role in defense against herbivores and pathogens. *Current Opinion in Plant Biology, 5*, 1–8.

Wittstock, U., Kliebenstein, D. J., Lambrix, V., Reichelt, M., & Gershenzon, J. (2003). Glucosinolate hydrolysis and its impact on generalist and specialist insect herbivores. In T. R. John (Ed.), *Recent advances in phytochemistry* (Vol. 37, pp. 101–125). Elsevier.

Woods, E. C., Hastings, A. P., Turley, N. E., Heard, S. B., & Agrawal, A. A. (2011). Adaptive geographical clines in the growth and defense of a native plant. *Ecological Monographs, 82*(2), 149–168.

Yuan, G., Wang, X., Guo, R., & Wang, Q. (2010). Effect of salt stress on phenolic compounds, glucosinolates, myrosinase and antioxidant activity in radish sprouts. *Food Chemistry, 121*(4), 1014–1019.

Zhang, J., Sun, X., Zhang, Z., Ni, Y., Zhang, Q., Liang, X., … Tokuhisa, J. G. (2011). Metabolite profiling of *Arabidopsis* seedlings in response to exogenous sinalbin and sulfur deficiency. *Phytochemistry*.

Zhao, Z., Zhang, W., Stanley, B. A., & Assmann, S. M. (2008). Functional proteomics of *Arabidopsis thaliana* guard cells uncovers new stomatal signaling pathways[W][OA]. *Plant Cell, 20*(12), 3210–3226.

Züst, T., Joseph, B., Shimizu, K. K., Kliebenstein, D. J., & Turnbull, L. A. (2011). Using knockout mutants to reveal the growth costs of defensive traits. *Proceedings of the Royal Society B, Biological Sciences, 278*(1718), 2598–2603.

CHAPTER THREE

Nonlinear Selection and a Blend of Convergent, Divergent and Parallel Evolution Shapes Natural Variation in Glucosinolates

D.J. Kliebenstein[*,§,1], **N.I. Cacho**[¶]
[*]University of California, Davis, Davis, CA, United States
[§]University of Copenhagen, Frederiksberg, Denmark
[¶]Universidad Nacional Autónoma de México, Mexico City, Mexico
[1]Corresponding author: E-mail: kliebenstein@ucdavis.edu

Contents

Abstract

The molecular mechanisms underlying organismal fitness in complex environments is just beginning to be illuminated. One of the pre-eminent model systems that span the molecular to field fitness chasm is the natural variation in glucosinolate defence metabolites within the Capparales. In this system, there have been extensive surveys of variation in the same phenotypes across a breadth of species and the molecular basis of

Advances in Botanical Research, Volume 80
ISSN 0065-2296
http://dx.doi.org/10.1016/bs.abr.2016.06.002

the major polymorphisms both within and between species is largely known. This is showing an extensive array of evolutionary processes that create and recreate specific polymorphisms and traits that is highly similar both within and between species. Further, early field trials with single gene recreations of natural variation are showing that selection is highly fluctuating both from site to site and from year to year within a location. This review goes into the specific ecological, evolutionary and molecular observations for each of the major loci controlling natural variation in glucosinolates.

1. INTRODUCTION

Natural phenotypic variation within or between species or populations is controlled by complex networks of genes and associated polymorphisms (Falconer & Mackay, 1996; Fisher, 1930; Lynch & Walsh, 1998). There is vastly less clarity about the effect of this variation on the organisms' fitness or on the selective or demographic forces that shape this variation (Bull, 1987; Hedrick, Ginevan, & Ewing, 1976; Levene, 1953; Prasad et al., 2012; Stahl, Dwyer, Mauricio, Kreitman, & Bergelson, 1999). While models based on mutation–selection balance theory and associated demographic evolution establish that selection acts to largely eliminate variation through persistent purifying natural selection, empirical studies often identify higher rates of allelic variation that cannot be explained solely by purifying/directional selection (Adler, Karban, & Strauss, 2001; Barton & Turelli, 2004; Hufford et al., 2012; Kimura, 1968; Korves et al., 2007; Milla, Escudero, & Iriondo, 2009; Mothershead & Marquis, 2000; Pyhajarvi, Hufford, Mezmouk, & Ross-Ibarra, 2013; Tian, Traw, Chen, Kreitman, & Bergelson, 2003; Turelli, 1984; Wentzell, Boeye, Zhang, & Kliebenstein, 2008; Wolf, Howe, & Hambrick, 2000). Thus, a key question remains within ecology and evolution surrounding the extent to which this genetic variation is neutral in origin or maintained through natural selection. And, if shaped by selection, what are the relative roles of purifying/directional selection and nonlinear selective processes such as fluctuating or balancing selection.

A particularly intriguing problem in evolution is to understand the genetic mechanisms at play when different species display the same phenotypic variation. Is this caused by parallel evolution whereby two species, starting from the same beginning genetic material, end with similar phenotypic variation independently [a case of a similar outcome upon 'replaying the tape of life' (Gould, 1990)]? Or is this caused by convergent evolution whereby the organisms attain the same phenotypic polymorphism from different genetic starting material and via different mechanisms? Advances in both high-throughput genotyping and phenotyping are enabling identification of the underlying causal genes necessary to begin answering this question. Common approaches to identify genes and gene networks include a combination of quantitative trait locus (QTL) mapping and genome wide association (GWA) mapping that have been extensively discussed elsewhere in the literature (Hirschhorn & Daly, 2005; Liu, 1998; Lynch & Walsh, 1998; Mauricio, 2001; Nordborg & Weigel, 2008; Weigel & Nordborg, 2005). Typically, the genes underlying particular phenotype(s) are identified in a specific model or crop system. This approach has infrequently extended to a phylogenetic perspective examining the role of selection and genetic mechanisms underlying a common phenotype (or its common variation) across species within a lineage. We argue that undertaking an explicit phylogenetic perspective on the evolution of a phenotype and associated causal mechanisms is necessary to ascertain general patterns of evolutionary mechanisms.

Limitations for undertaking a phylogenetic perspective to examine whether parallelism or convergence underlies the genetic basis of phenotypic variation include that surveys across species tend to be available only for few phenotypes with such variation; that even fewer systems exist for which the causal genes have been cloned to allow an informed comparison of underlying genetic mechanisms, and; that a robust phylogenetic framework might not be available for a given clade. The glucosinolate defence system in the Capparales is one of the few systems in which between and within species variation is being assessed at both the phenotypic and causal genetic levels [(Halkier & Gershenzon, 2006; Kerwin et al., 2015; Sønderby, Geu-Flores, & Halkier, 2010); reviewing the progress in other pathways, such as the anthocyanin pathway, is beyond the scope of this paper]. Glucosinolates show extensive within and between species genetic variation and have been surveyed across a wide array of Capparales (Daxenbichler et al., 1991; Rodman, 1980; Rodman, Kruckeberg, & Alshehbaz, 1981; Rodman, Soltis, Soltis, Sytsma, & Karol, 1998). The genes in the glucosinolate biosynthetic pathway have been nearly completely cloned, which has enabled a deep understanding of the roles of genome and gene duplication in the evolution of the various glucosinolate synthesis pathways (Edger et al., 2015; Kliebenstein, 2008; Sønderby, Geu-Flores, et al., 2010). Thus, the glucosinolate synthesis pathway is emerging as a unique model to study the selective and evolutionary processes underlying natural variation within and between species. In this review, we discuss the major causal loci for glucosinolate variation and how this variation is shaped within and between species. For the purpose of this review, when, from the same genetic starting material, two species (independently of their degree of relatedness) achieve the same phenotype (here variation in glucosinolate production) through the same genetic mechanisms, we use the term parallelism. We refer to all other cases where similar glucosinolate variation is achieved as convergence (that is, when the genetic starting point is different, or when different molecular mechanisms are involved).

2. LOCI

2.1 Species Sampling

The vast majority of studies that report genetic variation in glucosinolate accumulation and structure within the Capparales focus on within species variation in species of the mustard family (Brassicaceae) that are cultivated as vegetables (Gols, Bukovinszky, et al., 2008; Gols, Wagenaar, et al., 2008; Harper et al., 2012; Kabouw, Biere, van der Putten, & van Dam, 2010; Li, Chen, et al., 2014; Magrath et al., 1994; Magrath, Herron, Giamoustaris, & Mithen, 1993; Magrath & Mithen, 1993; Mithen, Clarke, Lister, & Dean, 1995; Mithen, Raybould, & Giamoustairs, 1995; Parkin et al., 1994; Poelman, van Dam, van Loon, Vet, & Dicke, 2009). Species beyond vegetable cultivars that have been sampled for within species variation include *Arabidopsis thaliana* and close relatives (Brachi et al., 2015; Chan, Rowe, & Kliebenstein, 2010; Clauss, Dietel, Schubert, & Mitchell-Olds, 2006; Kazemi-Dinan, Sauer, Stein, Kramer, & Muller, 2015; Kliebenstein, Kroymann, et al., 2001; Windsor et al., 2005), *Boechera stricta* (Manzaneda, Prasad, & Mitchell-Olds, 2010; Prasad et al., 2012; Schranz, Manzaneda, Windsor, Clauss, & Mitchell-Olds, 2009), species of *Streptanthus* and close relatives, including *Caulanthus*, *Guillenia* and *Stanleya* (Cacho, Kliebenstein, & Strauss, 2015; Rodman et al., 1981), and a few other scattered species (Al-Gendy, El-gindi, Hafez, & Ateya, 2010; Berhow et al., 2013; Daxenbichler, Spencer, & Schroeder, 1980; Rodman, 1980). In addition, a few papers have documented between species variation across Capparales (Cacho et al., 2015; Daxenbichler et al., 1991; Spencer & Daxenbichler, 1980). While still limited and not phylogenetically evenly distributed across Brassicaceae (and much less so across Capparales), the sampling to date allows us to begin exploring the evolutionary processes at play in shaping glucosinolate genetic variation. We conducted a survey of the phenotypic consequences of the major polymorphic loci that are known to control glucosinolate structural variation within the Capparales (Fig. 1). Our survey reveals that nearly all the assessed loci are present and polymorphic in the majority of lineages sampled to date (Fig. 1). Below, we discuss each locus and highlight its particular evolutionary trajectories.

2.2 AOP2

The presence or absence of alkenyl glucosinolates both within and across closely related species is one of the most widely spread polymorphisms for aliphatic glucosinolate accumulation in the Capparales (Fig. 1). QTL mapping in *Brassica* and *Arabidopsis* has revealed that this variation is caused by a single locus, even in polyploid species of *Brassica* (Gao, Li, Yang, McCombie, & Quiros, 2004; Kliebenstein, Lambrix, Reichelt, Gershenzon, & Mitchell-Olds, 2001; Kliebenstein, Pedersen, Barker, & Mitchell-Olds, 2002; Kliebenstein, Gershenzon, & Mitchell-Olds, 2001; Li & Quiros, 2003; Magrath et al., 1994; Mithen, Clarke, et al., 1995). The *AOP2* locus of *Arabidopsis* was one of the first QTLs to be cloned in any plant species, and natural variation in gene expression was key to identify the causal gene within the final genomic region (Kliebenstein, Lambrix, et al., 2001). The *AOP2* locus encodes a 2-oxoacid dependent dioxygenase that catalyses the conversion of a methylsulfinyl glucosinolate to an alkenyl glucosinolate (Kliebenstein, Lambrix, et al., 2001). Accessions that have a functional copy of this gene accumulate alkenyl glucosinolates while the accumulation of methylsulfinyl glucosinolates is linked to loss-of-function alleles in *AOP2*.

Data suggest that *AOP2* has undergone a loss of function within *Arabidopsis* at least two times (Chan et al., 2010). One loss-of-function event was via a small frameshift indel that leads to the expression of an AOP2 RNA that is unable to make a functional enzyme, thus leading to the accumulation of the methylsulfinyl glucosinolate precursor (Kliebenstein, Lambrix, et al., 2001). The other loss-of-function allele is generated by a local inversion as described below for the generation of the AOP3 locus (Chan et al., 2010).

Fine-scale mapping also allowed the rapid cloning and confirmation of the *Brassica* orthologue of the *Arabidopsis AOP2* (Gao et al., 2004; Li & Quiros, 2003). Similar to the variation in *Arabidopsis*, a loss of function in *Brassica* was due to a small frameshift indel in *AOP2* that allowed for a similar RNA unable to make a functional enzyme, suggesting parallelism. These data on the presence or absence of alkenyl glucosinolates in both *Brassica* and *Arabidopsis* associated with independent loss-of-function events within

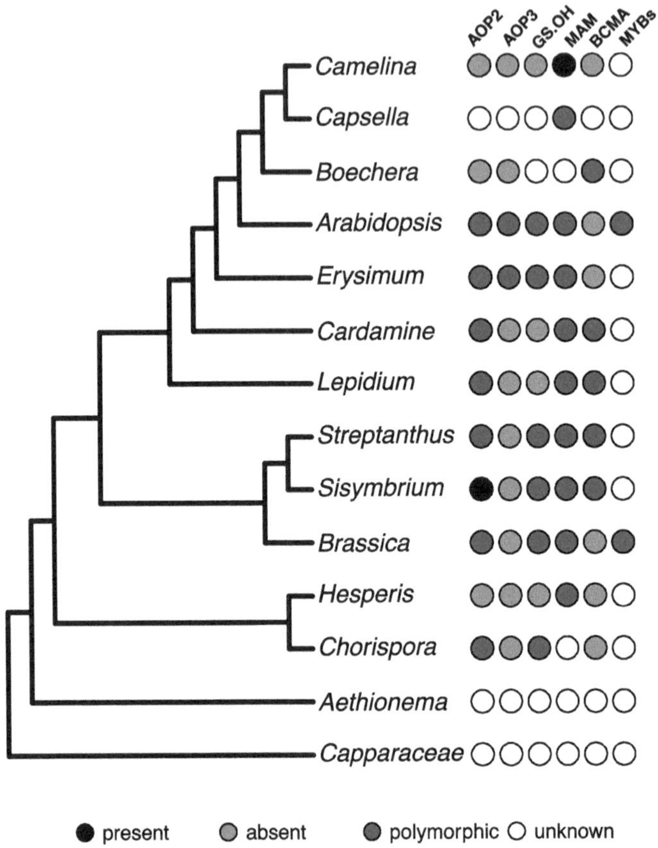

Figure 1 ***Natural variation in glucosinolate loci across lineages of Brassicaceae.*** Shown is a survey of publications looking at within or between species variation of aliphatic glucosinolate structure. The genetic state of each locus was classified as described for each locus in the text. For example, *AOP2* was classified as present if there were alkenyl glucosinolates present within a lineage and as polymorphic if there were lines with and lines without alkenyl glucosinolates either within a single species or across species within each lineage. A locus was called unknown if current information was insufficient to ascertain its status in a given lineage. For example, *GS-OH* status is only possible to call if there are 4C or longer alkenyl glucosinolates as these are the precursors for the enzyme. *The phylogenetic relationships were redrawn from Huang, C. H., Sun, R. R., Hu, Y., Zeng, L. P., Zhang, N., Cai, L. M.,… Ma, H. (2016). Resolution of Brassicaceae phylogeny using nuclear genes uncovers nested radiations and supports convergent morphological evolution.* Molecular Biology and Evolution, 33*(2), 394–412.*

AOP2 suggest that this polymorphism across this family may be caused by repeated loss of functions in the conserved *AOP2* locus (Fig. 1). Testing this hypothesis will require cloning the causal *AOP2* gene for alkenyl glucosinolate production in other lineages of Brassicaceae and, more broadly, of Capparales (Fig. 1).

2.3 AOP3

The production of 3-hydroxypropyl glucosinolate is very limited within the Capparales and so far, it appears to only be present in the *Arabidopsis* and *Erysimum* lineages (Clauss et al., 2006; Daxenbichler et al., 1980; Kliebenstein, Kroymann, et al., 2001) (Fig. 1). In *Arabidopsis*, 3-hydroxypropyl glucosinolate production depends on the AOP3 enzyme, which appears to have arisen via a gene duplication event followed by neo-functionalization of the *AOP2* gene (Chan et al., 2010; Kliebenstein, Lambrix, et al., 2001). It appears that 3-hydroxypropyl glucosinolate is only present in a fraction of the tested genotypes, both for *Arabidopsis* and *Erysimum* (Chan et al., 2010; Clauss et al., 2006; Kliebenstein, Lambrix, et al., 2001; Kliebenstein, Kroymann, et al., 2001). Interestingly, *AOP3* has also undergone a separate duplication event followed by subfunctionalization within *A. thaliana* (Chan et al., 2010). The *AOP3* gene underwent a local tandem triplication that was then followed by a consequent inversion with the tandem *AOP2*. The new *AOP3* copy acquired the *AOP2* promoter creating a localized expression in foliar tissues, while the ancestral *AOP3* copy maintained the original seed-specific expression pattern (Chan et al., 2010). This explains why in *A. thaliana*, 3-hydroxypropyl accumulates in the seeds of all accessions but in the leaves of only about a third of the accessions. A concomitant loss of *AOP2* caused by this inversion shifted the foliar glucosinolate profile from one dominated by alkenyls to one dominated by 3-hydroxypropyl glucosinolate. Current evidence suggests that this foliar specific *AOP3*/*AOP2* shift occurred at least once within the *A. thaliana* lineage, but additional similar shifts cannot be ruled out due to the limited number of accessions that have been examined with sequencing technologies sufficient to find this polymorphism (Chan et al., 2010). Similarly, it remains to be seen if 3-hydroxypropyl production in *Erysimum* is via an *AOP3* orthologue or through a different gene that has evolved the same function independently (Fig. 1).

2.4 Elong

The *Elong* locus controls the production of chain-elongated glucosinolates. *Elong* contains genes that encode methylthioalkylmalate (MAM) enzymes that function with two other enzymes that can cyclically add a carbon to methionine or elongated methionine's (Benderoth, Pfalz, & Kroymann, 2009; Li & Quiros, 2002; de Quiros et al., 2000; Textor et al., 2004; Textor, de Kraker, Hause, Gershenzon, & Tokuhisa, 2007). The length of the chain in aliphatic glucosinolates is a highly variable phenotype in *A. thaliana* and species of *Brassica*; most accessions accumulate glucosinolates with either three carbon (3C) or four carbon (4C) molecules. In *Arabidopsis* and *Brassica,* the production of 3C and 4C glucosinolates is largely controlled by two different MAM genes, *MAM1* and *MAM2* in *Arabidopsis* (Benderoth et al., 2009; Haubold, Kroymann, Ratzka, Mitchell-Olds, & Wiehe, 2002; Li & Quiros, 2002; de Quiros et al., 2000; Textor et al., 2004, 2007). Interestingly, the two MAM genes evolved the ability to stop at either 3C or 4C elongation independently in *Brassica* and *Arabidopsis.* Thus, the 3C MAM genes are different homologues in *Brassica* and *Arabidopsis.* The 4C enzymes also evolved separately from the same source MAM progenitor in the two lineages. In fact, these two independent evolutionary trajectories in *Arabidopsis* and *Brassica* are showcased by the fact that the *Arabidopsis* genes can produce further elongated products up to C7 and C8 while the *Brassica* genes stop at 4C.

In addition to the *Arabidopsis* and *Brassica* polymorphisms, other species contain a wide array of elongation chemistry that is not captured within these model systems. *Camelina* does not accumulate short-chain glucosinolates and instead the MAM enzyme in this species appears to have enabled the production of C10 and C11 glucosinolates (Berhow et al., 2013). In species of *Streptanthus* and close relatives, the polymorphism is tri-allelic, with accessions accumulating C4, C5 or C6 glucosinolates (Cacho et al., 2015). A pattern that emerges from comparison of *Streptanthus* with *Arabidopsis* and *Brassica* is that individual accessions tend to accumulate predominantly one chain length rather than a combination of glucosinolates of different chain lengths. In *Arabidopsis*, this is because the C4 and C3 alleles are unable to recombine with each other, thus behaving as a monogenic Mendelian locus even though there are multiple genes within the *Elong* locus (Haubold et al., 2002; Kroymann & Mitchell-Olds, 2005). Understanding the mechanisms leading to what appears to be a tri-allelic locus in *Streptanthus* and close relatives (Cacho et al., 2015), as well as exploration of chain length in other species might shed additional light on the *Elong* locus and how a diversity of chains in glucosinolates is encoded and manifested.

2.5 GS-OH

Another widely spread polymorphism is the ability to hydroxylate alkenyl glucosinolates (Fig. 1). This maps to a single locus in *Arabidopsis* and *Brassica* that is named *GS-OH* (van Doorn et al., 1998; van Doorn et al., 1999; Kliebenstein et al., 2002; Kliebenstein, Gershenzon, et al., 2001; Parkin et al., 1994; Wentzell et al., 2007). Cloning of *GS-OH* in *Arabidopsis* identified a 2-oxoacid-dependent dioxygenase as the causal gene (Hansen et al., 2008). This enzyme occurs as a single copy in *Arabidopsis* and is unrelated to the AOP2/AOP3 enzymes (Hansen et al., 2008). *Brassica* also contains a single locus polymorphism that controls the hydroxylation of alkenyl glucosinolates but this locus remains uncloned (Gao et al., 2007; Parkin et al., 1994). There is evidence suggesting that, unlike the *AOP* and *Elong* loci, the *GS-OH* locus in *Arabidopsis* and *Brassica* might not belong to a single conserved gene or gene family and are unrelated loci. Biochemically, the hydroxylation of alkenyl glucosinolates can give rise to two enantiomers, and there is variation across the phylogeny on which enantiomers are produced. *Brassica napus* makes only the 2*R* enantiomer, and *Crambe abyssinica* makes only the 2*S* enantiomer (Daubos et al., 1998; van Doorn et al., 1998, 1999). In contrast, the *GS-OH* from *Arabidopsis* leads to the production of both 2*S* and 2*R* enantiomers (Hansen et al., 2008). Variation in the stereochemistry of an enzyme suggests that each of these species likely utilizes a different enzyme to catalyse the same reaction (a case of convergence). Thus, it appears that the stereochemistry is unique to each of the three lineages in which the *GS-OH* polymorphism is present. Because changes in enzymatic stereochemistry are unlikely, this pattern in *GS-OH* suggests a more likely model of independent convergent evolution of the hydroxylation of alkenyl glucosinolates in each lineage (Fig. 1). Testing this hypothesis will require cloning the *GS-OH* locus from each of these lineages.

2.6 BCMA

The *BCMA* locus, which controls the production of branch-chain aliphatic glucosinolates, was largely overlooked until investigations on glucosinolate variation within *B. stricta* identified a single large effect polymorphism for the presence or absence of these glucosinolates (Prasad et al., 2012; Schranz et al., 2009). In *B. stricta*, the *BCMA* locus contains two genes, *BCMA2* and *BCMA3*, both of which encode CYP79F enzymes that convert amino acids into oximes in the first step of glucosinolate biosynthesis (Kahn, Fahrendorf, Halkier, & Moller, 1999). BCMA3 catalyses the conversion of valine to enable the production of branch-chain glucosinolates while BCMA2 catalyses the conversion of methionine. The *BCMA* locus changes the level of *BCMA3* relative to *BCMA2*, which enables the plant to genetically shift between the two glucosinolate profiles within *B. stricta*. Importantly, it has been shown that there are fitness effects of *BCMA* locus in the field mediated by its influence on plant chemistry and insect resistance (Lee, Anderson, & Mitchell-Olds, 2014; Prasad et al., 2012; Schranz et al., 2009). Using the information on the phenotypic function of variation at the *BCMA* locus, specifically a distinct shift from methionine- to valine-derived glucosinolate production, we reanalysed existing data and found evidence that the *BCMA* locus may be present in a number of species (Fig. 1). Interestingly, in species with more extensive sampling, as in some species of the *Streptanthus* clade, there is evidence that the *BCMA* locus is polymorphic both within and between species (Cacho et al., 2015). Also, *BCMA* is not present or polymorphic within the *Brassica* or *Arabidopsis* lineages, suggesting that these lineages may have lost this locus. Future work is required to identify and clone this locus in lineages other than *Boechera* to assess if this variation represents evolutionary parallelism or convergence on the same phenotype due to selection on this locus.

2.7 MYBs and Other Transcription Factors

Because the genetic loci described above alter the structure of the accumulating glucosinolates within a given sample, they are relatively easy to identify and characterize phenotypically (Magrath et al., 1993). In contrast, regulatory variation, which solely alters the accumulation of groups of glucosinolates and has not been linked to altering their structure, is harder to identify. Thus, the characterization of the genetic variation in aliphatic glucosinolate regulation depended on first identifying key MYB transcription factors responsible for regulating the aliphatic glucosinolate pathway, which was achieved through standard forward genetics approaches (Beekwilder et al., 2008; Gigolashvili, Berger, et al., 2007; Gigolashvili, Engqvist, Yatusevich, Muller, & Flugge, 2008; Gigolashvili, Yatusevich, Berger, Müller, & Flügge, 2007; Hirai et al., 2007; Sønderby, Burow, Rowe, Kliebenstein, & Halkier, 2010; Sønderby et al., 2007). Once MYB28, 29 and76 were identified as key aliphatic glucosinolate regulators, it was possible to reanalyse existing QTL and GWA data sets and show that these transcription factors underlie and are likely responsible for significant natural genetic variation within *A. thaliana* aliphatic glucosinolate content (Brachi et al., 2015; Chan et al., 2010; Sønderby, Burow, et al., 2010; Sønderby et al., 2007; Wentzell et al., 2007). In addition to *Arabidopsis*, natural variation within this MYB family also colocates with QTLs controlling genetic variation in aliphatic glucosinolates within species of *Brassica* (Bisht et al., 2009; Brown et al., 2015; Traka et al., 2013; Xu et al., 2015). Moreover, MYBs have been shown to have presence/absence variation within *B. napus* and other species of *Brassica* (Harper et al., 2012), which could suggest a potential causal role with respect to intraspecific aliphatic glucosinolate across *Brassica*. Further studies are needed to ascertain if these MYBs and their associated homologues are involved in controlling variation within and across other lineages in the Capparales.

In addition to the MYBs, there are a large number of transcription factors that have been linked to glucosinolate regulation (Levy, Wang, Kaspi, Parrella, & Abel, 2005; Li, Gaudinier, et al., 2014; Schweizer et al., 2013; Zhang, Li, Huai, Zhou, & Kliebenstein, 2015). These include a wide range of transcription factor families, from bHLHs to homeobox domain proteins to NACs (NAM, ATAF1 and −2, and CUC2) and beyond. These transcription factor families have, however, not been investigated as potentially controlling natural variation in aliphatic glucosinolates and remain as possible candidates for contributing to variation in uncharacterized loci.

2.8 Remaining Polygenicity for Aliphatic Glucosinolates

The *AOP2, AOP3, Elong, GS-OH* and *BCMA* loci (described in Sections 2.2−2.6) control the major detectable QTLs and the dominant fraction of measurable natural variation present in aliphatic glucosinolates in the species assessed to date. This has led to suggestions that these loci are all that is required to fully interpret natural variation in the glucosinolate defence system (Brachi et al., 2015). In support of such argument, it has been shown that it is possible to create an algebraic model using these loci alone that can explain >80% of the genetic variation observed in aliphatic glucosinolates (Brachi et al., 2015; Chan et al., 2010; Hageman, Kliebenstein, & Churchill, 2012; Wentzell et al., 2007). However, other studies have shown that natural selection on aliphatic glucosinolates can discriminate between lines that differ by 5% or less in glucosinolate content (Bidart-Bouzat & Kliebenstein, 2008; Kerwin et al., 2015; Lankau, 2007; Lankau & Kliebenstein, 2009; Lankau & Strauss, 2007, 2008; Mauricio, 1998; Mauricio & Rausher, 1997). By such argument, the ~20% of variation left unexplained by the algebraic argument raises as quite significant, and also it is implied that a genetic model would have to explain nearly 100% of aliphatic glucosinolate variation before it would be a reasonable predictor of how selective forces shape natural variation in glucosinolates.

Thus, the question becomes, if a six locus model can explain the vast majority of glucosinolate variation (~80%), how many additional loci would be required to achieve 100% of variation explained by such model? An answer to this question is beginning to come into focus from several GWA studies across accessions of *A. thaliana* (Brachi et al., 2015; Chan, Rowe, Corwin, Joseph, & Kliebenstein, 2011; Chan et al., 2010). These studies found most of the expected major causal loci, but they also showed that there was a vast set of additional genes with significant associations with aliphatic glucosinolate variation. Using insertional mutants, it has been shown that over 70% of such genes has a significant effect on glucosinolate content (Chan et al., 2011). And this was true even when using a moderate significance threshold to call potential candidate genes (Chan et al., 2011). Taken altogether, these studies suggest that there are potentially hundreds of genes required to create a causal model that fully explains natural variation in aliphatic glucosinolate accumulation.

2.9 Indolic Glucosinolates

The genes underlying qualitative and quantitative variation in tryptophan-derived indole glucosinolates have received much less attention. While QTL and GWA studies have identified a large number of loci that correlate with the accumulation of indolic glucosinolates, the causal genes are largely not yet understood (Kliebenstein, Kroymann, et al., 2001; Pfalz, Vogel, Mitchell-Olds, & Kroymann, 2007; Wentzell et al., 2007). One QTL locus, the *Indole Glucosinolate Modifier1* (*IGM1*) has been identified in relation to the accumulation of indole glucosinolates modified at the four carbon in *A. thaliana* (Pfalz, Vogel, & Kroymann, 2009). It has been shown that a single gene, *CYP81F2*, underlies the *IGM1* QTL. The enzyme encoded by *CYP81F2*, a cytochrome P450 monooxygenase, catalyses the conversion of I3M glucosinolate to 4OH-I3M glucosinolate and contains *cis*-expression polymorphism within *Arabidopsis* (Pfalz et al., 2009). The *IGM1* locus also contributes to defence against generalist aphids in the lab (but apparently not to lepidopterans) as well as to the formation of key non-host pathogen resistance mechanisms within *A. thaliana* (Bednarek et al., 2009; Clay, Adio, Denoux, Jander, & Ausubel, 2009; Kim & Jander, 2007; Mueller et al., 2010; Pfalz et al., 2009). That *IGM1* is implicated in non-host resistance posits a conundrum because non-host resistance is largely considered to be a core invariable function in plants (Jones & Dangl, 2006). Yet, for this one *A. thaliana* non-host resistance mechanism, the core of the pathway displays presence or absence variation, suggesting that non-host resistance is not shaped exclusively by core conserved mechanisms but potentially involves a net of overlapping mechanisms that while being individually polymorphic, could, in sum, present an image of conservation. While there is enticing evidence in other species for variation in this indolic structure (Cacho et al., 2015), whether this locus controls natural variation within indole glucosinolates in other species remains to be addressed.

2.10 Glucosinolate Activation Loci

A key aspect of the glucosinolate function, one with great ecological and evolutionary implications in biotic interactions, is their ability to be activated to create a suite of different activation products. A typical aliphatic glucosinolate can be activated to make the associated nitrile, isothiocyanate, thiocyanate or epithionitrile compounds, each with different biological activities (Burow & Wittstock, 2009). This differential activation had largely been thought to be driven by non-enzymatic properties such as pH or free-iron but has now been shown to be driven by the presence or absence of a family of enzymes, nitrile specifier proteins, that can drive the formation of nitriles, thiocyanates or epithionitriles (Bones & Rossiter, 2006; Burow et al., 2009; Burow & Wittstock, 2009; Burow et al., 2008; Foo et al., 2000; Lambrix, Reichelt, Mitchell-Olds, Kliebenstein, & Gershenzon, 2001; Mumm et al., 2008; Voelckel et al., 2010; Zabala et al., 2005). The absence of one of these proteins leads to the default isothiocyanate pathway (Burow & Wittstock, 2009). In *Arabidopsis* and *Brassica* foliar tissue, the presence or absence of the epithiospecifier protein encoded by the *ESP* locus is the dominant determinant of whether that plant makes nitriles or isothiocyanates upon tissue disruption (Jeffery et al., 2003; Lambrix et al., 2001; Matusheski et al., 2006). This polymorphism changes the resistance to various lepidopteran herbivores in the lab and unknown herbivores in the field (Jander, Cui, Nhan, Pierce, & Ausubel, 2001; Kerwin et al., 2015; Kliebenstein et al., 2002; Kliebenstein, Gershenzon, et al., 2001). In addition to the dominant *ESP* locus, there are additional polymorphisms within the other nitrile specifier proteins and associated proteins (Agee et al., 2010; Wentzell et al., 2008). Altogether, variation in the presence/absence and associated polymorphisms lead to a complex system in which the formation of nitriles versus isothiocyanates in *A. thaliana* is dependent on the interaction of genetic variation with the tissue being tested, the age of the plant being measured, and the biotic environment in which the plant is grown (Wentzell et al., 2008; Wentzell & Kliebenstein, 2008). This complexity and the role of environment surrounding whether nitriles or isothiocynates are produced contrasts to the glucosinolate synthesis structural loci described above, which are largely insensitive to the environment and age of the plant. This suggests that the genetic variation underlying glucosinolate activation could be plastic to the environment, and such plasticity may allow for fine-tuning of the glucosinolate structural variation to optimize fitness within a particular environment. Glucosinolate activation has not yet been broadly assayed outside of *Brassica*, *Arabidopsis* and New Zealand *Pachycladon* species, which all contain variation in *ESP*, limiting our ability to make broader conclusions that span the Capparales, or even the Brassicaceae lineage (Voelckel et al., 2010).

3. EVOLUTIONARY IMPLICATIONS

3.1 Evolutionary Mechanisms

The glucosinolate synthesis structural loci seem to follow a pattern of repeated evolution whereby the same phenotype arises reiteratively across the phylogeny. In the case of aliphatic glucosinolates, the evolution of similar phenotypes seems to be via a diversity of processes. The natural variation in alkenyl glucosinolates is through the parallel generation of nonfunctional alleles within the conserved *AOP2* locus. Loss-of-function variation is also present in MYBs within both the *Arabidopsis* and *Brassica* lineages. There is also the potential for convergent evolution as exemplified by the *GS-OH* locus where it is highly likely that three lineages evolved the production of the ability to hydroxylate alkenyl glucosinolates via co-opting different enzymes. The *Elong* locus is a complex combination of divergent and convergent evolution. This locus has a complex pattern whereby a gene family of MAMs underwent repeated duplication events and ensuing neofunctionalization resulting in variation in the number of carbon additions within the elongation cycle. This leads to a situation where the same length glucosinolates in *Brassica* and *Arabidopsis* are produced by MAM genes which evolved independently and thus are not orthologues.

Most importantly, all of the above polymorphisms and their associated evolution/variation exist both between and within species across nearly the entire Brassicaceae, and potentially Capparales, suggesting that there is selection for glucosinolate diversity associated with the *AOP/MAM/*

BCMA loci. Yet, that these genes are not orthologues conserved across the evolutionary history of Brassicaceae, suggests that eco-evolutionary forces oppose their survival through speciation events. And that subsequent processes allow for diversity to evolve (e.g., through gene duplication) and selection acts to preserve similar phenotypic variants throughout the phylogeny via either independent loss-of-function events or convergent evolution of chemical properties. Thus, the major evolutionary pressures on (at least) aliphatic glucosinolates are not lineage specific nor directional. An alternative model, one defining evolutionary pressures on glucosinolate synthesis as nonlinear and similar across Brassicaceae (or Capparales), emerges. Such a model is supported by glucosinolate loci frequently showing patterns of balancing selection within and between various plant species (Bakker, Traw, Toomajian, Kreitman, & Bergelson, 2008; Benderoth et al., 2006; Brachi et al., 2015; Wright, Lauga, & Charlesworth, 2002).

3.2 Field Trials and Selective Pressures

While genome sequencing and quantitative trait mapping can help unveil some aspects of the evolution of the aliphatic glucosinolate pathway, these methods cannot reveal the pressure or selective regime driving the observed genetic variation. Understanding the ecological and evolutionary forces enabling the generation and maintenance of variation in the glucosinolate pathways (whether aliphatic, indolic or aromatic) requires direct field trials, both in controlled conditions and in more natural settings. The null model to be tested would be that observed polymorphisms are evolving due to stochastic demographic processes of neutral loci, i.e., they have no effect on fitness or any other adaptive value. To date, there is vast evidence to reject such null model: studies have identified fitness effects of glucosinolate variation in both laboratory and field trials (Bidart-Bouzat & Kliebenstein, 2008; Brachi et al., 2015; Kerwin et al., 2015; Kroymann & Mitchell-Olds, 2005; Lankau, 2007; Lankau & Kliebenstein, 2009; Lankau & Strauss, 2007, 2008; Mauricio & Rausher, 1997; Mauricio, Rausher, & Burdick, 1997; Prasad et al., 2012; Züst et al., 2012). Moreover, evidence points at glucosinolate variation being non-neutral and having fitness effects. Thus, the question becomes what type(s) of selection, or more comprehensively, what is the selection landscape shaping the evolution of glucosinolate variation?

Studies to date suggest that in field conditions, selection operating on glucosinolate variation is largely nonlinear and shaped by the changing complexity of the environments occupied by plants. Balancing selective pressures have been observed on the *MAM* locus and likely the *MYB* locus. The *MAM* locus controls differential sensitivity to two different aphids, with the 3C and 4C alleles of this locus being optimal for one but not the other aphid species (Züst et al., 2012). These alleles show a geographic distribution across Europe that agrees with the presence or absence of the respective aphid. Critically, in central Europe, where the two alleles and the two aphids co-occur, there is a pattern suggestive of balancing or fluctuating selection occurring on the *MAM* locus, where alleles occurrence track the relative abundance of the two aphid species (Kroymann & Mitchell-Olds, 2005; Züst et al., 2012). Similarly, a study using field trials to assess the segregation of the *BCMA* locus in two different sites found a significant *BCMA* × site interaction, suggesting local adaptation at the *BCMA* locus (Prasad et al., 2012). Evidence for variation in the biotic environment driving balancing selection was found in *Brassica*, where the production of glucosinolates and associated fitness effects were genetically polymorphic and responded to variation in aphids and lepidopterans (Lankau, 2007; Lankau & Kliebenstein, 2009; Lankau & Strauss, 2007, 2008). In addition to insects, the local plant community can impose selection on glucosinolate variation. A study examining the effects of glucosinolate content on competitive ability and community composition using *Brassica* lines of high and low glucosinolate production found that the high glucosinolate *Brassica* lines outcompeted other plant species, the low glucosinolate *Brassicas* outcompeted the high glucosinolate ones (suggesting tradeoffs in defence and competitive ability) and other plant species outcompeted low glucosinolate *Brassica* lines (Lankau & Strauss, 2007). This established a setting where genetic variation in glucosinolate content was maintained by competition both within and between species.

That variation in glucosinolate profiles differential affects fitness in various biotic environments suggests that the glucosinolate pathway may be continuously being shaped by highly fluctuating selective pressures that rapidly shift across the landscape at a given time but also across time within a single location. Both local insect and plant populations are dynamic, and demographic fluctuations can be quite dramatic (for example, in years marked by drastic differences in temperature and precipitation). A direct test of this fluctuating selection concept utilized lines that represented natural glucosinolate variation within *Arabidopsis* via single gene manipulations in addition to their combinations. Planting these defined genetic variants in multiple sites showed that the major selective pressure was fluctuations both between sites and between years at the same site

(Kerwin et al., 2015). These fluctuations of the fitness value for individual loci were not small and could reach nearly a 200% difference in fitness. Further, no allele was optimal across all field trials, suggesting that fluctuating selection, at least within *Arabidopsis*, could drive the formation and the maintenance of genetic variation in aliphatic glucosinolates (Kerwin et al., 2015). Given that most Brassicaceae are attacked by the same or similar generalist and specialist herbivores and pathogens, including lepidopterans, aphids, fungi and other organisms, it is possible that there is some constancy in the selective pressure imposed by their enemies that different species in this family are experiencing within a given environment. Such constancy could explain that the same loci are showing blends of parallel, convergent and divergent evolution across Brassicaceae that gives rise to similar phenotypic diversity (Fig. 1).

4. FUTURE PERSPECTIVES FOR UNDERSTANDING GLUCOSINOLATE VARIATION

Significant progress has been made in the last decade or two in making the glucosinolate pathway a model system in ecology and evolution to identify causal genes underlying natural variation and testing the fitness consequences of variation at such loci under field conditions. This has led to the identification of nearly all the major causal genes in a couple of lineages (Fig. 1) and direct assessment of their field consequences in limited field trials. This, however, represents only the first dabble into the true complexity of this system. To fully understand the evolution of glucosinolates in the Capparales will require the cloning and validation of the major causal genes in one or more species within each major lineage across the order. Only then we will be able to ascertain the relative contribution of parallel loss-of-function events versus independent evolution of new genes for the same enzymatic process in shaping the evolution of the major loci underlying glucosinolate diversity. Also, because fluctuating selection could be playing a key role in shaping the evolution of the glucosinolate pathway, long-term field trials with single gene manipulants at each of the causal genes of interest might shed valuable insights on how complexity in the glucosinolate pathway originates and is maintained. Further, this would be optimal if conducted in multiple species that co-exist within the same environment. Recent advances in genome engineering suggest such an approach might be feasible in the near future. One could thus, begin to envision decade-long field trials across the native range of wild species (e.g., species of

Streptanthus across the Sierra Nevada), where glucosinolate polymorphisms evolved naturally, using lines for which glucosinolate polymorphisms were re-engineered. Insights from such long-term experiments would allow us to understand and characterize the role of fluctuating selective regimes (and their main players – competitors, herbivores, pathogens, etc.) in shaping variation in glucosinolates.

ACKNOWLEDGEMENTS

Funding for this work was provided by the NSF awards IOS 1339125 and MCB 1330337 to DJK, the USDA National Institute of Food and Agriculture, Hatch project number CA-D-PLS-7033-H to DJK, the Danish National Research Foundation (DNRF99) grant to DJK and a PAPIIT-UNAM grant (IA201516) to NIC.

REFERENCES

Adler, L. S., Karban, R., & Strauss, S. Y. (2001). Direct and indirect effects of alkaloids on plant fitness via herbivory and pollination. *Ecology, 82*(7), 2032–2044.

Agee, A. E., Surpin, M., Sohn, E. J., Girke, T., Rosado, A., Kram, B. W., ... Raikhel, N. V. (2010). MODIFIED VACUOLE PHENOTYPE1 is an *Arabidopsis* myrosinase-associated protein involved in endomembrane protein trafficking. *Plant Physiology, 152*(1), 120–132.

Al-Gendy, A. A., El-gindi, O. D., Hafez, A. S., & Ateya, A. M. (2010). Glucosinolates, volatile constituents and biological activities of *Erysimum corinthium Boiss.* (Brassicaceae). *Food Chemistry, 118*(3), 519–524.

Bakker, E. G., Traw, M. B., Toomajian, C., Kreitman, M., & Bergelson, J. (2008). Low levels of polymorphism in genes that control the activation of defense response in *Arabidopsis thaliana. Genetics, 178*(4), 2031–2043.

Barton, N. H., & Turelli, M. (2004). Effects of genetic drift on variance components under a general model of epistasis. *Evolution: International Journal of Organic Evolution, 58*(10), 2111–2132.

Bednarek, P., Pislewska-Bednarek, M., Svatos, A., Schneider, B., Doubsky, J., Mansurova, M., ... Schulze-Lefert, P. (2009). A glucosinolate metabolism pathway in living plant cells mediates broad-spectrum antifungal defense. *Science, 323*(5910), 101–106.

Beekwilder, J., van Leeuwen, W., van Dam, N. M., Bertossi, M., Grandi, V., Mizzi, L., ... Bovy, A. (2008). The impact of the absence of aliphatic glucosinolates on insect herbivory in *Arabidopsis. PLoS One, 3*(4), e2068.

Benderoth, M., Pfalz, M., & Kroymann, J. (2009). Methylthioalkylmalate synthases: genetics, ecology & evolution. *Phytochemistry Reviews, 8*(1), 255–268.

Benderoth, M., Textor, S., Windsor, A. J., Mitchell-Olds, T., Gershenzon, J., & Kroymann, J. (2006). Positive selection driving diversification in plant secondary metabolism. *Proceedings of the National Academy of Sciences of the United States of America, 103*(24), 9118–9123.

Berhow, M. A., Polat, U., Glinski, J. A., Glensk, M., Vaughn, S. F., Isbell, T., ... Gardner, C. (2013). Optimized analysis and quantification of glucosinolates from *Camelina sativa* seeds by reverse-phase liquid chromatography. *Industrial Crops and Products, 43*, 119–125.

Bidart-Bouzat, M. G., & Kliebenstein, D. J. (2008). Differential levels of insect herbivory in the field associated with genotypic variation in glucosinolates in *Arabidopsis thaliana. Journal of Chemical Ecology, 34*(8), 1026–1037.

Bisht, N. C., Gupta, V., Ramchiary, N., Sodhi, Y. S., Mukhopadhyay, A., Arumugam, N., … Pradhan, A. K. (2009). Fine mapping of loci involved with glucosinolate biosynthesis in oilseed mustard (*Brassica juncea*) using genomic information from allied species. *TAG. Theoretical and Applied Genetics. Theoretische und angewandte Genetik, 118*(3), 413–421.

Bones, A. M., & Rossiter, J. T. (2006). The enzymic and chemically induced decomposition of glucosinolates. *Phytochemistry, 67*(11), 1053–1067.

Brachi, B., Meyer, C. G., Villoutreix, R., Platt, A., Morton, T. C., Roux, F., & Bergelson, J. (2015). Coselected genes determine adaptive variation in herbivore resistance throughout the native range of *Arabidopsis thaliana. Proceedings of the National Academy of Sciences of the United States of America, 112*(13), 4032–4037.

Brown, A. F., Yousef, G. G., Reid, R. W., Chebrolu, K. K., Thomas, A., Krueger, C., … Juvik, J. A. (2015). Genetic analysis of glucosinolate variability in broccoli florets using genome-anchored single nucleotide polymorphisms. *TAG. Theoretical and Applied Genetics. Theoretische und angewandte Genetik, 128*(7), 1431–1447.

Bull, J. J. (1987). Evolution of phenotypic variance. *Evolution: International Journal of Organic Evolution, 41*(2), 303–315.

Burow, M., Losansky, A., Muller, R., Plock, A., Kliebenstein, D. J., & Wittstock, U. (2009). The genetic basis of constitutive and herbivore-induced ESP-independent nitrile formation in *Arabidopsis. Plant Physiology, 149*(1), 561–574.

Burow, M., & Wittstock, U. (2009). Regulation and function of specifier proteins in plants. *Phytochemistry Reviews, 8*, 87–99.

Burow, M., Zhang, Z. Y., Ober, J. A., Lambrix, V. M., Wittstock, U., Gershenzon, J., & Kliebenstein, D. J. (2008). ESP and ESM1 mediate indol-3-acetonitrile production from indol-3-ylmethyl glucosinolate in *Arabidopsis. Phytochemistry, 69*(3), 663–671.

Cacho, N. I., Kliebenstein, D. J., & Strauss, S. Y. (2015). Macroevolutionary patterns of glucosinolate defense and tests of defense-escalation and resource availability hypotheses. *New Phytologist, 208*(3), 915–927.

Chan, E. K., Rowe, H. C., Corwin, J. A., Joseph, B., & Kliebenstein, D. J. (2011). Combining genome-wide association mapping and transcriptional networks to identify novel genes controlling glucosinolates in *Arabidopsis thaliana. PLoS Biology, 9*(8), e1001125.

Chan, E. K. F., Rowe, H. C., & Kliebenstein, D. J. (2010). Understanding the evolution of defense metabolites in *Arabidopsis thaliana* using genome-wide association mapping. *Genetics, 185*(3), 991–1007.

Clauss, M. J., Dietel, S., Schubert, G., & Mitchell-Olds, T. (2006). Glucosinolate and trichome defenses in a natural *Arabidopsis lyrata* population. *Journal of Chemical Ecology, 32*(11), 2351–2373.

Clay, N. K., Adio, A. M., Denoux, C., Jander, G., & Ausubel, F. M. (2009). Glucosinolate metabolites required for an *Arabidopsis* innate immune response. *Science, 323*(5910), 95–101.

Daubos, P., Grumel, V., Iori, R., Leoni, O., Palmieri, S., & Rollin, P. (1998). *Crambe abyssinica* meal as starting material for the production of enantiomerically pure fine chemicals. *Industrial Crops and Products, 7*(2–3), 187–193.

Daxenbichler, M. E., Spencer, G. F., Carlson, D. G., Rose, G. B., Brinkler, A. M., & Powell, R. G. (1991). Glucosinolate composition of seeds from 297 species of wild plants. *Phytochemistry, 30*, 2623–2638.

Daxenbichler, M. E., Spencer, G. F., & Schroeder, W. P. (1980). 3-hydroxypropylglucosinolate, a new glucosinolate in seeds of erysimum-hieracifolium and malcolmia-maritima. *Phytochemistry, 19*(5), 813–815.

van Doorn, H. E., van der Kruk, G. C., van Holst, G. J., Raaijmakers-Ruijs, N., Postma, E., Groeneweg, B., & Jongen, W. H. F. (1998). The glucosinolates sinigrin and progoitrin

are important determinants for taste preference and bitterness of Brussels sprouts. *Journal of the Science of Food and Agriculture, 78*(1), 30–38.
van Doorn, J. E., van der Kruk, G. C., van Holst, G. J., Schoofs, M., Broer, J. B., & de Nijs, J. J. M. (1999). Quantitative inheritance of the progoitrin and sinigrin content in Brussels sprouts. *Euphytica, 108*(1), 41–52.
Edger, P. P., Heidel-Fischer, H. M., Bekaert, M., Rota, J., Gloeckner, G., Platts, A. E., … Wheat, C. W. (2015). The butterfly plant arms-race escalated by gene and genome duplications. *Proceedings of the National Academy of Sciences of the United States of America, 112*(27), 8362–8366.
Falconer, D. S., & Mackay, T. F. C. (1996). *Introduction to quantitative genetics*. Longman, Harlow: Essex.
Fisher, R. A. (1930). *The genetical theory of natural selection*. Oxford: Clarendon.
Foo, H. L., Gronning, L. M., Goodenough, L., Bones, A. M., Danielsen, B.-E., Whiting, D. A., & Rossiter, J. T. (2000). Purification and characterisation of epithiospecifier protein from *Brassica napus*: enzymic intramolecular sulphur addition within alkenyl thiohydroximates derived from alkenyl glucosinolate hydrolysis. *FEBS Letters, 468*(2–3), 243–246.
Gao, M. Q., Li, G. Y., Yang, B., McCombie, W. R., & Quiros, C. F. (2004). Comparative analysis of a Brassica BAC clone containing several major aliphatic glucosinolate genes with its corresponding *Arabidopsis* sequence. *Genome, 47*(4), 666–679.
Gao, M. Q., Li, G. Y., Yang, B., Qiu, D., Farnham, M., & Quiros, C. (2007). High-density *Brassica oleracea* linkage map: identification of useful new linkages. *TAG. Theoretical and Applied Genetics. Theoretische und angewandte Genetik, 115*(2), 277–287.
Gigolashvili, T., Berger, B., Mock, H. P., Müller, C., Weisshaar, B., & Flügge, U. I. (2007). The transcription factor HIG1/MYB51 regulates indolic glucosinolate biosynthesis in *Arabidopsis thaliana. Plant Journal, 50*(5), 886–901.
Gigolashvili, T., Engqvist, M., Yatusevich, R., Muller, C., & Flugge, U. I. (2008). HAG2/MYB76 and HAG3/MYB29 exert a specific and coordinated control on the regulation of aliphatic glucosinolate biosynthesis in *Arabidopsis thaliana. New Phytologist, 177*(3), 627–642.
Gigolashvili, T., Yatusevich, R., Berger, B., Müller, C., & Flügge, U. I. (2007). The R2R3-MYB transcription factor HAG1/MYB28 is a regulator of methionine-derived glucosinolate biosynthesis in *Arabidopsis thaliana. Plant Journal, 51*(2), 247–261.
Gols, R., Bukovinszky, T., van Dam, N. M., Dicke, M., Bullock, J. M., & Harvey, J. A. (2008). Performance of generalist and specialist herbivores and their endoparasitoids differs on cultivated and wild *Brassica* populations. *Journal of Chemical Ecology, 34*(2), 132–143.
Gols, R., Wagenaar, R., Bukovinszky, T., van Dam, N. M., Dicke, M., Bullock, J. M., & Harvey, J. A. (2008). Genetic variation in defense chemistry in wild cabbages affects herbivores and their endoparasitoids. *Ecology, 89*(6), 1616–1626.
Gould, S. (1990). *Wonderful life: The burgess shale and the nature of history*. New York: Norton and Company.
Hageman, R. B., Kliebenstein, D. J., & Churchill, G. A. (2012). What can causal networks tell us about metabolic pathways? *PLoS Computational Biology, 8*(4).
Halkier, B. A., & Gershenzon, J. (2006). Biology and biochemistry of glucosinolates. *Annual Review of Plant Biology, 57*, 303–333.
Hansen, B. G., Kerwin, R. E., Ober, J. A., Lambrix, V. M., Mitchell-Olds, T., Gershenzon, J., … Kliebenstein, D. J. (2008). A novel 2-oxoacid-dependent dioxygenase involved in the formation of the goiterogenic 2-hydroxybut-3-enyl glucosinolate and generalist insect resistance in *Arabidopsis. Plant Physiology, 148*(4), 2096–2108.
Harper, A. L., Trick, M., Higgins, J., Fraser, F., Clissold, L., Wells, R., … Bancroft, I. (2012). Associative transcriptomics of traits in the polyploid crop species *Brassica napus. Nature Biotechnology, 30*(8), 798–802.

Haubold, B., Kroymann, J., Ratzka, A., Mitchell-Olds, T., & Wiehe, T. (2002). Recombination and gene conversion in a 170-kb genomic region of *Arabidopsis thaliana*. *Genetics, 161*(3), 1269–1278.

Hedrick, P. W., Ginevan, M. E., & Ewing, E. P. (1976). Genetic polymorphism in heterogenous environments. *Annual Review of Ecology and Systematics, 7*, 1–32.

Hirai, M., Sugiyama, K., Sawada, Y., Tohge, T., Obayashi, T., Suzuki, A., ... Saito, K. (2007). Omics-based identification of *Arabidopsis* Myb transcription factors regulating aliphatic glucosinolate biosynthesis. *Proceedings of the National Academy of Sciences of the United States of America, 104*, 6478–6483.

Hirschhorn, J. N., & Daly, M. J. (2005). Genome-wide association studies for common diseases and complex traits. *Nature Reviews Genetics, 6*(2), 95–108.

Huang, C. H., Sun, R. R., Hu, Y., Zeng, L. P., Zhang, N., Cai, L. M., ... Ma, H. (2016). Resolution of Brassicaceae phylogeny using nuclear genes uncovers nested radiations and supports convergent morphological evolution. *Molecular Biology and Evolution, 33*(2), 394–412.

Hufford, M. B., Xu, X., van Heerwaarden, J., Pyhajarvi, T., Chia, J. M., Cartwright, R. A., ... Ross-Ibarra, J. (2012). Comparative population genomics of maize domestication and improvement. *Nature Genetics, 44*(7), 808–811.

Jander, G., Cui, J., Nhan, B., Pierce, N. E., & Ausubel, F. M. (2001). The TASTY locus on chromosome 1 of *Arabidopsis* affects feeding of the insect herbivore *Trichoplusia ni*. *Plant Physiology, 126*(2), 890–898.

Jeffery, E. H., Brown, A. F., Kurilich, A. C., Keck, A. S., Matusheski, N., Klein, B. P., & Juvik, J. A. (2003). Variation in content of bioactive components in broccoli. *Journal of Food Composition and Analysis, 16*(3), 323–330.

Jones, J. D. G., & Dangl, J. L. (2006). The plant immune system. *Nature, 444*(7117), 323–329.

Kabouw, P., Biere, A., van der Putten, W. H., & van Dam, N. M. (2010). Intra-specific differences in root and shoot glucosinolate profiles among white cabbage (*Brassica oleracea* var. *capitata*) cultivars. *Journal of Agricultural and Food Chemistry, 58*(1), 411–417.

Kahn, R. A., Fahrendorf, T., Halkier, B. A., & Moller, B. L. (1999). Substrate specificity of the cytochrome P450 enzymes CYP79A1 and CYP71E1 involved in the biosynthesis of the cyanogenic glucoside dhurrin in *Sorghum bicolor* (L.) *Moench*. *Archives of Biochemistry and Biophysics, 363*(1), 9–18.

Kazemi-Dinan, A., Sauer, J., Stein, R. J., Kramer, U., & Muller, C. (2015). Is there a trade-off between glucosinolate-based organic and inorganic defences in a metal hyperaccumulator in the field? *Oecologia, 178*(2), 369–378.

Kerwin, R., Feusier, J., Corwin, J., Rubin, M., Lin, C., Muok, A., ... Kliebenstein, D. J. (2015). Natural genetic variation in *Arabidopsis thaliana* defense metabolism genes modulates field fitness. *eLife, 4*.

Kim, J. H., & Jander, G. (2007). *Myzus persicae* (green peach aphid) feeding on *Arabidopsis* induces the formation of a deterrent indole glucosinolate. *Plant Journal, 49*(6), 1008–1019.

Kimura, M. (1968). Evolutionary rate at the molecular level. *Nature, 217*, 624–626.

Kliebenstein, D., Lambrix, V., Reichelt, M., Gershenzon, J., & Mitchell-Olds, T. (2001). Gene duplication and the diversification of secondary metabolism: side chain modification of glucosinolates in *Arabidopsis thaliana*. *Plant Cell, 13*, 681–693.

Kliebenstein, D., Pedersen, D., Barker, B., & Mitchell-Olds, T. (2002). Comparative analysis of quantitative trait loci controlling glucosinolates, myrosinase and insect resistance in *Arabidopsis thaliana*. *Genetics, 161*(1), 325–332.

Kliebenstein, D. J. (2008). A role for gene duplication and natural variation of gene expression in the evolution of metabolism. *PLoS One, 3*(3), e1838.

Kliebenstein, D. J., Gershenzon, J., & Mitchell-Olds, T. (2001). Comparative quantitative trait loci mapping of aliphatic, indolic and benzylic glucosinolate production in *Arabidopsis thaliana* leaves and seeds. *Genetics, 159*(1), 359–370.

Kliebenstein, D. J., Kroymann, J., Brown, P., Figuth, A., Pedersen, D., Gershenzon, J., & Mitchell-Olds, T. (2001). Genetic control of natural variation in *Arabidopsis thaliana* glucosinolate accumulation. *Plant Physiology, 126*(2), 811–825.

Korves, T. M., Schmid, K. J., Caicedo, A. L., Mays, C., Stinchcombe, J. R., Purugganan, M. D., & Schmitt, J. (2007). Fitness effects associated with the major flowering time gene *FRIGIDA* in *Arabidopsis thaliana* in the field. *American Naturalist, 169*(5), E141–E157.

Kroymann, J., & Mitchell-Olds, T. (2005). Epistasis and balanced polymorphism influencing complex trait variation. *Nature, 435*(7038), 95–98.

Lambrix, V., Reichelt, M., Mitchell-Olds, T., Kliebenstein, D. J., & Gershenzon, J. (2001). The *Arabidopsis* epithiospecifier protein promotes the hydrolysis of glucosinolates to nitriles and influences *Trichoplusia ni* herbivory. *Plant Cell, 13*(12), 2793–2807.

Lankau, R. A. (2007). Specialist and generalist herbivores exert opposing selection on a chemical defense. *New Phytologist, 175*(1), 176–184.

Lankau, R. A., & Kliebenstein, D. J. (2009). Competition, herbivory and genetics interact to determine the accumulation and fitness consequences of a defence metabolite. *Journal of Ecology, 97*(1), 78–88.

Lankau, R. A., & Strauss, S. Y. (2007). Mutual feedbacks maintain both genetic and species diversity in a plant community. *Science, 317*, 1561–1563.

Lankau, R. A., & Strauss, S. Y. (2008). Community complexity drives patterns of natural selection on a chemical defense of *Brassica nigra*. *American Naturalist, 171*(2), 150–161.

Lee, C. R., Anderson, J. T., & Mitchell-Olds, T. (2014). Unifying genetic canalization, genetic constraint, and genotype-by-environment interaction: QTL by genomic background by environment interaction of flowering time in *Boechera stricta*. *PLoS Genetics, 10*(10).

Levene, H. (1953). Genetic equilibrium when more than one ecological niche is available. *American Society of Naturalists, 87*(836), 331–333.

Levy, M., Wang, Q. M., Kaspi, R., Parrella, M. P., & Abel, S. (2005). *Arabidopsis* IQD1, a novel calmodulin-binding nuclear protein, stimulates glucosinolate accumulation and plant defense. *Plant Journal, 43*(1), 79–96.

Li, B., Gaudinier, A., Taylor-Teeples, M., Nham, N. T., Ghaffari, C., Benson, D. S., … Kliebenstein, D. J. (2014). Promoter based integration in plant defense regulation. *Plant Physiology, 166*(4), 1803–1820.

Li, F., Chen, B. Y., Xu, K., Wu, J. F., Song, W. L., Bancroft, I., … Wu, X. M. (2014). Genome-wide association study dissects the genetic architecture of seed weight and seed quality in rapeseed (*Brassica napus* L.). *DNA Research, 21*(4), 355–367.

Li, G., & Quiros, C. F. (2003). In planta side-chain glucosinolate modification in *Arabidopsis* by introduction of dioxygenase *Brassica* homolog BoGSL-ALK. *TAG. Theoretical and Applied Genetics. Theoretische und angewandte Genetik, 106*(6), 1116–1121.

Li, G. Y., & Quiros, C. F. (2002). Genetic analysis, expression and molecular characterization of BoGSL-ELONG, a major gene involved in the aliphatic glucosinolate pathway of Brassica species. *Genetics, 162*(4), 1937–1943.

Liu, B.-H. (1998). *Statistical genomics: Linkage, mapping and QTL analysis*. Boca Raton, FL: CRC Press.

Lynch, M., & Walsh, B. (1998). *Genetics and analysis of quantitative traits*. Sunderland, MA: Sinauer Associates, Inc.

Magrath, R., Bano, F., Morgner, M., Parkin, I., Sharpe, A., Lister, C., … Mithen, R. (1994). Genetics of aliphatic glucosinolates. I. Side chain elongation in *Brassica napus* and *Arabidopsis thaliana*. *Heredity, 72*, 290–299.

Magrath, R., Herron, C., Giamoustaris, A., & Mithen, R. (1993). The inheritance of aliphatic glucosinolates in *Brassica napus*. *Plant Breeding, 111*, 55–72.

Magrath, R., & Mithen, R. (1993). Maternal effects on the expression of individual aliphatic glucosinolates in seeds and seedlings of *Brassica napus*. *Plant Breeding, 111*, 249–252.

Manzaneda, A. J., Prasad, K. V. S. K., & Mitchell-Olds, T. (2010). Variation and fitness costs for tolerance to different types of herbivore damage in *Boechera stricta* genotypes with contrasting glucosinolate structures. *New Phytologist, 188*(2), 464–477.

Matusheski, N. V., Swarup, R., Juvik, J. A., Mithen, R., Bennett, M., & Jeffery, E. H. (2006). Epithiospecifier protein from broccoli (*Brassica oleracea* L. ssp *italica*) inhibits formation of the anticancer agent sulforaphane. *Journal of Agricultural and Food Chemistry, 54*(6), 2069–2076.

Mauricio, R. (1998). Costs of resistance to natural enemies in field populations of the annual plant *Arabidopsis thaliana*. *American Naturalist, 151*(1), 20–28.

Mauricio, R. (2001). Mapping quantitative trait loci in plants: uses and caveats for evolutionary biology. *Nature Reviews Genetics, 2*(5), 370–381.

Mauricio, R., & Rausher, M. D. (1997). Experimental manipulation of putative selective agents provides evidence for the role of natural enemies in the evolution of plant defense. *Evolution: International Journal of Organic Evolution, 51*(5), 1435–1444.

Mauricio, R., Rausher, M. D., & Burdick, D. S. (1997). Variation in the defense strategies of plants: are resistance and tolerance mutually exclusive? *Ecology, 78*(5), 1301–1311.

Milla, R., Escudero, A., & Iriondo, J. M. (2009). Inherited variability in multiple traits determines fitness in populations of an annual legume from contrasting latitudinal origins. *Annals of Botany, 103*(8), 1279–1289.

Mithen, R., Clarke, J., Lister, C., & Dean, C. (1995). Genetics of aliphatic glucosinolates.III. Side-chain structure of aliphatic glucosinolates in *Arabidopsis thaliana*. *Heredity, 74*(Part 2), 210–215.

Mithen, R., Raybould, A. F., & Giamoustairs, A. (1995). Divergent selection for secondary metabolites between wild populations of *Brassica oleracea* and its implications for plant-herbivore interactions. *Heredity, 75*, 472–484.

Mothershead, K., & Marquis, R. J. (2000). Fitness impacts of herbivory through indirect effects on plant-pollinator interactions in *Oenthera macrocarpa*. *Ecology, 81*(1), 30–40.

Mueller, R., de Vos, M., Sun, J. Y., Sonderby, I. E., Halkier, B. A., Wittstock, U., & Jander, G. (2010). Differential effects of indole and aliphatic glucosinolates on lepidopteran herbivores. *Journal of Chemical Ecology, 36*(8), 905–913.

Mumm, R., Burow, M., Bukovinszkine'Kiss, G., Kazantzidou, E., Wittstock, U., Dicke, M., & Gershenzon, J. (2008). Formation of simple nitriles upon glucosinolate hydrolysis affects direct and indirect defense against the specialist herbivore, *Pieris rapae*. *Journal of Chemical Ecology, 34*(10), 1311–1321.

Nordborg, M., & Weigel, D. (2008). Next-generation genetics in plants. *Nature, 456*(7223), 720–723.

Parkin, I., Magrath, R., Keith, D., Sharpe, A., Mithen, R., & Lydiate, D. (1994). Genetics of aliphatic glucosinolates.II. Hydroxylation of alkenyl glucosinolates in *Brassica napus*. *Heredity, 72*, 594–598.

Pfalz, M., Vogel, H., & Kroymann, J. (2009). The gene controlling the indole glucosinolate modifier1 quantitative trait locus alters indole glucosinolate structures and aphid resistance in *Arabidopsis*. *Plant Cell, 21*(3), 985–999.

Pfalz, M., Vogel, H., Mitchell-Olds, T., & Kroymann, J. (2007). Mapping of QTL for resistance against the crucifer specialist herbivore *Pieris brassicae* in a new *Arabidopsis* inbred line population, Da(1)-12×Ei-2. *PLoS One, 2*(6), e578.

Poelman, E. H., van Dam, N. M., van Loon, J. J. A., Vet, L. E. M., & Dicke, M. (2009). Chemical diversity in *Brassica oleracea* affects biodiversity of insect herbivores. *Ecology, 90*(7), 1863–1877.

Prasad, K. V. S. K., Song, B.-H., Olson-Manning, C., Anderson, J. T., Lee, C.-R., Schranz, M. E., … Mitchell-Olds, T. (2012). A gain-of-function polymorphism controlling complex traits and fitness in nature. *Science, 337*(6098), 1081–1084.

Pyhajarvi, T., Hufford, M. B., Mezmouk, S., & Ross-Ibarra, J. (2013). Complex patterns of local adaptation in teosinte. *Genome Biology and Evolution, 5*(9), 1594–1609.

de Quiros, H. C., Magrath, R., McCallum, D., Kroymann, J., Schnabelrauch, D., Mitchell-Olds, T., & Mithen, R. (2000). α-Keto acid elongation and glucosinolate biosynthesis in *Arabidopsis thaliana. TAG. Theoretical and Applied Genetics. Theoretische und angewandte Genetik, 101*, 429–437.

Rodman, J. (1980). Population variation and hybridization in sea-rockets (*Cakile*, Cruciferae): seed glucosinolate characters. *American Journal of Botany, 67*, 1145–1159.

Rodman, J. E., Kruckeberg, A. R., & Alshehbaz, I. A. (1981). Chemotaxonomic diversity and complexity in seed glucosinolates of *Caulanthus* and *Streptanthus* (Cruciferae). *Systematic Botany, 6*(3), 197–222.

Rodman, J. E., Soltis, P. S., Soltis, D. E., Sytsma, K. J., & Karol, K. G. (1998). Parallel evolution of glucosinolate biosynthesis inferred from congruent nuclear and plastid gene phylogenies. *American Journal of Botany, 85*(7), 997–1006.

Schranz, M. E., Manzaneda, A. J., Windsor, A. J., Clauss, M. J., & Mitchell-Olds, T. (2009). Ecological genomics of *Boechera stricta*: identification of a QTL controlling the allocation of methionine- vs branched-chain amino acid-derived glucosinolates and levels of insect herbivory. *Heredity, 102*(5), 465–474.

Schweizer, F., Fernandez-Calvo, P., Zander, M., Diez-Diaz, M., Fonseca, S., Glauser, G., … Reymond, P. (2013). *Arabidopsis* basic helix-loop-helix transcription factors MYC2, MYC3, and MYC4 regulate glucosinolate biosynthesis, insect performance, and feeding behavior. *Plant Cell, 25*(8), 3117–3132.

Sønderby, I. E., Burow, M., Rowe, H. C., Kliebenstein, D. J., & Halkier, B. A. (2010). A complex interplay of three R2R3 MYB transcription factors determines the profile of aliphatic glucosinolates in *Arabidopsis. Plant Physiology, 153*(1), 348–363.

Sønderby, I. E., Geu-Flores, F., & Halkier, B. A. (2010). Biosynthesis of glucosinolates – gene discovery and beyond. *Trends in Plant Science, 15*(5), 283–290.

Sønderby, I. E., Hansen, B. G., Bjarnholt, N., Ticconi, C., Halkier, B. A., & Kliebenstein, D. J. (2007). A systems biology approach identifies a R2R3 MYB gene subfamily with distinct and overlapping functions in regulation of aliphatic glucosinolates. *PLoS One, 2*(12), e1322.

Spencer, G. F., & Daxenbichler, M. E. (1980). Gas chromatography–mass spectrometry of nitriles, isothiocyanates and oxazolidinethiones derived from cruciferous glucosinolates. *Journal of the Science of Food and Agriculture, 31*, 359–367.

Stahl, E. A., Dwyer, G., Mauricio, R., Kreitman, M., & Bergelson, J. (1999). Dynamics of disease resistance polymorphism at the *Rpm1* locus of *Arabidopsis. Nature, 400*, 667–671.

Textor, S., Bartram, S., Kroymann, J., Falk, K. L., Hick, A., Pickett, J. A., & Gershenzon, J. (2004). Biosynthesis of methionine-derived glucosinolates in *Arabidopsis thaliana*: recombinant expression and characterization of methylthioalkylmalate synthase, the condensing enzyme of the chain-elongation cycle. *Planta, 218*(6), 1026–1035.

Textor, S., de Kraker, J. W., Hause, B., Gershenzon, J., & Tokuhisa, J. G. (2007). MAM3 catalyzes the formation of all aliphatic glucosinolate chain lengths in *Arabidopsis. Plant Physiology, 144*(1), 60–71.

Tian, D., Traw, M. B., Chen, H., Kreitman, M., & Bergelson, J. (2003). Fitness costs of R-gene-mediated resistance in *Arabidopsis thaliana. Nature, 423*(6935), 74–77.

Traka, M. H., Saha, S., Huseby, S., Kopriva, S., Walley, P. G., Barker, G. C., … Mithen, R. F. (2013). Genetic regulation of glucoraphanin accumulation in Beneforte (R) broccoli. *New Phytologist, 198*(4), 1085–1095.

Turelli, M. (1984). Heritable genetic variation via mutation-selection balance: Lerch's zeta meets the abdominal bristle. *Theoretical Population Theory, 25*, 138–193.

Voelckel, C., Mirzaei, M., Reichelt, M., Luo, Z., Pascovici, D., Heenan, P. B., … Lockhart, P. J. (2010). Transcript and protein profiling identify candidate gene sets of potential adaptive significance in New Zealand Pachycladon. *BMC Evolutionary Biology, 10.*

Weigel, D., & Nordborg, M. (2005). Natural variation in *Arabidopsis*. How do we find the causal genes? *Plant Physiology, 138*(2), 567–568.

Wentzell, A. M., Boeye, I., Zhang, Z. Y., & Kliebenstein, D. J. (2008). Genetic networks controlling structural outcome of glucosinolate activation across development. *PLoS Genetics, 4*(10), e1000234.

Wentzell, A. M., & Kliebenstein, D. J. (2008). Genotype, age, tissue, and environment regulate the structural outcome of glucosinolate activation. *Plant Physiology, 147*(1), 415–428.

Wentzell, A. M., Rowe, H. C., Hansen, B. G., Ticconi, C., Halkier, B. A., & Kliebenstein, D. J. (2007). Linking metabolic QTLs with network and cis-eQTLs controlling biosynthetic pathways. *PLoS Genetics, 3*(9), 1687–1701.

Windsor, A. J., Reichelt, M., Figuth, A., Svatos, A., Kroymann, J., Kliebenstein, D. J., … Mitchell-Olds, T. (2005). Geographic and evolutionary diversification of glucosinolates among near relatives of *Arabidopsis thaliana* (Brassicaceae). *Phytochemistry, 66*(11), 1321–1333.

Wolf, A. T., Howe, R. W., & Hambrick, J. L. (2000). Genetic diversity and population structure of the serpentine endemic *Calystegia collina* (Convolvulaceae) in Northern California. *American Journal of Botany, 87*(8), 1138–1146.

Wright, S. I., Lauga, B., & Charlesworth, D. (2002). Rates and patterns of molecular evolution in inbred and outbred *Arabidopsis*. *Molecular Biology and Evolution, 19*(9), 1407–1420.

Xu, J. F., Long, Y., Wu, J. G., Xu, H. M., Zhao, Z. G., Wen, J., … Shi, C. H. (2015). QTL identification on two genetic systems for rapeseed glucosinolate and erucic acid contents over two seasons. *Euphytica, 205*(3), 647–657.

Zabala, M. D., Grant, M., Bones, A. M., Bennett, R., Lim, Y. S., Kissen, R., & Rossiter, J. T. (2005). Characterisation of recombinant epithiospecifier protein and its over-expression in *Arabidopsis thaliana*. *Phytochemistry, 66*(8), 859–867.

Zhang, Y. Y., Li, B. H., Huai, D. X., Zhou, Y. M., & Kliebenstein, D. J. (2015). The conserved transcription factors, MYB115 and MYB118, control expression of the newly evolved benzoyloxy glucosinolate pathway in *Arabidopsis thaliana*. *Frontiers in Plant Science, 6.*

Züst, T., Heichinger, C., Grossniklaus, U., Harrington, R., Kliebenstein, D. J., & Turnbull, L. A. (2012). Natural enemies drive geographic variation in plant defenses. *Science, 338*(6103), 116–119.

CHAPTER FOUR

Glucosinolate Regulation in a Complex Relationship – MYC and MYB – No One Can Act Without Each Other

H. Frerigmann
University of Cologne, Cologne, Germany
E-mail: henning.frerigmann@uni-koeln.de

Contents

Abstract

The well-studied pathway of glucosinolate (GSL) biosynthesis in *Arabidopsis thaliana* serves as a model system for secondary metabolites in plants. Most biosynthetic genes are known and were used in systematic co-expression approaches to identify their transcriptional regulators. Two groups of homologous MYB and bHLH transcription factors (TFs) interact with each other and fulfil the main role of direct regulation. The MYB factors are crucial for the specificity of the pathway and control genes in the core-structure

Advances in Botanical Research, Volume 80
ISSN 0065-2296
http://dx.doi.org/10.1016/bs.abr.2016.06.005

pathway of indolic and aliphatic GSL, respectively, as well as the genes in the primary metabolism for the necessary substrates. The MYC-bHLH TFs are crucial for all types of GSL and act together with both types of MYB factors; furthermore they integrate phytohormone signals specifically from the jasmonate signal transduction to adjust the production of this important defence compound class. The regulation of the modification reactions of GSL is under control of a diverse set of other TFs, which have the potential to fine-tune also the core-structure pathway of the glucosinolate biosynthesis. However, most of these regulators still need to be identified and characterized.

Although many aspects and components involved in this transcriptional regulation are known, still not all known glucosinolate profiles can be explained by the activity of these modulators and point to the important role of transport and metabolism in determining accumulation pattern of GSL.

1. INTRODUCTION

Glucosinolates (GSL) are sulphur and nitrogen containing secondary metabolites which occur mainly in the order Capparales in the family Brassicaceae. Due to their presence in *Arabidopsis thaliana* this class became a model system for secondary metabolites and various aspects of GSL, e.g., biosynthesis and in vivo function have been addressed in the last decades. The GSL are well known for their function in plant defence against diverse pest like nematodes (Buskov, Serra, Rosa, Sorensen, & Sorensen, 2002; Lazzeri, Curto, Leoni, & Dallavalles, 2004), insects (Beekwilder et al., 2008; Zang, Kim, Park, & Hong, 2009), fungi (Bednarek et al., 2009; Sarwar, Kirkegaard, Wong, & Desmarchelier, 1998) and bacteria (Clay, Adio, Denoux, Jander, & Ausubel, 2009; Smith & Kirkegaard, 2002), but they serve also as an oviposition signal for specialist herbivores (Renwick, Haribal, Gouinguene, & Städler, 2006). To achieve this important function in plant defence GSL occur as phytoanticipins in basal amounts in the whole plant and contribute thereby to the innate immunity, but they are also induced by diverse signalling cascades to repel a specific attack (Cipollini, Enright, Traw, & Bergelson, 2004). This signal-dependent induction of GSL biosynthesis upon an attack has two reasons: (1) on the one hand it is important for the plant to produce a specific set of GSL, since different types affect the control of different pest (Müller et al., 2010; Pfalz, Vogel, & Kroymann, 2009); (2) on the other hand, GSL production may increase the photosynthetic requirements by at least 15% (Bekaert, Edger, Hudson, Pires, & Conant, 2012); therefore it is important to regulate the biosynthesis by the specific needs and save energy for growth if possible. While many of the molecular mechanisms underlying growth and defence trade-offs remain to be elucidated, hormone crosstalk has emerged as a major player in regulating trade-offs needed to achieve a balance (Huot, Yao, Montgomery, & He, 2014). These hormonal signals interact with each other and converge on the level of transcription factors (TFs) which translate these signals into the corresponding activation/repression of GSL biosynthetic genes (Table 1).

More than 1500 TFs were found in the genome of *A. thaliana* and thereby more than 6% of all genes (Riechmann et al., 2000), indicating the importance of transcriptional regulation in general. About 45% of the TF families like WRKY, NAC or the big AP2/EREBP (APETALA2/Ethylene responsive element binding protein) family are plant specific. This higher number of plant TFs can be probably explained by the sessile lifestyle of plants and the associated much more complex response to exogenous cues (Riechmann et al., 2000). The two biggest TF-families in *A. thaliana* are the MYB and the bHLH (basic-Helix-loop-Helix) family (Heim et al., 2003; Stracke, Werber, & Weisshaar, 2001), members of both families were shown to be responsible for the largest effects on GSL regulation (see the following sections). Understanding of the complex interplay of these widespread occurring components might provide a template of how to study and manipulate transcriptional regulation of other secondary metabolites in important agricultural plants. This review gives an update on the achieved progress made on transcriptional regulation of GSL biosynthesis in the last years.

2. SUBGROUP 12 R2R3-MYB TRANSCRIPTION FACTORS IN GLUCOSINOLATE REGULATION

2.1 General Introduction Into MYB Transcription Factors

The name of the MYB protein family originates from the sequence similarity to the product of the viral oncogene v-MYB the 'vian myeloblastosis virus'. Members of this TF family occur in many diverse eukaryotic organisms. In animal organisms, there are only a few members present, which all have three imperfect sequence repeats (R1, R2, R3) each about 53 amino acids long. The plant MYB-TFs could be arranged by the number of sequence repeats in three groups: 1R-MYB proteins (one sequence repeat), R2R3-MYB proteins (two sequence repeats) and the MYB-3R proteins (three sequence repeats), whereas the *At*MYB4R with four sequence repeats is an exceptional case. While in animals, only 3R-type

Table 1 List of Transcription Factors Reviewed in This Chapter in Terms of GLS Regulation

Main Name and Other Names	AGI Code	Main References in Terms of GLS Regulation
MYB34/ATR1/HIG3	AT5G60890	Celenza et al. (2005), Frerigmann and Gigolashvili (2014a)
MYB51/HIG1	AT1G18570	Gigolashvili, Berger, et al. (2007), Frerigmann and Gigolashvili (2014a)
MYB122/HIG2	AT1G74080	Gigolashvili, Berger, et al. (2007), Frerigmann and Gigolashvili (2014a)
MYB28/HAG1/PMG1	AT5G61420	Hirai et al. (2007), Sønderby et al. (2007), Gigolashvili, Yatusevich, et al. (2007)
MYB29/HAG3/PMG2	AT5G07690	Hirai et al. (2007), Sønderby et al. (2007), Gigolashvili et al. (2008)
MYB76/HAG2	AT5G07700	Sønderby et al. (2007), Gigolashvili et al. (2008), Sønderby et al. (2010)
bHLH04/MYC4	AT4G17880	Schweizer et al. (2013), Frerigmann et al. (2014)
bHLH05/ATR2/MYC3	AT5G46760	Schweizer et al. (2013), Frerigmann et al. (2014)
bHLH06/MYC2/JAI1/JIN1/rd22BP1	AT1G32640	Dombrecht et al. (2007), Schweizer et al. (2013), Frerigmann et al. (2014)
bHLH28/MYC5/NIG1	AT5G46830	Frerigmann et al. (2014)
ERF6	AT4G17490	Xu et al. (2016)
MYB115/LAG1	AT5G40360	Zhang et al. (2015)
MYB118/PGA37/LAG2	AT3G27785	Zhang et al. (2015)
IQD	AT3G09710	Levy et al. (2005)
ATDof1/OBP2	AT1G07640	Skirycz et al. (2006)
SLIM1	AT1G73730	Maruyama-Nakashita et al. (2006)
WRKY18	AT4G31800	Schön et al. (2013)
WRKY40	AT1G80840	Schön et al. (2013)

MYBs occur, the R2R3-MYBs represent with 125 members the biggest subgroup in plants (Stracke et al., 2001).

Each of the sequence repeats consist of three α-helices, from which the last two build a helix-turn-helix motif. The third helix is in contact with the

DNA and the respective other third helix (Ogata et al., 1994). Although some MYBs possess three sequence repeats, the R2 and R3 domains are sufficient for DNA binding (Gabrielsen, Sentenac, & Fromageot, 1991). The MYB domain is not only necessary for DNA binding, but also for protein—protein interactions, for example, with bHLH TFs (Goff, Cone, & Chandler, 1992). Actually, regulation of many metabolic pathways takes place in cooperation with bHLH TFs (Zimmermann, Heim, Weisshaar, & Uhrig, 2004). Outside of the MYB domain the proteins have a very diverse structure, but from the sequence homology of the MYB domain and other motifs at C-terminal part the R2R3-MYBs in Arabidopsis could be classified into 25 subgroups. Within a subgroup, paralogs often control the same metabolic pathway in different cell types as a result of differences in expression patterns (Dubos et al., 2010). Besides the subgroup motifs with unknown function, all MYBs possess at the C-terminus conserved serine and threonine residues which might represent possible target motifs for serine/threonine kinases. Such posttranslational modifications were already been shown for B-MYB (Petrovas, Jeay, Lewis, & Sonenshein, 2003).

The best characterized regulators for GSL biosynthesis are a group of six homologous R2R3-MYB-TFs. These six regulators are close homologs belonging to subgroup 12 (Fig. 1) and can be distinguished from other MYB factors by the same motif [L/F]LN[K/R]VA (Stracke et al., 2001). These MYBs are unique to the GSL containing Brassicales and arose via whole-genome and tandem duplication with ensuing neo-functionalization to focus on one branch of the GSL pathway (Bekaert et al., 2012). Thus, from the functional perspective they can be subdivided in two groups which regulate the aliphatic glucosinolate (AG) or the indolic glucosinolate (IG) biosynthesis (for details, see later sections). Members of the subgroup regulating AG biosynthesis MYB28, MYB29 and MYB76 are referred to as HAG-MYBs (high aliphatic GSL) whereas regulators of IG MYB34, MYB51 and MYB122 are called HIG-MYBs (high indolic GSL), whenever generalized (Fig. 1).

2.2 HAG-MYB Factors in the Regulation of Aliphatic Glucosinolates

GLS incorporate sulphur from glutathione (GSH) and during sulphation by 3′-phosphoadenosine 5′-phosphosulfate (PAPS). Some AG contain 1 molecule extra sulphur provided by methionine, but they do not have to contain that additional sulphur, due to subsequent loss during biosynthesis. Accordingly Hirai and coworkers could show in a gene-to-metabolite network analysis

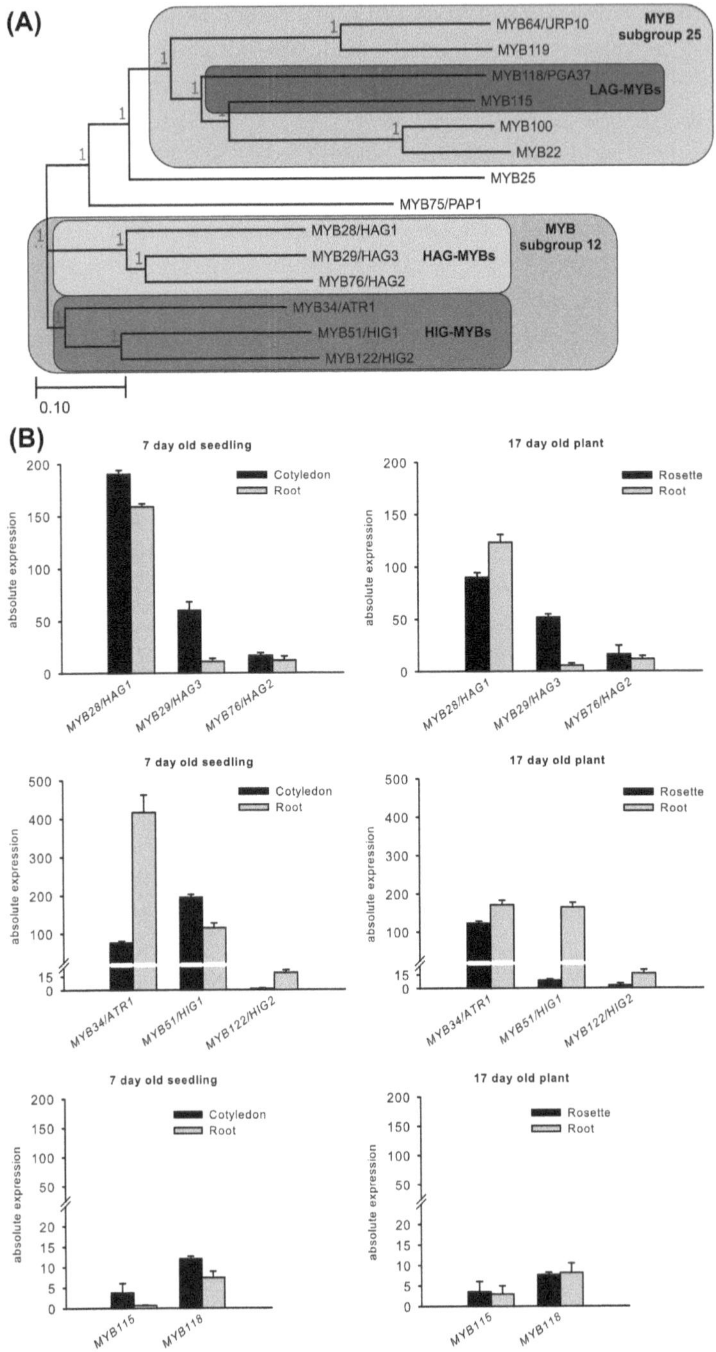
(A)
MYB64/URP10
MYB119
MYB118/PGA37
MYB115
MYB100
MYB22
MYB25
MYB75/PAP1
MYB subgroup 25
LAG-MYBs
MYB28/HAG1
MYB29/HAG3
MYB76/HAG2
MYB34/ATR1
MYB51/HIG1
MYB122/HIG2
HAG-MYBs
HIG-MYBs
MYB subgroup 12
0.10
(B)
7 day old seedling
17 day old plant
Cotyledon
Rosette
Root
absolute expression
MYB28/HAG1
MYB29/HAG3
MYB76/HAG2
MYB34/ATR1
MYB51/HIG1
MYB122/HIG2
MYB115
MYB118

that the genes and metabolites involved in GLS metabolism are co-ordinately modulated upon sulphur deficiency (Hirai et al., 2004). Such high correlation of AG gene expression enabled identification of new genes in the biosynthesis of AG (Hirai et al., 2005). Similarly, co-expression analysis of publicly available, transcriptome data identified MYB28/HAG1/PMG1 (production of methionine-derived glucosinolate 1; in the following MYB28/HAG1) and MYB29/PMG2 as candidates for positive regulators of AG (Hirai et al., 2007).

Together with MYB76/HAG2, these MYBs were shown to affect AG production (Gigolashvili, Engqvist, Yatusevich, Müller, & Flügge, 2008; Gigolashvili, Yatusevich, Berger, Müller, & Flügge, 2007; Hirai et al., 2007; Sønderby et al., 2007). MYB28/HAG1 is the main regulator and affects short- and long-chained AG, whereas MYB29 and MYB76 affect only short-chained AG and knockouts contain only slightly reduced AG level. *MYB29* is in contrast to *MYB28*, which is strongly inducible by methyl-jasmonate (MeJA) and thereby plays potentially a role in response to MeJA signalling (Hirai et al., 2007). Furthermore, although only *MYB28/HAG1* is considered as the major regulator of long-chained AG, constitutive (*Pro35S*) expression of all three *HAGs* positively affects the expression of both *MAM1* and *MAM3* (Gigolashvili, Engqvist, et al., 2008; Gigolashvili, Yatusevich, et al., 2007), which results for *MYB76/HAG2* and *MYB29/HAG3* in even higher level of 8MSOO, a long-chained AG (Gigolashvili et al., 2008). Additionally, a *trans*-activation assay shows that the three HAG-MYBs can similarly induce the promoters of many AG-genes (Gigolashvili et al., 2008). In contrast to the *Pro35S*-triggered expression, only *ProMYB28:MYB28/HAG1* was able to induce the biosynthesis of

Figure 1 ***Phylogenetic tree of R2R3-MYB transcription factors (TFs) in glucosinolate (GSL) regulation.*** (A) The relationship of selected R2R3-MYB TFs is shown. The six well-characterized positive regulators of GSL of MYB subgroup 12 can be further subdivided in the two functional groups HIG-MYBs and HAG-MYBs. Two members of subgroup 25 (LAG-MYBs – low aliphatic glucosinolates) are shown to be negative regulators of aliphatic glucosinolate (AG). The amino acid sequences of MYB75/PAP1 (subgroup 6) and MYB25 (subgroup 23) were used as outgroup members for the phylogenetic tree. Subgroup classification is presented according to Dubos et al. (2010). Complete protein sequences were used for Clustal omega (Sievers et al., 2011) based calculation of the tree and the phylogenetic tree (newick) viewer (Huerta-Cepas et al., 2010) was used for preparation of the tree. (B) Absolute expression level of the six HAG- and HIG-MYBs and the two LAG-MYBs in cotyledons/rosettes and roots of 7- and 17-day-old wild type (Col-0) plants grown under continuous light on soil. Expression data were retrieved via eFP browser (Winter et al., 2007). *From Schmid et al. (2005).*

long-chained AG in seeds (Li et al., 2013), indicating that the differential observed effects are mainly due to differential expression of these TFs in different organs (Fig. 1) and cell types.

Thus *Pro35S* overexpression might give artificial results and presumably, knockout mutants reflect the real in vivo function better. The double knockout *myb28 myb29* confirmed the redundant function of both regulators; this line contains only very low amounts of total AG including a slightly more reduced amount of the long-chained AG in comparison to *myb28* alone (Sønderby et al., 2007). The single knockout *myb76* has only weak effects on AG gene expression and accumulation of AG (Gigolashvili et al., 2008), thus MYB76/HAG2 was imputed to have only an accessory role on GLS regulation, mainly due to its low expression (Fig. 1). However, the double mutant *myb28 myb76* exhibited a stronger reduction of the short-chained AG in comparison to the single mutants, since MYB76/HAG2 bears a positive effect on the expression on *MYB29/HAG3* and thereby on the short-chained AG (Sønderby, Burow, Rowe, Kliebenstein, & Halkier, 2010).

However, in all the *myb* knockout mutants, there is a lack of correlation between total level of biosynthetic transcripts and GLS chemotypes. As such, the metabolic chemotype is uncoupled from the total level of biosynthetic gene transcripts, which could indicate that a decrease in one or a few biosynthetic transcripts is enough to change total AG accumulation (Sønderby et al., 2010). This is strongly articulated in *myb28* which shows very low transcript levels for many AG genes, but still accumulates two-thirds of AG. The double mutant *myb28 myb29* has very similar transcript levels of the biosynthetic genes but nearly no AG left. Thus, the epistatic effect of MYB28/HAG1 and MYB29/HAG3 shown at the GLS level is not mirrored as an epistatic effect on the total level of biosynthetic transcripts but only on individual transcripts (Sønderby et al., 2010). Though loss of *MYB29* which has only weak impact on AG accumulation, caused an increase of *MYB28/HAG1* mRNA levels, which might be the reason why only a few AG genes are decreased due to lack of *MYB29/HAG3*. This systemic analysis of metabolite and transcript levels revealed a complex regulatory network within the MYBs, in which MYB76/HAG2 is also responsible for the distribution of the AG within the leaf and thus probably involved in the regulation of GLS transport within the leaf (Sønderby et al., 2010).

Thus these different approaches with endogenous and constitutive *35S* promoters indicate that the spatial distribution of GLS within the wild type leaf is at least partially due to the spatial expression pattern of the

MYBs. Correspondingly, all three HAGs are able to induce genes for both short- and long-chained AG if they are expressed in such specific cell types. The spatial expression pattern of the MYBs and the putative different interaction partners (such as bHLH TFs, see sections later) in these cells might thereby cause the observed differences. It can be furthermore assumed that these different TF-complexes regulate additionally GLS transport and breakdown processes, which could explain the uncoupling of AG biosynthetic genes and AG accumulation. Thus different biotic and abiotic challenges are integrated and induce a specific induction of the HAG-MYBs which in turn trigger a specific set of genes for biosynthesis, transport and metabolism in these distinct cells or tissues.

2.3 MYB-Factors of Indolic Glucosinolate Regulation

The other three members (HIG-MYBs) of the R2R3-subgroup 12, MYB34/HIG3/ATR1 (altered tryptophan regulation (ATR) 1) (in the following MYB34/ATR1), MYB51/HIG1 and MYB122/HIG2 regulate the tryptophan-dependent synthesis of GSL. MYB34/ATR1 was first described, due to the strong impact of the dominant mutant *atr1-D* on tryptophan biosynthesis. A base pair exchange in the 5′-region of *MYB34/ATR1* leads to strong induction of *MYB34/ATR1* and thereby to enhanced expression in *ASA1*, which encodes for the first enzyme in the production of indolic metabolites in the shikimate pathway (Bender & Fink, 1998). The reason for the high induction of *MYB34/ATR1* is still not completely understood, but the authors stated that the *MYB34/ATR1* transcript also includes a 5′-regulatory region with three short open reading frame (ORF), one of which is prematurely terminated by the *atr1-D* mutation. Some years later, Celenza et al. (2005) could show that MYB34/ATR1 has also a positive effect on the IG biosynthesis. The T-DNA insertion line *myb34* has lower IG level, whereas the overexpression of *MYB34/ATR1* leads to more IG, but also an increase in free indole-3-acetic-acid (IAA) level.

It is worth to mention that the strong *Pro35S*-triggered overexpression of *MYB34/ATR1* could not completely reflect the *atr1D* phenotype. Although the expression of *MYB34/ATR1* and the transcript level of IG biosynthesis genes are even higher in plants with *Pro35S*-controlled overexpression, the effect on IG content is weaker. An explanation for this could be that the *MYB34/ATR1* transcripts might not be produced in the appropriate spatial and temporal patterns to optimally affect the metabolic pathway

(Celenza et al., 2005), thus necessary interaction partners such as bHLH TFs might not be present (see sections later).

MYB51/HIG1 was identified in an High-performance liquid chromatography (HPLC)-based screen of a transposon-based activation-tagging population for high phenolic compounds (Schneider et al., 2005). Mass spectrometry-based analysis could clarify that the dominant mutant *HIG1-D* accumulates high level of indol-3-ylmethyl glucosinolate (I3M), while the knockout mutant *myb51* shows reduced content of this defence compound (Gigolashvili, Berger, et al., 2007). This reduction could be partially complemented by *Pro35S:MYB34*, however, leading to a strong auxin phenotype. An overexpression of *MYB122/HIG2* was not able to fully complement the IG reduction in *myb51*, but the *Pro35S*-triggered overexpression in the wild type had a positive effect on IG and IAA level. Thus all three HIG-MYBs have a positive impact on IG and IAA amount, since they activate in vivo and in vitro many genes of the tryptophan and IG-core-structure pathway (Gigolashvili, Berger, et al., 2007). Transcriptome and metabolome analyses of HIG- and HAG-MYB overexpression and micro-RNA-knockdown lines revealed that these TFs activate not only the IG- and AG-core-structure pathway, but also genes in the primary metabolism (Malitsky et al., 2008). Yatusevich et al. (2010) could verify with *trans*-activation assays that both classes of GSL-MYB TFs can activate the pathways to the necessary GSL substrates GSH and PAPS (Fig. 2).

The basal level of IG is completely abolished in *myb34 myb51 myb122* triple knockout (Frerigmann & Gigolashvili, 2014a). However, different organ-specific expression (Fig. 1 and (Frerigmann & Gigolashvili, 2014a)) reveals distinct impacts of these three genes on the accumulation of IG. While MYB34/ATR1 constitutes the main regulator of IG in roots and is crucial for the accumulation of 1MO-I3M, MYB51/HIG1 shows a dominant role in the shoot and for the modified indolic glucosinolate 4MO-I3M, while MYB122/HIG2 has an accessory role with only a low impact on IG biosynthesis (Frerigmann & Gigolashvili, 2014a). The influence of the MYBs on the accumulation of IGs after treatment with different hormones such as abscisic acid (ABA), jasmonic acid (JA), salicylic acid (SA) and ethylene (ET) confirms the varying regulation of the modified IGs by these factors (Fig. 3). The expression of *MYB51/HIG1* correlates with *CYP81F2* in the databases (ATTED II (Obayashi, Hayashi, Saeki, Ohta, & Kinoshita, 2009) and expression angler (Toufighi, Brady, Austin, Ly, & Provart, 2005)); both genes are induced by ET, SA and other signals such as flagellin or oligogalacturonides, which are associated with microbial infections

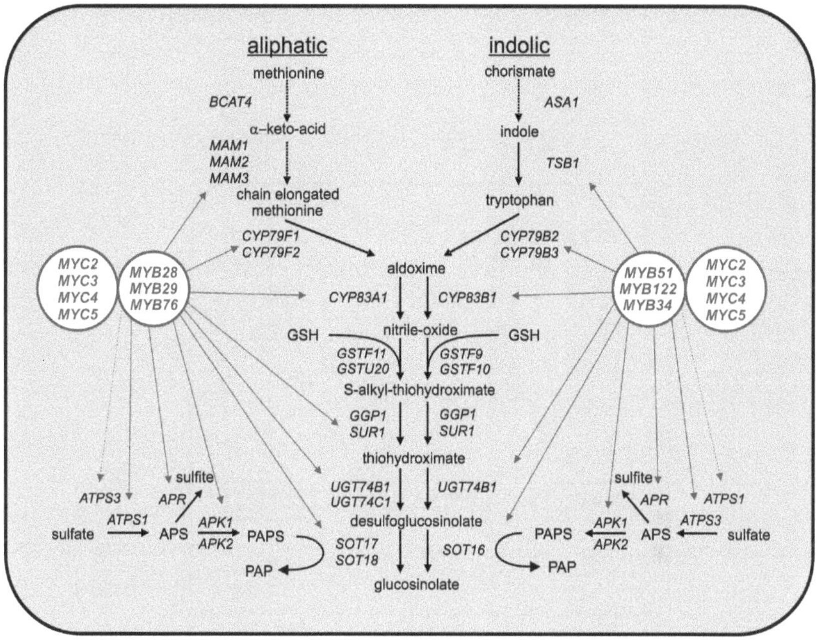

Figure 2 ***Transcriptional activation of aliphatic and indolic glucosinolate genes by MYB and bHLH transcription factors (TFs).*** The MYB-MYC complexes regulate probably all depicted GLS biosynthesis genes. The microarray data for bhlh04 bhlh05 bhlh06 (Schweizer et al., 2013) indicate that these homologous MYC-bHLH TFs regulate all mentioned genes (except CYP79F1) in the glucosinolate (GSL) pathway positively ($p < .05$). Additionally most of the aliphatic glucosinolate (AG)-related genes (including CYP79F1) were downregulated in myb28 myb29 (FDR adjusted *p*-values; $p < .05$) (Sønderby et al., 2010). The displayed *arrows* present the proved *trans*-activation capacity of the respective HIG- or HAG-MYB TFs (Yatusevich et al., 2010). *Figure adopted from Yatusevich, R., Mugford, S. G., Matthewman, C., Gigolashvili, T., Frerigmann, H., Delaney, S.,... Kopriva, S. (2010) Genes of primary sulfate assimilation are part of the glucosinolate biosynthetic network in* Arabidopsis thaliana. Plant Journal, *62, 1–11.*

(Clay et al., 2009; Ferrari et al., 2007; Segarra, Mir, Martínez, & León, 2010; Winter et al., 2007). Correspondingly, Frerigmann and Gigolashvili (2014a) could show that both, ET and SA, induce the production of 4MO-I3M; however, this induction is lacking in *myb51*. The expression of *MYB34/ATR1* is antagonistically regulated to *MYB51/HIG1* and correlated in the databases with *CYP81F4*, which is important for the production of 1MO-I3M (Pfalz et al., 2011) (Fig. 6). The production of 1MO-I3M occurs likewise antagonistically to 4MO-I3M and is strongly induced by the phytohormones JA and ABA (Frerigmann & Gigolashvili, 2014a; Mikkelsen et al., 2003). This induction is strongly impaired in *myb34* and also partially

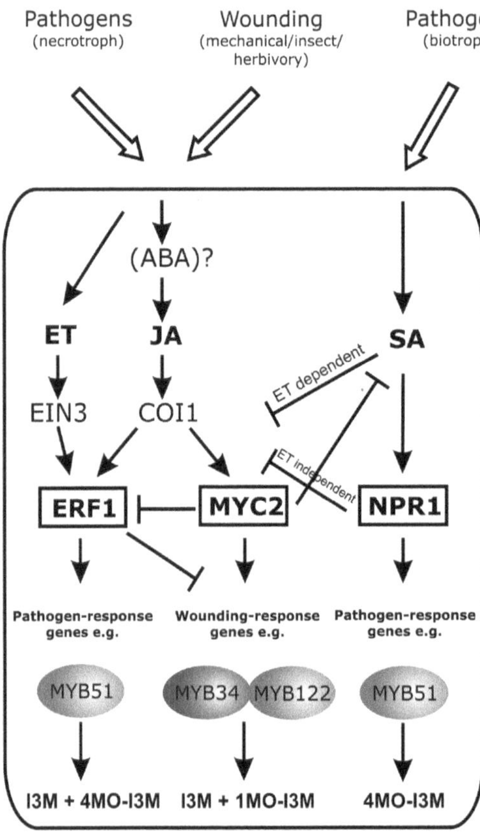

Figure 3 ***Signalling model for indolic glucosinolate regulation in plant defence.*** The figure depicts the postulated simplified hormonal signalling cascades in *Arabidopsis thaliana* in response to different pathogens on the regulation of the HIG-MYBs and the corresponding production of IG. *Modified after Lorenzo, O., Chico, J. M., Sanchez-Serrano, J. J., Solano, R. (2004). JASMONATE-INSENSITIVE1 encodes a MYC transcription factor essential to discriminate between different jasmonate-regulated defense responses in* Arabidopsis. *Plant Cell, 16, 1938–1950 and Mikkelsen M. D., Petersen B. L., Glawischnig E., Jensen A. B., Andreasson E., Halkier B. A. (2003). Modulation of CYP79 genes and glucosinolate profiles in* Arabidopsis *by defense signaling pathways.* Plant Physiology, *131, 298–308 and Frerigmann, H., Gigolashvili, T. (2014a). MYB34, MYB51 and MYB122 distinctly regulate indolic glucosinolate biosynthesis in* Arabidopsis thaliana. Molecular Plant, *7, 814–828.*

in *myb122*, indicating that both genes are involved in ABA- and MeJA-dependent regulation of 1MO-I3M and I3M.

While the JA signalling pathway is supposed to act in the defence response to mechanical wounding, insects and herbivory, the phytohormones SA and

ET act together with JA in response to microbial and fungal pathogens (Fig. 3, reviewed in Lorenzo, Chico, Sanchez-Serrano, & Solano, 2004; Mikkelsen et al., 2003). ABA is mostly associated with responses to drought stress; however, ABA is induced by necrotrophic pathogens and mutants involved in ABA signalling or biosynthesis are more susceptible to some necrotrophic fungi (Adie et al., 2007). These different signalling pathways intersect with each other, as a result of such complex interactions, different sets of genes and metabolites are activated as shown for IG in Fig. 3. However, although *MYB34* and *MYB51* are correlated with *CYP81F4* and *CYP81F2*, and important for the MeJA/ABA- and ET/SA-dependent induction of 1MO-I3M and 4MO-I3M, respectively, the MYBs do not directly regulate these genes for the modification. Thus it is hypothesized that additional regulators of *CYP81F4* or *CYP81F2* exist. These yet unknown regulators might be either (1) upstream of the respective HIG-MYB and activate in parallel the MYB TF and CYP81F gene in the same cell/tissue or (2) both are controlled by the same upstream regulator to facilitate a concerted expression for specific induction of the modified IG in specific cells/tissues (Frerigmann & Gigolashvili, 2014a).

3. SUBGROUP IIIe bHLH TRANSCRIPTION FACTORS IN GLUCOSINOLATE REGULATION

3.1 General Introduction Into bHLH Transcription Factors

The bHLH-TF family can be found ubiquitously in all three eukaryotic kingdoms of life (Heim et al., 2003). This family is defined by an about 60 amino acid long bHLH domain, which consists of two regions with different functions. The N-terminal part of this domain consists of 15 basic amino acids which are involved in DNA-binding. The C-terminal part of this domain forms a helix-loop-helix (HLH) structure which is important for protein–protein interactions (Murre et al., 1994, 1989). The HLH domain is especially important for homo- and heterodimerizations between bHLH proteins, leading to different DNA-binding specificities (Ellenberger, Fass, Arnaud, & Harrison, 1994; Shimizu et al., 1997). Therefore the basic regions of both interacting proteins are necessary for the sequence-specific interaction with the DNA (Heim et al., 2003).

The amino acids at position 9, 13 and 17 in this basic region are crucial for the ability of DNA binding. bHLH TFs which contain histidine, glutamine and arginine (H-E-R) at these positions can bind variants of the E-Box

(CANNTG) (Ledent & Vervoort, 2001). However, since bHLH115 can activate in a *trans*-activation experiment the promoter *AtCHS*min (with CA**TC**TG) and the promoter *AtDFR* (with CA**CG**TG), but not the promoter *AtBAN* (likewise with CA**CG**TG), shows that, beside the H-E-R-motif and the corresponding *cis*-element in the promoter, also other factors like coactivating interaction partners, repressors and the position of the *cis*-element play a role (Heim et al., 2003). MYB TFs act frequently as co-activating interaction partners for bHLH factors in the regulation of biosynthetic pathways (Zimmermann et al., 2004). bHLH proteins without an H-E-R-motif, which are mostly not able to bind DNA, but due to the highly conserved interaction domain can build inactive bHLH complexes by heterodimerization and act as repressors (Heim et al., 2003).

The 133 bHLHs from *A. thaliana* were grouped by their structural similarity in to 12 subgroups. The members of these subgroups often displayed high functional redundancy (Heim et al., 2003).

3.2 MYC-bHLH Transcription Factors Are Well-Known Jasmonic Acid Signalling Components

The bHLH subgroup IIIe consists of four members the so-called MYC-bHLH TFs: bHLH04/MYC4, bHLH05/MYC3/ATR2 (in the following bHLH05/ATR2), bHLH06/MYC2/JIN1 (in the following bHLH06/MYC2) and bHLH28/MYC5/NIG1 (in the following bHLH28/MYC5) (Fig. 4). This subgroup was shown to be crucial for GLS regulation in cooperation with the six MYB TF from subgroup 12 (Frerigmann, Berger, & Gigolashvili, 2014; Schweizer et al., 2013).

bHLH05/ATR2 was originally found similarly to MYB34/ATR1 and CYP83B1/ATR4 in a screen for ATR (Smolen, Pawlowski, Wilensky, & Bender, 2002). The *atr2D* mutant is resistant against the toxic tryptophan analogue 5-MT by activating *ASA1*, a key enzyme in the tryptophan biosynthesis (Smolen et al., 2002). However, a simple overexpression of bHLH05/ATR2 did not mimic all phenotypes, such as activation of many stress genes. The mutant *atr2D* has a point mutation at a conserved amino acid position (D94N) in the N-terminal part of the protein, and an overexpression of this mutated protein and other variants with uncharged amino acids at this position confirmed the stress phenotype of *atr2D*. Thus *atr2D* was likely to be an altered-function allele that identifies a key regulatory site in this bHLH factor. The combination of *atr2D* with *atr1D* showed additive effects on the tryptophan- and the tryptophan-metabolism-pathway; however, at this point the authors could not confirm

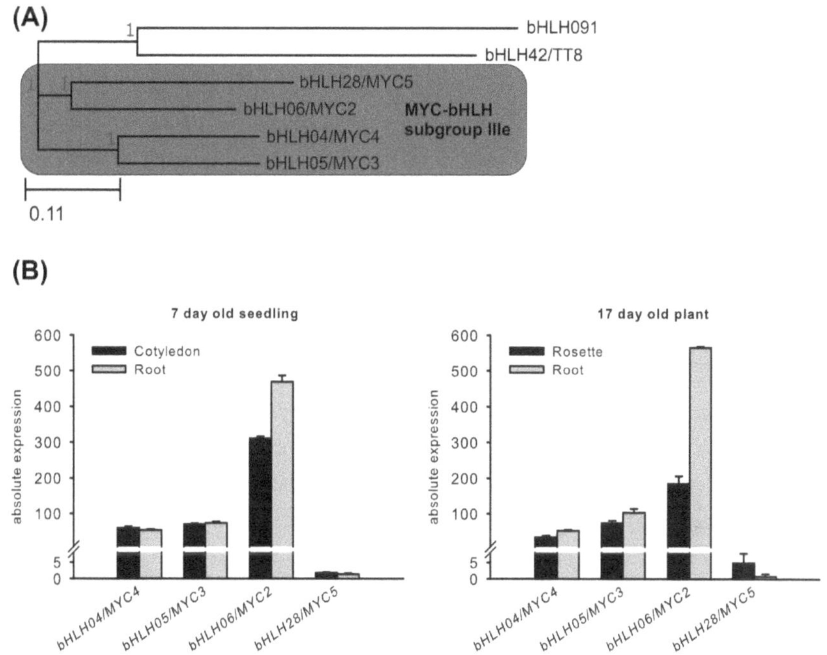

Figure 4 ***MYC-bHLH transcription factors (TFs) in* Arabidopsis thaliana.** (A) The relationship of the four MYC-bHLH TFs from subgroup IIIe together with two outgroup members (bHLH42/TT8 (subgroup IIIf) and bHLH091 (subgroup II)). The six complete protein sequences were used for Clustal omega (Sievers et al., 2011) based calculation of the tree and the phylogenetic tree (newick) viewer (Huerta-Cepas et al., 2010) was used for preparation of the tree. (B) Absolute expression level of the four MYC-bHLH TFs in cotyledons/rosettes and roots of 7- and 17-day-old wild type (Col-0) plants grown under continuous light on soil. Expression data were retrieved via eFP browser (Winter et al., 2007). *From Schmid et al. (2005).*

a physical interaction between the bHLH and MYB TFs, therefore it was not clear if the normal bHLH05/ATR2 variant acts also in tryptophan regulation.

The second bHLH TF of this subgroup bHLH06/MYC2 is an intensively analysed TF with extensive importance for transcriptional reprogramming upon ABA and JA treatment (Abe et al., 2003, 1997; Dombrecht et al., 2007; Lorenzo et al., 2004). In the ABA signal transduction pathway, bHLH06/MYC2 acts in cooperation with MYB TFs and regulates gene expression of drought stress and ABA-inducible-genes (Abe et al., 2003). The overexpression leads, therefore, to an enhanced ABA sensitivity. In the JA-signalling pathway, bHLH06/MYC2 acts as well positively on the transcription of many genes which are connected with the wounding response (Fig. 3). On the other hand, the microarray data from Dombrecht et al. (2007) indicate a negative impact of bHLH06/MYC2 on the JA-dependent expression of many tryptophan and IG genes. Correspondingly, *bhlh06* accumulates more IG upon JA treatment and for many years bHLH06/MYC2 was believed to be a negative regulator of IG (Dombrecht et al., 2007); however, many genes for IG biosynthesis such as *CYP79B2* and *CYP79B3* are co-regulated with bHLH06/MYC2 in the ATTED II database www.atted.jp (Obayashi et al., 2009). Similarly bHLH04/MYC4, which is the closest homolog to bHLH05/MYC3 (Fig. 4), is co-regulated with several genes for AG biosynthesis. The fourth homolog of this subgroup, bHLH28/MYC5/NIG1 (*Na*Cl *i*nducible *g*ene1), is by far the lowest expressed homolog (Fig. 4). This gene was first described in a study for TF which are differentially regulated upon NaCl (Kim & Kim, 2006), and therefore similar as bHLH06/MYC2 in an interrelation with the phytohormone ABA, correspondingly both mutants *bhlh28* and *bhlh06* are more sensitive to ABA and NaCl (Dombrecht et al., 2007; Kim & Kim, 2006).

bHLH06/MYC2 is known for a long time as an important component of the JA signal transduction pathway, and in the last decade the molecular mechanism were elucidated. *COI1* (Coronantine Insensitive 1) encodes for a central protein in the JA signalling and *coi1* lacks the JA-dependent effects. This F-box protein belongs to the SCF-complex and acts together with the JAZ proteins as a receptor for the biological active JA jasmonate-isoleucine (JA-Ile) (Sheard et al., 2010) (Fig. 5). bHLH06/MYC2 interacts with the JAZ proteins and is transcriptional inactive due to the recruitment of further repressor proteins (NINJA (novel interactor of JAZ)), TPL (topless) and TPR (TPL related) (Pauwels et al., 2010). In the presence of JA-Ile the JAZ proteins are ubiquitinated and degraded by the 26S-proteasome and thereupon bHLH06/MYC2 retains its transcriptional activity (Fig. 5) (reviewed by Chini, Boter, & Solano 2009). But *bhlh06* is in comparison to *coi1* not completely JA-insensitive, it was therefore suspected that other proteins act in the JAZ-dependent signal transduction of JA. Indeed, several studies showed in the last years that amongst other TFs mainly several homologous bHLH TFs interact with the JAZ proteins. Members of the closely related bHLH subgroup IIId (Sasaki-Sekimoto et al., 2013; Song et al., 2013), subgroup IIIf (Qi et al., 2011) and also the other close homologs of bHLH06/MYC2 in subgroup

IIIe, except for bHLH28/MYC5, were shown to interact with JAZ proteins and affect thereby the JA signal transduction (Cheng et al., 2011; Fernández-Calvo et al., 2011; Niu & Figueroa, 2011). Although two studies failed to prove physical interaction of bHLH28/MYC5 with any of the JAZ proteins, Figueroa and Browse (2015) could show in a *Daucus carota* protoplast system that bHLH28/MYC5 can similarly to bHLH06/MYC2, bHLH05/MYC3 and bHLH04/MYC4 activate JAZ promoters, which can be similarly inhibited by JAZ1ΔJas repressor, indicating a JAZ-MYC5 interaction in vivo.

With the help of multiple *myc-bhlh* mutants, Fernández-Calvo et al. (2011) identified a redundant function of bHLH04/MYC4, bHLH05/ATR2 and bHLH06/MYC2 on some JA-dependent effects. The JA-dependent induction of *JAZ10* and *VSP2*, the reduced root length upon JA and the susceptibility to the coronatine producing bacterial hemibiotrophic pathogen *Pseudomonas syringae* pv *tomato* DC3000 are already strongly impaired in *bhlh06* alone, but double knockouts with its homologs and particularly the triple mutant *bhlh04 bhlh05 bhlh06* exhibit clearly additive function in JA signalling of the three MYCs, with bHLH06/MYC2 as the main regulator. It is noteworthy that *coi1* was in all respects even more insensitive to JA (Fernández-Calvo et al., 2011).

But there are also differences between these MYCs: while *Pro35S:bHLH06/MYC2* and *Pro35S:bHLH04/MYC4* (with and without JA) exhibit a negative impact on *PDF1* expression, the overexpression of *bHLH05*/ATR2 shows the opposite (Niu & Figueroa, 2011). *PDF1* is regulated by ERF1 (Fig. 3), the integrator of ET and JA signalling pathways (Berrocal-Lobo & Molina, 2004). However, the different *myc-bhlh* knockouts do not show the opposite effect on *PDF1* expression upon JA (Fernández-Calvo et al., 2011). This indicates that the effects are not solely dependent on a specific MYC-bHLH, but due to overexpression or knockout different bHLH homo- and heterodimers can occur, which leads to varying DNA-binding characteristics. The three MYCs have the ability to form homo- and heterodimer with each other, however, with different affinity (Fernández-Calvo et al., 2011). Since all three MYCs bind to the same G-box and only bHLH04/MYC4 displays a slight weaker affinity to the G-box variants (Fernández-Calvo et al., 2011; Godoy et al., 2011), it can be assumed that the differences between these TFs are due to tissue-specific expression and the differences in the interaction capacities with bHLH, JAZs, MYBs and other interaction partners.

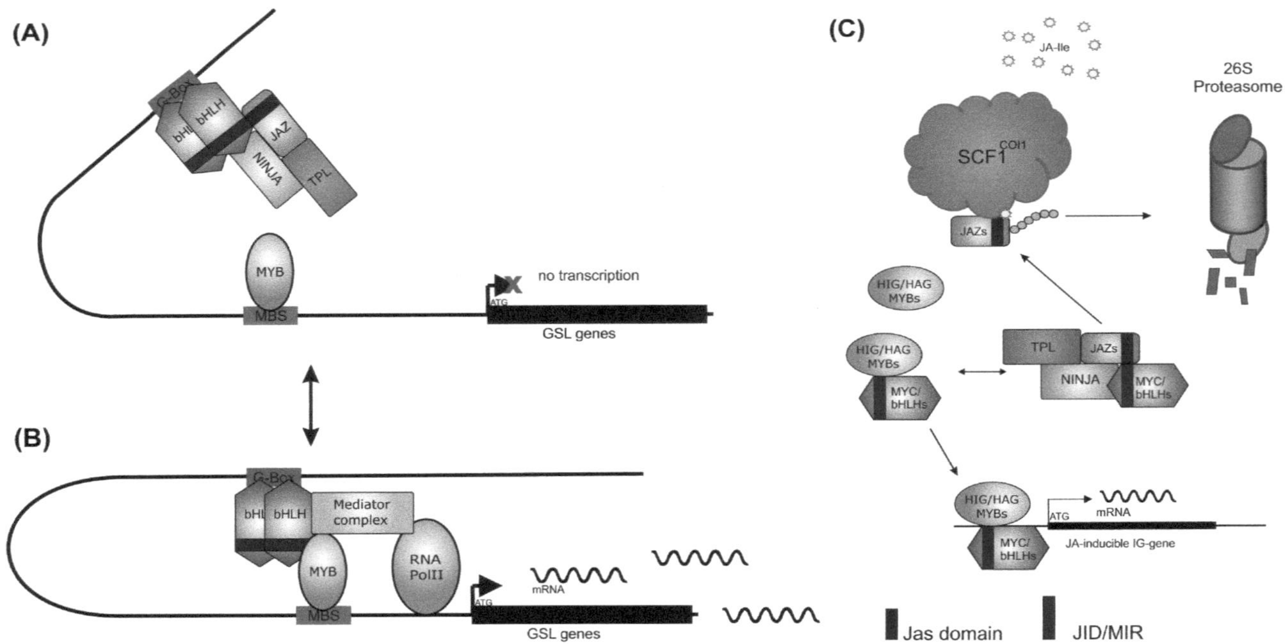

Figure 5 ***Model for regulation of glucosinolate (GSL) by MYB and bHLH in* Arabidopsis.** In the absence of JA, JAZ repressors bind to bHLH and inhibit the interaction between the bHLH proteins and R2R3-MYBs (A), attenuating the potential of the MYB-bHLH complex to activate the promoters of GSL pathway genes (B). Both MYB- and JAZ-binding is mediated via the JID/MIR domain in the bHLH proteins (C). Competitive binding to bHLH proteins by JAZs and MYBs as well as preferential binding of some bHLH proteins to MYBs rather than to JAZ allows the basal transcriptional activity of the bHLH-MYB complex and allows the production of the moderate level of GSL present in wild type plants. In the presence of JA, the JAZ proteins build together with the SCF1-COI1 complex the receptor for bioactive JA-Ile and are directed to 26S-

proteasome and thereby release the bHLH proteins from this transcriptional inactive complex. The binding of bHLH-TFs and JAZ is mediated via the Jas domain in the JAZ proteins, which is also necessary for the interaction with COI1. Thus upon JA the bHLH proteins form more protein complexes with MYBs. Once the MYBs and bHLH are simultaneously recruited to the promoter of GSL genes, they activate the transcription of GSL biosynthesis genes. MYBs binds to MBS boxes present in GSL genes, whereas bHLH proteins bind to G boxes present in the same genes and the transcription of GSL genes is activated. The function of bHLHs is fulfilled by the tethering of the mediator complex, and probably the chromatin modifying factors to DNA, which unwind chromatin, make it accessible to the MYBs and RNA polymerase II. *The figure is adopted from Frerigmann, H., Berger, B., Gigolashvili, T. (2014). bHLH05 is an interaction partner of MYB51 and a novel regulator of glucosinolate biosynthesis in* Arabidopsis. Plant Physiology, *166, 349–369; copyright American Society of Plant Biologists (www.plantphysiol.org).*

4. THE MYC-bHLH Interact With the GSL-MYBs AND ARE THEREBY CRUCIAL REGULATORS OF IG AND AG

The *bhlh06* mutant exhibited the main effects on JA-dependent gene expression; however, the growth of the generalist herbivore *Spodoptera littoralis* was much faster in *bhlh04 bhlh05* in comparison to other single and double mutants with *bhlh06*; and surprisingly *S. littoralis* gained even more weight in *bhlh04 bhlh05 bhlh06* than in *coi1* (Fernández-Calvo et al., 2011). To gain more insights into the underlying mechanism and specifically MYC-regulated genes Schweizer et al. (2013) analysed the transcriptome of the triple mutant *bhlh04 bhlh05 bhlh06*. Surprisingly, they found that out of the 50 most deregulated genes the majority were known genes for GLS biosynthesis. Analysis of important genes for AG and IG biosynthesis showed that these genes are heavily and some completely downregulated in *bhlh04 bhlh05 bhlh06* while these genes were not as strong downregulated in *coi1*. Correspondingly, there were nearly no GSLs detectable in the *myc-bhlh* triple mutant, but still considerable amounts in *coi1* (Schweizer et al., 2013). However, the well-known HIG- and HAG-MYB regulators for GSL biosynthesis were not downregulated, indicating that the MYC-bHLH TFs act together with these regulators.

The MYC-bHLHs bind via an N-terminal domain (the so-called JID-domain (JAZ interaction domain)) to the Jas-domain of the JAZ (Fig. 5) (Chini et al., 2009; Fernández-Calvo et al., 2011). The interaction of the JAZs with COI1 occurs similarly via this Jas-domain, but this interaction is dependent on JA-Ile and JA-Ile-dose-dependent (Chini et al., 2009). The JID domain is also known from other bHLH TFs as the MYB-interaction-region (MIR) (Hichri et al., 2011). Therefore Schweizer et al. (2013) tested successfully the interaction of all six HIG- and HAG-MYBs with the JID-domain of the three MYCs in a yeast-2-hybrid (Y2H) assay. To proof if the MYCs also directly regulate GLS-gene expression together with the HIG-/HAG-MYBs the authors analysed after JA treatment via ChIP-seq if GLS promoters were bound by bHLH06/MYC2. From the 27 deregulated GLS genes in *bhlh04 bhlh05 bhlh06* the promoters of 14 genes showed enriched reads with bHLH06/MYC2:FLAG (Schweizer et al., 2013).

Frerigmann et al. (2014) identified bHLH05/ATR2 in an Y2H screen as an interaction partner of MYB51/HIG1. With SPLIT-YFP and pull-down assay they could proof with full-length bHLH proteins that all MYCs (including bHLH28/MYC5) interact with all six MYB TFs of the subgroup 12. The quantitative pulldown analysis exhibited also a varying interaction capability of different MYB-MYC combinations (Frerigmann et al., 2014). However, this differential interaction capability does not lead to fundamentally different GSL amounts in leaves of the single *myc-bhlh* mutants. Only *bhlh05* had significantly reduced level of AG, but the *bhlh04 bhlh05* double mutant with about 80% reduction of AG and IG demonstrate that these two MYCs act functional redundant and are most important for basal GSL regulation. However upon JA treatment the remaining bHLH06/MYC2 is sufficient to complement the majority of IG (Frerigmann et al., 2014). This might be explained by the lower interaction affinity of bHLH05/ATR2 (Cheng et al., 2011) and bHLH04/MYC4 (Niu & Figueroa, 2011) to JAZ proteins than bHLH06/MYC2; thus without JA treatment most of the bHLH06/MYC2 are caught inactive in MYC-JAZ complexes, while more bHLH05/ATR2 and bHLH04/MYC4 proteins are available to interact with MYB proteins. Upon JA treatment bHLH06/MYC2 proteins are released from the inactive complexes in a high number and are sufficient to substitute most of the missing MYCs in *bhlh04 bhlh05*. The MYCs act in a hierarchical higher role and are normally not rate limiting, this explains why single mutants show no obvious GSL phenotype and additionally an MYC overexpression does not lead to higher GSL level. Corroborating this hypothesis, Frerigmann et al. (2014) showed that the D94N amino acid exchange in *atr2D* impairs the interaction capability with JAZ1 but favours the interaction with MYB34/ATR1. Goossens, Swinnen, Vanden Bossche, Pauwels, and Goossens (2015) could show that this mutation blocks the interaction to all 12 JAZ proteins, only JAZ1 and JAZ10 show slight remaining interaction capability due to the occurrence of a second Jas-like domain. This cryptic MYC2-interacting domain (CMID) lacks conserved residues that have been shown to be essential for COI1 interaction and can thus be considered more specific for interaction with MYC TFs (Moreno et al., 2013). Mutation of the conserved homologous amino acid in MYC2 features the same interaction capacities, thus leading to a constitutively active MYC protein with even stronger effect on stress gene expression like *LOX2* and *TAT3*, but not to induction of *PDF1* (Goossens et al., 2015).

However, increase in MYC/bHLHs alone does not trigger higher GSL biosynthesis if there is not in parallel a higher abundance of the MYB partner. The combination of *atr1D* with *atr2D* leads therefore not only to higher induction of *ASA1* (Smolen et al., 2002), but to a further doubling of IG accumulation and IG gene induction in comparison to *atr1D* alone (Frerigmann et al., 2014). The MYB/MYC complexes regulate a vast number of genes in the GSL core-structure pathway but also in the supplying primary metabolism (Malitsky et al., 2008; Schweizer et al., 2013; Yatusevich et al., 2010), but the most severe effects can be observed on the first genes *CYP79B2, CYP79B3* and *CYP79F1, CYP79F2* of the respective pathway.

The triple mutant *bhlh04 bhlh05 bhlh06* revealed still a slight induction of IG upon JA. Since *myb34/51/122* shows the same induction upon JA this rest IG biosynthesis is presumably not due to the remaining homolog bHLH28/MYC5 but due to regulation by other TFs which also regulate *CYP79B2* and *CYP79B3* for the biosynthesis of other indolic phytoalexins (Frerigmann et al., 2016).

The fourth MYC homolog of subgroup IIIe bHLH28/MYC5 is very low expressed (Fig. 4) and the single mutant *bhlh28* shows no aberrant GSL level. Thus although it interacts with the GSL-regulating MYBs it is not clear if it really affects GSL biosynthesis (Frerigmann et al., 2014). Figueroa and Browse (2015) generated transgenic lines expressing bHLH28/MYC5 fused to an SRDX (SUPERMAN repressive domain X). With this technique the transcriptional activator is converted into a dominant repressor. They found that this bHLH28/MYC5:SRDX fusion phenocopied mutants defective in JA signalling, thus bHLH28/MYC5 activates normally probably specific JA-inducible MYB TFs for pollen development; however, the effect on GSL biosynthesis was not studied.

Thus the four MYC-bHLH TFs connect the signalling of the important defence hormone JA directly with the regulation of the major defence compounds in Arabidopsis. Due to this connection the plant is able to fine-tune the biosynthesis on several levels upon diverse situations. For example, low red:far-red light ratios that occur in shade conditions, such as dense canopies, promote plant growth to outcompete neighbours but have a repressive effect on JA-dependent defences. The MYC-bHLH TFs are destabilized in shade (Chico et al., 2014), and thereby the induction of GSL genes by MeJA or *Botrytis cinerea* is much lower and the plants become more susceptible against *B. cinerea* (Cargnel, Demkura, & Ballaré, 2014).

At present it is unclear if orthologous MYC-bHLH TFs utilize the same mechanism in a complex with other MYB-TFs for different defence compounds in GLS-free species. Lenka et al. (2015) hypothesized that the MeJA inducible biosynthesis of paclitaxel (Taxol) whose corresponding genes contained a large number of E-boxes in the promoters, might be under control of orthologous MYCs. The overexpression of three homologous bHLH from *Taxus cuspidata* in cultured *T. cuspidata* cell culture negatively regulated genes for paclitaxel; however, overexpression of the MYC-bHLH alone did also not induce GLS biosynthesis, but even repress the function of MYB51 (Frerigmann et al., 2014). More research is needed to verify if similar MYC-MYB complexes are similarly crucial for regulation of other secondary metabolites in different species in the plant kingdom.

5. HIG-MYB-DEPENDENT EFFECTS ON IG-RELATED INDOLIC COMPOUNDS

The first intermediate from tryptophan in IG biosynthesis – indole-3-acetaldoxime (IAOx) – is a central metabolite which is used for biosynthesis of IG, IAA, and different indolic phytoalexins (Böttcher et al., 2014; Rajniak, Barco, Clay, & Sattely, 2015; Zhao et al., 2002). Two homologous cytochrome P450 monooxygenases CYP71A12 and CYP71A13 convert IAOx to indole-3-actonitrile (IAN), indole-3-carboxyaldehyde (ICHO) and in presence of cysteine also to CysIAN (Klein, Anarat-Cappillino, & Sattely, 2013; Nafisi et al., 2007) as precursors for camalexin, indole carboxylic acids and cyanogenic phytoalexins (Müller et al., 2015; Rajniak et al., 2015). The resulting IAN might also serve as an intermediate for auxin biosynthesis, suggesting that either the different IAN-metabolizing enzymes are competing for the substrate or IAN is highly channelled within each pathway. However, since these genes are strongly linked with camalexin biosynthesis they are supposed to act primarily in phytoalexin production (Nafisi et al., 2007). *MYB51/HIG1* and *MYB122/HIG2* are also strongly co-regulated with the genes *CYP71A12*, *CYP71A13* and *CYP71B15/PAD3* and the *myb34 myb51 myb122* triple mutant is partially impaired in UV-dependent camalexin accumulation (Frerigmann, Glawischnig, & Gigolashvili, 2015) and Flg22-induced indole carboxylic acids production (Frerigmann et al., 2016). However, Frerigmann et al. (2015) showed that the HIG-MYBs do not directly regulate *CYP71A13*, but positively regulate tryptophan genes and *CYP79B2/3* to provide IAOx as a substrate (Fig. 6).

Overexpression of HIG-MYB factors especially of *MYB34/ATR1* lead to an increase of free IAA (Celenza et al., 2005; Gigolashvili, Berger, et al., 2007) and depending on the overexpression strength even to a high auxin phenotype (Malitsky et al., 2008). Beside the mentioned IAN production by *CYP71A12/A13* homologs, there is another metabolic link between IG and IAA. In the presence of nitrile-specifier proteins (NSPs) I3M can be degraded by myrosinases into IAN (Kong, Kissen, & Bones,

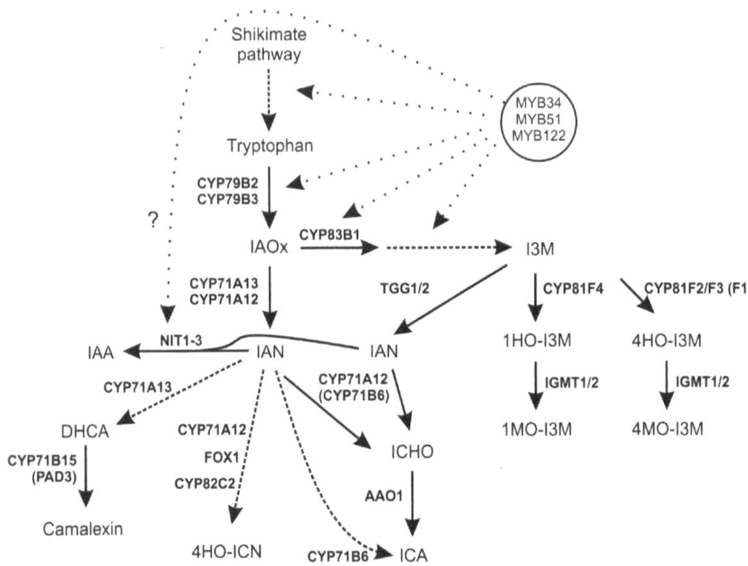

Figure 6 ***HIG-MYB-dependent effect on IG-related indolic compounds.*** The figure illustrates the direct regulatory impact of the three HIG-MYB transcription factors (TFs) (*dotted lines*) on the biosynthesis of related indolic compounds (Gigolashvili, Berger, et al., 2007; Frerigmann & Gigolashvili, 2014a; Frerigmann et al., 2015). *Solid lines* indicate single enzymatic steps, whereas *dashed arrows* stand for several enzymatic steps. *IAOx*, indole-3-acetaldoxime; *IAN*, indole-3-acetonitrile; *IAA*, indole-3-acetic acid; *ICHO*, indole-3-carbaldehyde; *ICA*, indole-3-carboxylic acid; *DHCA*, R-dihydro-camalexic acid; *4HO-ICN*, 4-hydroxyindole-3-carbonyl nitrile; *I3M*, indole-3-methyl-GSL; *1HO-I3M*, 1-hydroxy-I3M; *1MO-I3M*, 1-methoxy-I3M; *4HO-I3M*, 4-hydroxy-I3M; *4MO-I3M*, 4-methoxy-I3M.

2012), which in turn can be hydrolysed by nitrilases (NITs) into IAA (Vorwerk et al., 2001). *NIT1-3* are co-expressed in the databases with *MYB34/ATR1* (Obayashi et al., 2009) and the high expression of *NIT2* in *Pro35S-MYB34* plants but not in plants overexpressing *MYB51/HIG1D* might explain the high-IAA chemotype of *MYB34/ATR1* overexpressing lines (Gigolashvili, Berger, et al., 2007), similarly *NIT3* is more induced by overexpression of *MYB34/ATR1* (Malitsky et al., 2008). However, it is not clear if the *NITs* are directly regulated by HIG-MYBs or secondary effects like GSL-breakdown products induce their expression.

6. GLUCOSINOLATE-FEEDBACK MECHANISM

GLS represent important compounds in plant defence, as signalling compounds but can also serve as storage for nutrients, therefore it seems important for the plant to sense GSL amounts and trigger the biosynthesis if necessary. Such a possible feedback mechanism was first described in the *cyp83b1* (*atr4-1*) mutant (Smolen & Bender, 2002). This mutation leads to a lack of the first enzyme in the IG-core-structure biosynthesis (Naur et al., 2003), which results in very low amounts of IG, since the homologous enzyme CYP83A1 (Fig. 2) mainly takes methionine-derived acetaldoximes as substrates (Hemm, Ruegger, & Chapple, 2003). In addition to the low IG level *cyp83b1* has also elevated transcript level of IG-biosynthesis genes (Smolen & Bender, 2002). This increased gene expression is thought to be a GSL-feedback mechanism which activates *MYB34/ATR1* due to the low IG amounts (Halkier & Gershenzon, 2006). However, the *cyp83b1 myb34* double mutant still exhibits higher expression of *CYP79B2, CYP79B3* and *CYP83B1* than the wild type (Celenza et al., 2005), it seems therefore likely that MYB51/HIG1 and MYB122/HIG2 play also a role in the IG-feedback mechanism. Similarly, Celenza et al. (2005) showed that the IG-free double mutant *cyp79b2 cyp79b3* has elevated transcript level of *MYB34/ATR1* and *CYP83B1*.

This GSL-feedback mechanism is not only restricted to IG, but was also observed in AG. PAPS is the universal sulphate donor and is used for the last step in the GSL core-structure biosynthesis in the cytosol. The block of PAPS transport from the chloroplast to the cytosol in *papst1* (PAPS-TRANSPORTER1) leads to accumulation of IG and AG precursors (desulphoglucosinolates) and reduced accumulation of AG. This comes along with higher transcript level of many AG and IG biosynthetic genes. Amongst them also *APK1* and *APK2* show strong induction (Gigolashvili et al., 2012), both are regulated by HIG- and HAG-MYB TFs (Yatusevich et al., 2010), whereas the not MYB-controlled homologs *APK3* and *APK4* are not induced. The lack of these most important isoforms for GSL biosynthesis, *APK1* and *APK2*, results in a general reduction of GSL down to 20% of wild type level. A microarray and qPCR analysis with *apk1 apk2* revealed an overall induction of HAG- and HIG-MYB regulated AG- and IG-biosynthetic genes and also of all six *HAG*- and *HIG-MYBs* (Mugford et al., 2009). Noteworthy, no other known transcriptional regulator like *MYC-bHLH* TFs nor *OBP2*, *IQD1* or *TFL2* are upregulated in *apk1 apk2* (unpublished suppl. data Mugford et al., 2009). This indicates that the feedback mechanism most likely functions via the six MYBs of subgroup 12.

This previous observations were all made with artificial mutant models which do not occur normally, thus to analyse the impact in a potential naturally occurring situation in which the plants need to adapt their metabolism, Frerigmann and Gigolashvili (2014b) used a long-lasting sulphate deficiency situation. Upon this condition, GSL amounts are dramatically decreased, probably to use the sulphur which is stored in GSL for vitally important functions. The low GSL levels induced also upon this long-lasting sulphate deficiency a GSL-feedback signalling which triggered the expression of *MYB51/HIG1, MYB122/HIG2* and *MYB28/HAG1*, whereas *MYB29/HAG3* and *MYB76/HAG2* were strongly downregulated. However, the exclusive induction of these MYBs could be specific pattern upon sulphate deficiency since the key regulator of sulphate deficiency response Sulphate Limitation 1 (SLIM1) negatively controls *HIG-* and *HAG-MYB* expression at the same time (Frerigmann & Gigolashvili, 2014b). Thus the MYBs represent presumably the central element to integrate various signals to determine the desired GSL production.

In the last three examples, desulphoglucosinolates accumulate as premature intermediates due to the lack of sufficient PAPS for sulphation, but the first two examples block the biosynthesis much earlier. Therefore it is unlikely that intermediates of the GLS core-structure pathway form a signal for this feedback signal. Thus, presumably, readily formed GSLs or rather their breakdown products act as the signal, as shown in three examples: (1) the PEN2-mediated metabolism products of 4MO-I3M upon Flg22 were shown to be crucial as a signal for callose deposition (Clay et al., 2009) and furthermore important to boost the IG and indolic phytoalexin production (Frerigmann et al., 2016); (2) similarly Johansson et al. (2014) and Andersson et al. (2015) could show that IG- and AG-dependent breakdown products are involved in the formation of the hypersensitive response upon pathogen attack; (3) another study reported that the well-known GSL-degrading myrosinase TTG1 is important for ABA-mediated stomatal closure (Zhao, Zhang, Stanley, & Assmann, 2008).

Andersson et al. (2015) concluded that the reduction of the total GSH pool due to the conjugation with GSL-derived isothyocyanates and the corresponding increase in the redox potential represent the signal for the hypersensitive response, but more research is needed to find the real signal and/or proteins involved in the GSL-feedback mechanism cascade to activate *HIG-* and *HAG-MYB* expression for further increase of GSL biosynthesis.

7. HIG-HAG CROSSTALK – POSSIBLE REASONS FOR THE RECIPROCAL REGULATION OF ALIPHATIC AND INDOLIC GLUCOSINOLATES

The deeper analyses of HIG- and HAG-MYBs knockout and overexpression lines indicated a reciprocal negative 'crosstalk' (Gigolashvili, Berger, et al., 2007, 2008). This crosstalk was peculiar between AG and IG but also for some other tryptophan-derived metabolites such as camalexin (Malitsky et al., 2008). However, the regulation of the different GSL classes by the concurrence of the HIG- and HAG-MYBs does not result in clear picture. Whereas *trans*-activation assays in *Nicotiana benthamiana* showed a distinct negative impact of the HAG-TFs on *HIG-MYB* expression (Gigolashvili et al., 2008), the overexpression of the HAG-MYBs in *A. thaliana* did not lead to reduced *HIG*-expression (Malitsky et al., 2008; Sønderby et al., 2007).

In addition to that, the HAG-TFs display a variable impact on the accumulation of IG. While Beekwilder et al. (2008) reported higher IG level in *myb28 myb29*, Sønderby et al. (2007) and Müller et al. (2010) observed no prominent higher IG level, but rather lower IG amounts in the seeds. The overexpression of the HAG-MYBs led to differentially reduced amounts of I3M and 1MO-I3M, but mostly to an induction of 4MO-I3M (Malitsky et al., 2008; Sønderby et al., 2007). Likewise overexpression of the HIG-MYBs did not mandatorily lead to fewer AG, in fact Malitsky et al. (2008) showed that overexpression lines of *MYB34/ATR1* and *MYB51/HIG1* under the '*650*'-promoter contained reduced AG amounts, whereas the use of *35S*-promoter showed only for *MYB51/HIG1* this negative impact, but not for *MYB34/ATR1* and *MYB122/HIG2* (Gigolashvili, Berger, et al., 2007). Thus, it seems that particularly strong expression of the HIGs or HAGs may cause a negative effect on the respective other GLS class, while moderate overexpression has no negative effect.

Interestingly, overexpression of *MYB29/HAG3* and *MYB76/HAG2* did not alter the expression of the *HIGs*, but there is still a strong reduction of IG-biosynthetic genes and of I3M/1MO-I3M (Malitsky et al., 2008). This points to the involvement of other transcriptional acting factors, which have an impact on IG-gene expression due to high expression of the *HAGs*. The reason for that might be the competition of HAG- and HIG-MYBs for interacting bHLH-TFs which are necessary for promoter binding (see previous section).

Correspondingly, it was shown that the *myb34 myb51 myb122* triple mutant has higher AG level in the roots (Frerigmann & Gigolashvili, 2014a). However, this AG accumulation is not caused by a reciprocal induction of *HAG*-gene expression (unpublished data, Frerigmann et al.). The *myb28 myb29* mutant shows also no general induction of the HIGs, it can therefore be ruled out that the HAG- and HIG-MYBs negatively regulate each other. The reciprocal amounts of AG and IG in the corresponding knockout lines are rather caused by the lack of competition for intermediates in the GSL biosynthesis such as GSH, PAPS and commonly used enzymes like the C-S lyase or UGT74B1. Likewise, the competition for electrons for the numerous cytochrome-P450-monooxygenases in these biosynthetic pathways might play a role (Grubb & Abel, 2006).

In the overexpression lines of the *HAGs* and *HIGs*, there is additionally the already mentioned competition for interacting TFs and the GSL-feedback system present, which has at least for the *HAG-MYBs* a negative impact on specific IG. Aspects in this favour are that *HAG*-overexpression lines, which contain, despite a generally lower IG amount, more 4MO-I3M (Sønderby et al., 2007; Malitsky et al., 2008) or fewer 4MO-I3M in *myb28 myb29* and in respective *microRNA-HAG*-lines (Beekwilder et al., 2008; Malitsky et al., 2008). Upon hormone treatment AG and 1MO-I3M reciprocally accumulate in comparison to 4MO-I3M (Frerigmann & Gigolashvili, 2014a). Furthermore, *MYB34/ATR1* (the important regulator for 1MO-I3M accumulation) is co-regulated with the *HAG-MYBs* (Frerigmann et al., 2015; Malitsky et al., 2008; Obayashi et al., 2009). Thus it seems that 1MO-I3M and the (especially short-chained) AG have a similar function, contrary to 4MO-I3M. Without a specific stress condition there is a distinct ratio between these GSL. In *myb28 myb29* this low amount of AG is sensed and the plant tries to restore this ratio by induced expression of *MYB34/ATR1* and reduced expression of *MYB51/HIG1*. This leads to lower level of 4MO-I3M, although in total more IG accumulate (Beekwilder et al., 2008; Sønderby et al., 2007).

8. FURTHER REGULATORY PROTEINS IN GLUCOSINOLATE REGULATION

Many important transcriptional regulators have been identified and characterized in the last decade. However, a recent study showed that the promoters of AG genes are targeted by a large number of TFs unrelated to the MYBs and bHLHs (Li et al., 2014). Strangely, only a few interactions with bHLH06/MYC2 were found, but not with the other MYC-bHLHs or MYB51/HIG1 and MYB122/HIG2 although the promoter of *CYP83B1* was included. However, this systematic yeast-1-hybrid (Y1H) showed that the AG promoters are not only enriched for binding sites for the two known classes of TFs (MYB and bHLH) associated with regulating this pathway (see previous sections), but there was also a strong enrichment in basic leucine zipper domain (bZIP) and homeodomain DNA binding elements and a weaker enrichment in binding elements for the MADs, Auxin response Factor (ARF), and Apetala2 (AP2)/Ethylene-response factor (ERF) TFs (Li et al., 2014). Many of these TFs which bind to the promoters of *AG-genes* bind also to the promoters of the *HAG-MYBs* indicating a feed forward loop.

Li et al. (2014) further analysed in detail candidates, which were found to interact with at least three of the 21 promoters. More than 75% of tested TF mutants had significantly altered accumulation of at least some GLS. However, most of the mutants in this study led to higher GLS in the absence of biotic stresses, this increase disappeared in the presence of pest, suggesting that they are TFs that repress AG accumulation in the absence of biotic or abiotic stress (Li et al., 2014). Although the study was based on the TFs binding to AG-related promoters also a significant influence of the TFs on IG chemotypes was found. Thus regulation of AG is interconnected with IG biosynthesis, probably in an indirect manner (see Section 7).

The HIG/HAG-MYBs and MYC-bHLH TFs mainly regulate genes responsible for the first biosynthetic step (Frerigmann et al., 2014; Frerigmann & Gigolashvili, 2014a; Schweizer et al., 2013; Sønderby et al., 2010). However, this study indicates that there might be many undescribed TFs which allow the plant to fine-tune the regulation of GSL biosynthesis. No relationship was found between pathway position and TF-promoter binding behaviour for AG biosynthetic pathway genes. Thus, their analyses demonstrated that TFs are likely to interact with promoters in nonlinear modules of the pathway (Li et al., 2014).

Several other studies reported TFs which affect GSL regulation. One of this, the Dof-TF OBP2 was described to take part in the transcriptional network of GSLs (Skirycz et al., 2006). The expression of *OBP2* is triggered by MeJA, auxin, mechanical wounding and feeding by generalist herbivores. An overexpression of *OBP2* induces the accumulation of short-chained AG, IG and IAA by two- to threefolds. Quantitative real-time PCR analysis indicated that the IG accumulation takes place due to transcriptional activation of *CYP79B2*, *CYP79B3* and *CYP83B1* which induces the IAOx pool in the plant (Skirycz et al., 2006). However, the microarray data of the *Pro35S:OBP2* plants showed an induction of *MYB34/ATR1* along with all three *HAG-MYBs*. It might be therefore supposed that OBP2 itself is not able to induce all GSL-genes itself, but conducts this activation via regulation of specific HIG- and HAG-MYB.

Many GSL genes were identified by the strong linkage of GSL biosynthesis with sulphur metabolism and gene regulation upon sulphur deficiency. Upon sulphur deficiency SLIM1 positively regulates genes for sulphur assimilation and putative GLS breakdown, at the same time genes for GLS biosynthesis are repressed (Maruyama-Nakashita, Nakamura, Tohge, Saito, & Takahashi, 2006). The microarray data indicated that SLIM1 might have an impact on the expression of *MYB34/ATR1*, and indeed overexpression of SLIM1 in *A. thaliana* cell culture strongly represses *HIG-* and *HAG-MYB* expression (Frerigmann & Gigolashvili, 2014b). These results indicate that at least a part of the negative effects on GSL biosynthesis is mediated by repression of the *HIG-* and *HAG-MYBs*.

A screen of activation-tagged lines for altered GSL showed that the overexpression of IQD1 leads to doubled amounts of AG and IG (Levy, Wang, Kaspi, Parrella, & Abel, 2005). The expression of this calmodulin-binding protein seems to be independent of the classical hormone signalling cascades, but mechanical stimuli and aphid feeding induce slightly *IQD1* transcript. The authors assume that IQD1 decodes the Ca^{2+}-signatures, which are triggered by biotic and abiotic cues. Thus IQD1 could integrate alterations in the cytoplasmic Ca^{2+} concentrations caused by early wounding reactions or pathogens and thereby fine regulate many defence reactions such as the GSL biosynthesis. Interestingly only genes for IG-genes are upregulated by overexpression of *IQD1* but not for AG-genes or the myrosinase TGG1, which are strongly downregulated. Besides this repression of AG biosynthesis, AG accumulate in the same way as IG (Levy et al., 2005), which might arise due to a specific turnover of GLS.

Further analysis showed that IQD1 interacts with kinesin light chain-related protein-1 (KLCR1) and is subcellular localized to microtubules and the cell nucleus (Bürstenbinder et al., 2013). Thus it is hypothesized that IQD1 and related proteins provide Ca2/CaM-regulated scaffolds for facilitating cellular transport of specific cargo along microtubular tracks via kinesin motor proteins. Such intracellular transport processes associated with GLS are, for example, described in the PEN2-dependent innate immune response (Fuchs et al., 2015; Lipka et al., 2005). Fuchs et al. (2015) showed that pathogen-induced CYP81F2 (important for the production of the PEN2-substrate 4MO-I3M) is localized to the ER. Furthermore, they revealed a tight association of the PEN2-containing ER and mitochondria at sites of fungal invasion attempts, thus bringing together all the elements of a highly localized toxin synthesis, activation and export machinery. This turnover of 4MO-I3M by PEN2 is known to induce the genes *CYP79B2*, *CYP79B3* and *CYP83B1* (Frerigmann et al., 2016), which are similarly induced by *IQD1* overexpression. It can be hypothesized that the PEN2-dependent breakdown of GSL is more specific to IG and thereby the AG accumulate in spite of reduced AG gene expression. In addition *TGG1* is also downregulated (Levy et al., 2005), this myrosinase is associated with classical GSL breakdown; however, at some developmental stage *tgg1 tgg2* double mutant accumulates AG (Barth & Jander, 2006), thus repression of AG turnover might lead to this AG increase in *IQD1* overexpression lines. In summary, IQD1 might be involved in the activation of GLS transport processes which induce in a secondary phase the transcriptional activation of IG biosynthesis. It needs to be tested if the PEN2-dependent breakdown is also responsible for the observed negative impact on AG biosynthesis and turnover.

Recently, Zhang, Li, Huai, Zhou, and Kliebenstein (2015) described two new members of the R2R3-MYB TFs to be involved in the regulation of AG. *MYB115* and *MYB118*, two homologous MYBs from subgroup 25 (LAG-MYBs – LOW ALIPHATIC GSL, Fig. 1), were identified due to co-expression with key enzyme encoding genes in the newly evolved benzoyloxy GLS pathway. Y1H analyses revealed that these TFs interacted with the promoters of the biosynthetic genes for the AG modification reactions, but also with genes in the side chain elongation and core-structure pathway of AG (Zhang et al., 2015). Although these MYBs were previously described as positive regulators of embryogenesis (Wang et al., 2009; Zhang, Cao, Qu, & Gu, 2009), the authors observed a negative effect on the expression on the AG genes (Zhang et al., 2015). It was shown for other TFs that they can be activators or repressors depending on the position of the *cis*-element, but they might also interfere with the interaction of HAG-MYBs with MYCs (see previous sections).

The pathway to the benzoyloxy GLS is transcriptionally limited to the developing seeds and expression of *MYB115*, and *MYB118* is similarly mainly found there and only marginally in seedlings and developing plants (Fig. 1 (Winter et al., 2007)). The *myb118* single mutant exhibited higher expression of the modification pathway genes and *CYP83A1* in developing seeds, whereas *myb115* showed an effect on genes in side chain elongation.

However, the double mutant *myb115 myb118* revealed a redundant role on the modification pathway with even higher expression, but not on the side chain elongation (Zhang et al., 2015). Corresponding effects were observed for GSL amounts in the developing seeds: *myb118* had more AG with methylsulphinylalkyl but fewer methylthioalkyl side chains whereas *myb115* showed weaker effects and only more 4MTB (with methylthioalkyl side chain) and slightly reduced 4BZO (benzoyloxy GLS) level. Similar to the observed transcriptional effects *myb115 myb118* showed even stronger effects than *myb118* with even higher amounts of (1) short-chained AG and specifically more (2) AG with methylsulphinylalkyl side chains but lower level of BZO-dependent AG.

Although expressed mainly in seeds MYB115 and MYB118 affect production of AG throughout the whole plant with similar effect. The authors speculated that this might be a secondary effect. There may be an unknown cell type in which they are expressed and affect global GLS patterning by transporting GSL to foliar tissue. Otherwise, it might be a maternal effect, since altering GLS accumulation can influence broad regulatory patterns (Zhang et al., 2015). The activators for the BZ-GLS related genes are still not known, but the authors hypothesized that the homologs in subgroup 25 (Fig. 1) are potential candidates. Data from the co-expression database atted.jp (Obayashi et al., 2009) indicate that at least the two close homologs *MYB22* and *MYB100* are co-regulated with *MYB115* and *MYB118* leaving the opportunity that they act somehow in this transcriptional network.

Although *MYB51/HIG1* and *MYB34/ATR1* are strongly co-regulated with the genes for I3M modification *CYP81F2* and *CYP81F4*, respectively, they are not the direct regulators (Frerigmann & Gigolashvili, 2014a). Schön et al. (2013) found that the IG biosynthesis in *wrky18 wrky40* is critical for *Golovinomyces orontii* resistance. Detailed qPCR analysis revealed a strong increase in *CYP81F2* transcript in the resistant double mutant during early *G. orontii* infection, which results in 4MO-I3M accumulation. Correspondingly, *wrky18 wrky40 cyp81f2* triple reconstituted wild type-like susceptible phenotype. The promoter region of *CYP81F2* harbours 19 W-box and W-box-like motifs partly clustering in the 4000-bp upstream region of the gene, pointing to a direct regulation of this gene by WRKY TFs. However, no direct association of WRKY18 or WRKY40 to the *CYP81F2* promoter could be shown till now (Schön et al., 2013). Instead of this Pandey, Roccaro, Schön, Logemann, and Somssich (2010) showed with ChIP analysis direct binding of WRKY40 to promoter regions of *JAZ8, EDS1* and of the AP2/ERF-type TF-designated *RRTF1*. Thus, due to lack of binding of the *JAZ8* promoter the JA signalling is impaired in *wrky18 wrky40* upon *G. orontii* infection. Furthermore, microarray data from Bartsch et al. (2006) show that infected *eds1* knockout lines have reduced level of *CYP81F2* expression. Thus, the lack of *EDS1* repression in *wrky18 wrky40* can explain the high *CYP81F2* expression.

A recent study identified ERF6 as a substrate of the MPK3/MPK6 cascade important for regulation of IG (Xu et al., 2016). Expression of phospho-mimicking $ERF6^{4D}$ plants show higher expression of important genes for 4MO-I3M biosynthesis – *CYP81F2, IGMT1* and *IGMT2*. Correspondingly, these lines accumulate high levels of 4MO-I3M, while expression of the dominant-negative ERF6-EAR results in opposite phenotypes. ChIP qPCR analysis verified that ERF6 binds to GCC-box in the promoter of these genes (Xu et al., 2016). Activated ERF6 is also able to induce the expression *MYB51/HIG1* and *MYB122/HIG2*, but it is not yet clear if ERF6 is the direct regulator of the MYBs and thereby regulator of both components important for the production of the antimicrobial IG 4MO-I3M.

So far no direct regulator is known to activate or repress expression of *CYP81F4*, which is necessary for the hydroxylation of I3M to form 1HO-I3M and in turn 1MO-I3M.

9. SUMMARY AND FUTURE PERSPECTIVES

The subgroup 12 of R2R3-MYB TFs and the subgroup IIIe of bHLH TFs constitute the main transcriptional regulators of IG and AG biosynthesis in *A. thaliana*. In this regard, the MYB TFs are most important for the specificity and the magnitude of this regulation, since the MYC-bHLH TFs normally not represent a bottleneck for this pathway. The above-mentioned results imply that the three HIG-MYB proteins have mostly similar functional capacities, but due to their specific spatial and developmental expression they affect differentially expression and thereby accumulation of IG and related compounds. Thus the three *HIGs* are induced by various biotic and abiotic signals like defence-related hormones and integrate the information into a specific regulation of IG biosynthesis. Same is valid for the regulation of AG by the three HAG-TFs. Thus in the future the transcriptional regulation of these modulators and their control on the protein level needs to be addressed in depth.

Further work is needed to analyse the particular protein complexes formed in different cell types to regulate specific parts of the GSL pathway.

MYB and bHLH TFs are known from other pathways to act in a trimeric complex with a WD40 protein (Gonzalez, Zhao, Leavitt, & Lloyd, 2008; Pesch et al., 2015; Qi et al., 2011), but till now no candidate WD40 protein was described for GSL biosynthesis. However a systemic Y1H with promoters of the AG pathway indicated that a vast amount of undescribed TFs are involved in the regulation of GLS, further experiments need to dissect if they work in addition to the MYB/bHLH complex or independent of these master regulators to fine-tune the GSL biosynthesis.

ACKNOWLEDGEMENTS

This work was supported by the Deutsche Forschungsgemeinschaft (Project Reference Number: EXC 1028). I would like to thank Richard Jacoby and Tamara Gigolashvili for critical reading of the manuscript.

REFERENCES

Abe, H., Urao, T., Ito, T., Seki, M., Shinozaki, K., & Yamaguchi-Shinozaki, K. (2003). *Arabidopsis At*MYC2 (bHLH) and *At*MYB2 (MYB) function as transcriptional activators in abscisic acid signaling. *Plant Cell, 15*, 63.

Abe, H., Yamaguchi-Shinozaki, K., Urao, T., Iwasaki, T., Hosokawa, D., & Shinozaki, K. (1997). Role of *Arabidopsis* MYC and MYB homologs in drought-and abscisic acid-regulated gene expression. *Plant Cell, 9*, 1859.

Adie, B. A., Pérez-Pérez, J., Pérez-Pérez, M. M., Godoy, M., Sánchez-Serrano, J.-J., Schmelz, E. A., & Solano, R. (2007). ABA is an essential signal for plant resistance to pathogens affecting JA biosynthesis and the activation of defenses in *Arabidopsis*. *Plant Cell, 19*, 1665–1681.

Andersson, M. X., Nilsson, A. K., Johansson, O. N., Boztaş, G., Adolfsson, L. E., Pinosa, F., … Tör, M. (2015). Involvement of the electrophilic isothiocyanate sulforaphane in *Arabidopsis* local defense responses. *Plant Physiology, 167*, 251–261.

Barth, C., & Jander, G. (2006). *Arabidopsis* myrosinases TGG1 and TGG2 have redundant function in glucosinolate breakdown and insect defense. *Plant Journal, 46*, 549–562.

Bartsch, M., Gobbato, E., Bednarek, P., Debey, S., Schultze, J. L., Bautor, J., & Parker, J. E. (2006). Salicylic acid–independent ENHANCED DISEASE SUSCEPTIBILITY1 signaling in *Arabidopsis* immunity and cell death is regulated by the monooxygenase FMO1 and the nudix hydrolase NUDT7. *Plant Cell, 18*, 1038–1051.

Bednarek, P., Pislewska-Bednarek, M., Svatos, A., Schneider, B., Doubsky, J., Mansurova, M., … Schulze-Lefert, P. (2009). A glucosinolate metabolism pathway in living plant cells mediates broad-spectrum antifungal defense. *Science, 323*, 101–106.

Beekwilder, J., van Leeuwen, W., van Dam, N. M., Bertossi, M., Grandi, V., Mizzi, L., … Schipper, B. (2008). The impact of the absence of aliphatic glucosinolates on insect herbivory in *Arabidopsis*. *PLoS One, 3*.

Bekaert, M., Edger, P. P., Hudson, C. M., Pires, J. C., & Conant, G. C. (2012). Metabolic and evolutionary costs of herbivory defense: systems biology of glucosinolate synthesis. *New Phytologist, 196*, 596–605.

Bender, J., & Fink, G. R. (1998). A Myb homologue, ATR1, activates tryptophan gene expression in *Arabidopsis*. *Proceedings of the National Academy of Sciences of the United States of America, 95*, 5655–5660.

Berrocal-Lobo, M., & Molina, A. (2004). Ethylene response factor 1 mediates *Arabidopsis* resistance to the soilborne fungus *Fusarium oxysporum*. *Molecular Plant-Microbe Interactions, 17*, 763–770.

Böttcher, C., Chapman, A., Fellermeier, F., Choudhary, M., Scheel, D., & Glawischnig, E. (2014). The biosynthetic pathway of indole-3-carbaldehyde and indole-3-carboxylic acid derivatives in *Arabidopsis*. *Plant Physiology, 165*, 841–853.

Bürstenbinder, K., Savchenko, T., Müller, J., Adamson, A. W., Stamm, G., Kwong, R., ... Abel, S. (2013). *Arabidopsis* calmodulin-binding protein IQ67-domain 1 localizes to microtubules and interacts with kinesin light chain-related protein-1. *Journal of Biological Chemistry, 288*, 1871–1882.

Buskov, S., Serra, B., Rosa, E., Sorensen, H., & Sorensen, J. C. (2002). Effects of intact glucosinolates and products produced from glucosinolates in myrosinase-catalyzed hydrolysis on the potato cyst nematode (*Globodera rostochiensis* cv. Woll). *Journal of Agricultural and Food Chemistry, 50*, 690–695.

Cargnel, M. D., Demkura, P. V., & Ballaré, C. L. (2014). Linking phytochrome to plant immunity: low red: far:red ratios increase *Arabidopsis* susceptibility to *Botrytis cinerea* by reducing the biosynthesis of indolic glucosinolates and camalexin. *New Phytologist, 204*, 342–354.

Celenza, J. L., Quiel, J. A., Smolen, G. A., Merrikh, H., Silvestro, A. R., Normanly, J., & Bender, J. (2005). The *Arabidopsis* ATR1 MYB transcription factor controls indolic glucosinolate homeostasis. *Plant Physiology, 137*, 253–262.

Cheng, Z., Sun, L., Qi, T., Zhang, B., Peng, W., Liu, Y., & Xie, D. (2011). The bHLH transcription factor MYC3 interacts with the jasmonate ZIM-domain proteins to mediate jasmonate response in *Arabidopsis*. *Molecular Plant, 4*, 279–288.

Chico, J.-M., Fernández-Barbero, G., Chini, A., Fernández-Calvo, P., Díez-Díaz, M., & Solano, R. (2014). Repression of jasmonate-dependent defenses by shade involves differential regulation of protein stability of myc transcription factors and their JAZ repressors in *Arabidopsis*. *Plant Cell, 26*, 1967–1980.

Chini, A., Boter, M., & Solano, R. (2009). Plant oxylipins: COI1/JAZs/MYC2 as the core jasmonic acid-signalling module. *FEBS Journal, 276*, 4682–4692.

Cipollini, D., Enright, S., Traw, M. B., & Bergelson, J. (2004). Salicylic acid inhibits jasmonic acid-induced resistance of *Arabidopsis thaliana* to *Spodoptera exigua*. *Molecular Ecology, 13*, 1643–1653.

Clay, N. K., Adio, A. M., Denoux, C., Jander, G., & Ausubel, F. M. (2009). Glucosinolate metabolites required for an *Arabidopsis* innate immune response. *Science, 323*, 95–101.

Dombrecht, B., Xue, G. P., Sprague, S. J., Kirkegaard, J. A., Ross, J. J., Reid, J. B., ... Manners, J. M. (2007). MYC2 differentially modulates diverse jasmonate-dependent functions in *Arabidopsis*. *Plant Cell, 19*, 2225–2245.

Dubos, C., Stracke, R., Grotewold, E., Weisshaar, B., Martin, C., & Lepiniec, L. (2010). MYB transcription factors in *Arabidopsis*. *Trends in Plant Science, 15*, 573–581.

Ellenberger, T., Fass, D., Arnaud, M., & Harrison, S. C. (1994). Crystal structure of transcription factor E47: E-box recognition by a basic region helix-loop-helix dimer. *Genes & Development, 8*, 970–980.

Fernández-Calvo, P., Chini, A., Fernández-Barbero, G., Chico, J.-M., Gimenez-Ibanez, S., Geerinck, J., ... Franco-Zorrilla, J. M. (2011). The *Arabidopsis* bHLH transcription factors MYC3 and MYC4 are targets of JAZ repressors and act additively with MYC2 in the activation of jasmonate responses. *Plant Cell, 23*, 701–715.

Ferrari, S., Galletti, R., Denoux, C., De Lorenzo, G., Ausubel, F. M., & Dewdney, J. (2007). Resistance to *Botrytis cinerea* induced in *Arabidopsis* by elicitors is independent of salicylic acid, ethylene, or jasmonate signaling but requires PHYTOALEXIN DEFICIENT3. *Plant Physiology, 144*, 367–379.

Figueroa, P., & Browse, J. (2015). Male sterility in *Arabidopsis* induced by overexpression of a MYC5-SRDX chimeric repressor. *Plant Journal, 81*, 849–860.
Frerigmann, H., Berger, B., & Gigolashvili, T. (2014). bHLH05 is an interaction partner of MYB51 and a novel regulator of glucosinolate biosynthesis in *Arabidopsis. Plant Physiology, 166*, 349–369.
Frerigmann, H., & Gigolashvili, T. (2014a). MYB34, MYB51 and MYB122 distinctly regulate indolic glucosinolate biosynthesis in *Arabidopsis thaliana. Molecular Plant, 7*, 814–828.
Frerigmann, H., & Gigolashvili, T. (2014b). Update on the role of R2R3-MYBs in the regulation of glucosinolates upon sulfur deficiency. *Frontiers of Plant Science, 5*, 1–10.
Frerigmann, H., Glawischnig, E., & Gigolashvili, T. (2015). On the role of MYB34, MYB51 and MYB122 in the regulation of camalexin biosynthesis in *Arabidopsis thaliana. Frontiers of Plant Science, 6*, 1–11.
Frerigmann, H., Piślewska-Bednarek, M., Molina, A., Sánchez-Vallet, A., Glawischnig, E., Gigolashvili, T., & Bednarek, P. (2016). Regulation of pathogen triggered tryptophan metabolism in *Arabidopsis thaliana* by MYB transcription factors and indole glucosinolate conversion products. *Molecular Plant, 9*, 682–695.
Fuchs, R., Kopischke, M., Klapprodt, C., Hause, G., Meyer, A. J., Schwarzländer, M., … Lipka, V. (2015). Immobilized subpopulations of leaf epidermal mitochondria mediate PEN2-dependent pathogen entry control in *Arabidopsis. Plant Cell.* TPC2015-00887-RA.
Gabrielsen, O. S., Sentenac, A., & Fromageot, P. (1991). Specific DNA binding by c-Myb: evidence for a double helix-turn-helix-related motif. *Science, 253*, 1140–1143.
Gigolashvili, T., Berger, B., Mock, H. P., Müller, C., Weisshaar, B., & Flügge, U. I. (2007). The transcription factor HIG1/MYB51 regulates indolic glucosinolate biosynthesis in *Arabidopsis thaliana. Plant Journal, 50*, 886–901.
Gigolashvili, T., Engqvist, M., Yatusevich, M., Müller, C., & Flügge, U. I. (2008). HAG2/MYB76 and HAG3/MYB29 exert a specific and coordinated control on the regulation of aliphatic glucosinolate biosynthesis in *Arabidopsis thaliana. New Phytologist, 177*, 627–642.
Gigolashvili, T., Geier, M., Ashykhmina, N., Frerigmann, H., Wulfert, S., Krueger, S., … Flügge, U. I. (2012). The *Arabidopsis* thylakoid ADP/ATP carrier TAAC has an additional role in supplying plastidic phosphoadenosine 5′-phosphosulfate to the cytosol. *Plant Cell, 24*, 4187–4204.
Gigolashvili, T., Yatusevich, R., Berger, B., Müller, C., & Flügge, U. I. (2007). The R2R3-MYB transcription factor HAG1/MYB28 is a regulator of methionine-derived glucosinolate biosynthesis in *Arabidopsis thaliana. Plant Journal, 51*, 247–261.
Godoy, M., Franco-Zorrilla, J. M., Pérez-Pérez, J., Oliveros, J. C., Lorenzo, Ó., & Solano, R. (2011). Improved protein-binding microarrays for the identification of DNA-binding specificities of transcription factors. *Plant Journal, 66*, 700–711.
Goff, S. A., Cone, K. C., & Chandler, V. L. (1992). Functional analysis of the transcriptional activator encoded by the maize B gene: evidence for a direct functional interaction between two classes of regulatory proteins. *Genes & Development, 6*, 864–875.
Gonzalez, A., Zhao, M., Leavitt, J. M., & Lloyd, A. M. (2008). Regulation of the anthocyanin biosynthetic pathway by the TTG1/bHLH/Myb transcriptional complex in *Arabidopsis* seedlings. *Plant Journal, 53*, 814–827.
Goossens, J., Swinnen, G., Vanden Bossche, R., Pauwels, L., & Goossens, A. (2015). Change of a conserved amino acid in the MYC2 and MYC3 transcription factors leads to release of JAZ repression and increased activity. *New Phytologist, 206*, 1229–1237.
Grubb, C. D., & Abel, S. (2006). Glucosinolate metabolism and its control. *Trends in Plant Science, 11*, 89–100.
Halkier, B. A., & Gershenzon, J. (2006). Biology and biochemistry of glucosinolates. *Annual Review of Plant Biology, 57*, 303–333.

Heim, M. A., Jakoby, M., Werber, M., Martin, C., Weisshaar, B., & Bailey, P. C. (2003). The basic helix–loop–helix transcription factor family in plants: a genome-wide study of protein structure and functional diversity. *Molecular Biology and Evolution, 20*, 735–747.

Hemm, M. R., Ruegger, M. O., & Chapple, C. (2003). The *Arabidopsis ref2* mutant is defective in the gene encoding CYP83A1 and shows both phenylpropanoid and glucosinolate phenotypes. *Plant Cell, 15*, 179–194.

Hichri, I., Barrieu, F., Bogs, J., Kappel, C., Delrot, S., & Lauvergeat, V. (2011). Recent advances in the transcriptional regulation of the flavonoid biosynthetic pathway. *Journal of Experimental Botany*. erq442.

Hirai, M. Y., Klein, M., Fujikawa, Y., Yano, M., Goodenowe, D. B., Yamazaki, Y., … Suzuki, H. (2005). Elucidation of gene-to-gene and metabolite-to-gene networks in *Arabidopsis* by integration of metabolomics and transcriptomics. *Journal of Biological Chemistry, 280*, 25590–25595.

Hirai, M. Y., Sugiyama, K., Sawada, Y., Tohge, T., Obayashi, T., Suzuki, A., … Aoki, K. (2007). Omics-based identification of *Arabidopsis* MYB transcription factors regulating aliphatic glucosinolate biosynthesis. *Proceedings of the National Academy of Sciences of the United States of America, 104*, 6478–6483.

Hirai, M. Y., Yano, M., Goodenowe, D. B., Kanaya, S., Kimura, T., Awazuhara, M., … Saito, K. (2004). Integration of transcriptomics and metabolomics for understanding of global responses to nutritional stresses in *Arabidopsis thaliana*. *Proceedings of the National Academy of Sciences of the United States of America, 101*, 10205–10210.

Huerta-Cepas, J., Dopazo, J., & Gabaldón, T. (2010). ETE: a python Environment for Tree Exploration. *BMC Bioinformatics, 11*, 1.

Huot, B., Yao, J., Montgomery, B. L., & He, S. Y. (2014). Growth–defense tradeoffs in plants: a balancing act to optimize fitness. *Molecular Plant, 7*, 1267–1287.

Johansson, O. N., Fantozzi, E., Fahlberg, P., Nilsson, A. K., Buhot, N., Tör, M., & Andersson, M. X. (2014). Role of the penetration-resistance genes PEN1, PEN2 and PEN3 in the hypersensitive response and race-specific resistance in *Arabidopsis thaliana*. *Plant Journal, 79*, 466–476.

Kim, J., & Kim, H.-Y. (2006). Functional analysis of a calcium-binding transcription factor involved in plant salt stress signaling. *FEBS Letters, 580*, 5251–5256.

Klein, A. P., Anarat-Cappillino, G., & Sattely, E. S. (2013). Minimum set of cytochromes P450 for reconstituting the biosynthesis of camalexin, a major *Arabidopsis* antibiotic. *Angewandte Chemie International Edition, 52*, 13625–13628.

Kong, X. Y., Kissen, R., & Bones, A. M. (2012). Characterization of recombinant nitrile-specifier proteins (NSPs) of *Arabidopsis thaliana*: dependency on Fe (II) ions and the effect of glucosinolate substrate and reaction conditions. *Phytochemistry, 84*, 7–17.

Lazzeri, L., Curto, G., Leoni, O., & Dallavalles, E. (2004). Effects of glucosinolates and their enzymatic hydrolysis products via myrosinase on the root-knot nematode *Meloidogyne incognita* (Kofoid et White) Chitw. *Journal of Agricultural and Food Chemistry, 52*, 6703–6707.

Ledent, V., & Vervoort, M. (2001). The basic helix-loop-helix protein family: comparative genomics and phylogenetic analysis. *Genome Research, 11*, 754–770.

Lenka, S. K., Nims, N. E., Vongpaseuth, K., Boshar, R. A., Roberts, S. C., & Walker, E. L. (2015). Jasmonate-responsive expression of paclitaxel biosynthesis genes in *Taxus cuspidata* cultured cells is negatively regulated by the bHLH transcription factors *Tc*JAMYC1, *Tc*JAMYC2, and *Tc*JAMYC4. *Frontiers of Plant Science, 6*.

Levy, M., Wang, Q., Kaspi, R., Parrella, M. P., & Abel, S. (2005). *Arabidopsis* IQD1, a novel calmodulin-binding nuclear protein, stimulates glucosinolate accumulation and plant defense. *Plant Journal, 43*, 79–96.

Li, B., Gaudinier, A., Tang, M., Taylor-Teeples, M., Nham, N. T., Ghaffari, C., … Brady, S. M. (2014). Promoter-based integration in plant defense regulation. *Plant Physiology, 166*, 1803–1820.

Li, Y., Sawada, Y., Hirai, A., Sato, M., Kuwahara, A., Yan, X., & Hirai, M. Y. (2013). Novel insights into the function of *Arabidopsis* R2R3-MYB transcription factors regulating aliphatic glucosinolate biosynthesis. *Plant & Cell Physiology, 54*, 1335–1344.

Lipka, V., Dittgen, J., Bednarek, P., Bhat, R., Wiermer, M., Stein, M., ... Scheel, D. (2005). Pre-and postinvasion defenses both contribute to nonhost resistance in *Arabidopsis*. *Science, 310*, 1180–1183.

Lorenzo, O., Chico, J. M., Sanchez-Serrano, J. J., & Solano, R. (2004). JASMONATE-INSENSITIVE1 encodes a MYC transcription factor essential to discriminate between different jasmonate-regulated defense responses in *Arabidopsis*. *Plant Cell, 16*, 1938–1950.

Malitsky, S., Blum, E., Less, H., Venger, I., Elbaz, M., Morin, S., ... Aharoni, A. (2008). The transcript and metabolite networks affected by the two clades of *Arabidopsis* glucosinolate biosynthesis regulators. *Plant Physiology, 148*, 2021–2049.

Maruyama-Nakashita, A., Nakamura, Y., Tohge, T., Saito, K., & Takahashi, H. (2006). *Arabidopsis* SLIM1 is a central transcriptional regulator of plant sulfur response and metabolism. *Plant Cell, 18*, 3235–3251.

Mikkelsen, M. D., Petersen, B. L., Glawischnig, E., Jensen, A. B., Andreasson, E., & Halkier, B. A. (2003). Modulation of CYP79 genes and glucosinolate profiles in *Arabidopsis* by defense signaling pathways. *Plant Physiology, 131*, 298–308.

Moreno, J. E., Shyu, C., Campos, M. L., Patel, L. C., Chung, H. S., Yao, J., ... Howe, G. A. (2013). Negative feedback control of jasmonate signaling by an alternative splice variant of JAZ10. *Plant Physiology, 162*, 1006–1017.

Mugford, S. G., Yoshimoto, N., Reichelt, M., Wirtz, M., Hill, L., Mugford, S. T., ... Kramell, R. (2009). Disruption of adenosine-5′-phosphosulfate kinase in *Arabidopsis* reduces levels of sulfated secondary metabolites. *Plant Cell, 21*, 910–927.

Müller, R., de Vos, M., Sun, J. Y., Sønderby, I. E., Halkier, B. A., Wittstock, U., & Jander, G. (2010). Differential effects of indole and aliphatic glucosinolates on lepidopteran herbivores. *Journal of Chemical Ecology, 36*, 905–913.

Müller, T., Böttcher, C., Morbitzer, R., Götz, C., Lehmann, J., Lahaye, T., & Glawischnig, E. (2015). TALEN-mediated generation and metabolic analysis of camalexin-deficient *cyp71a12 cyp71a13* double knockout lines. *Plant Physiology*. pp. 00481.02015.

Murre, C., Bain, G., van Dijk, M. A., Engel, I., Furnari, B. A., Massari, M. E., ... Stuiver, M. H. (1994). Structure and function of helix-loop-helix proteins. *Biochimica et Biophysica Acta (BBA) – Gene Structure and Expression, 1218*, 129–135.

Murre, C., McCaw, P. S., Vaessin, H., Caudy, M., Jan, L., Jan, Y., ... Lassar, A. B. (1989). Interactions between heterologous helix-loop-helix proteins generate complexes that bind specifically to a common DNA sequence. *Cell, 58*, 537–544.

Nafisi, M., Goregaoker, S., Botanga, C. J., Glawischnig, E., Olsen, C. E., Halkier, B. A., & Glazebrook, J. (2007). *Arabidopsis* cytochrome P450 monooxygenase 71A13 catalyzes the conversion of indole-3-acetaldoxime in camalexin synthesis. *Plant Cell, 19*, 2039–2052.

Naur, P., Petersen, B. L., Mikkelsen, M. D., Bak, S., Rasmussen, H., Olsen, C. E., & Halkier, B. A. (2003). CYP83A1 and CYP83B1, two nonredundant cytochrome P450 enzymes metabolizing oximes in the biosynthesis of glucosinolates in *Arabidopsis*. *Plant Physiology, 133*, 63–72.

Niu, Y., & Figueroa, P. (2011). Characterization of JAZ-interacting bHLH transcription factors that regulate jasmonate responses in *Arabidopsis*. *Journal of Experimental Botany, 62*, 2143–2154.

Obayashi, T., Hayashi, S., Saeki, M., Ohta, H., & Kinoshita, K. (2009). ATTED-II provides coexpressed gene networks for *Arabidopsis*. *Nucleic Acids Research, 37*, D987–D991.

Ogata, K., Morikawa, S., Nakamura, H., Sekikawa, A., Inoue, T., Kanai, H., ... Nishimura, Y. (1994). Solution structure of a specific DNA complex of the Myb DNA-binding domain with cooperative recognition helices. *Cell, 79*, 639–648.

Pandey, S. P., Roccaro, M., Schön, M., Logemann, E., & Somssich, I. E. (2010). Transcriptional reprogramming regulated by WRKY18 and WRKY40 facilitates powdery mildew infection of *Arabidopsis*. *Plant Journal, 64*, 912–923.

Pauwels, L., Barbero, G. F., Geerinck, J., Tilleman, S., Grunewald, W., Pérez, A. C., ... Gil, E. (2010). NINJA connects the co-repressor TOPLESS to jasmonate signalling. *Nature, 464*, 788–791.

Pesch, M., Schultheiß, I., Klopffleisch, K., Uhrig, J. F., Koegl, M., Clemen, C. S., ... Hülskamp, M. (2015). TRANSPARENT TESTA GLABRA1 and GLABRA1 compete for binding to GLABRA3 in *Arabidopsis*. *Plant Physiology, 168*, 584–597.

Petrovas, C., Jeay, S., Lewis, R. E., & Sonenshein, G. E. (2003). B-Myb repressor function is regulated by cyclin A phosphorylation and sequences within the C-terminal domain. *Oncogene, 22*, 2011–2020.

Pfalz, M., Mikkelsen, M. D., Bednarek, P., Olsen, C. E., Halkier, B. A., & Kroymann, J. (2011). Metabolic engineering in *Nicotiana benthamiana* reveals key enzyme functions in *Arabidopsis* indole glucosinolate modification. *Plant Cell, 23*, 716–729.

Pfalz, M., Vogel, H., & Kroymann, J. (2009). The gene controlling the Indole Glucosinolate Modifier1 quantitative trait locus alters indole glucosinolate structures and aphid resistance in *Arabidopsis*. *Plant Cell, 21*, 985–999.

Qi, T., Song, S., Ren, Q., Wu, D., Huang, H., Chen, Y., ... Xie, D. (2011). The jasmonate-ZIM-domain proteins interact with the WD-repeat/bHLH/MYB complexes to regulate jasmonate-mediated anthocyanin accumulation and trichome initiation in *Arabidopsis thaliana*. *Plant Cell, 23*, 1795–1814.

Rajniak, J., Barco, B., Clay, N. K., & Sattely, E. S. (2015). A new cyanogenic metabolite in *Arabidopsis* required for inducible pathogen defence. *Nature, 525*, 376–379.

Renwick, J. A. A., Haribal, M., Gouinguene, S., & Städler, E. (2006). Isothiocyanates stimulating oviposition by the diamondback moth, *Plutella xylostella*. *Journal of Chemical Ecology, 32*, 755–766.

Riechmann, J. L., Heard, J., Martin, G., Reuber, L., Jiang, C.-Z., Keddie, J., ... Samaha, R. (2000). *Arabidopsis* transcription factors: genome-wide comparative analysis among eukaryotes. *Science, 290*, 2105–2110.

Sarwar, M., Kirkegaard, J. A., Wong, P. T. W., & Desmarchelier, J. M. (1998). Biofumigation potential of Brassicas. *Plant and Soil, 201*, 103–112.

Sasaki-Sekimoto, Y., Jikumaru, Y., Obayashi, T., Saito, H., Masuda, S., Kamiya, Y., ... Shirasu, K. (2013). bHLH transcription factors JA-ASSOCIATED MYC2-LIKE 1, JAM2 and JAM3 are negative regulators of jasmonate responses in *Arabidopsis*. *Plant Physiology, 163*, 291–304.

Schneider, A., Kirch, T., Gigolashvili, T., Mock, H. P., Sonnewald, U., Simon, R., ... Werr, W. (2005). A transposon-based activation-tagging population in *Arabidopsis thaliana* (TAMARA) and its application in the identification of dominant developmental and metabolic mutations. *FEBS Letters, 579*, 4622–4628.

Schön, M., Töller, A., Diezel, C., Roth, C., Westphal, L., Wiermer, M., & Somssich, I. E. (2013). Analyses of *wrky18 wrky40* plants reveal critical roles of SA/EDS1 signaling and indole-glucosinolate biosynthesis for *Golovinomyces orontii* resistance and a loss-of resistance towards *Pseudomonas syringae* pv. *tomato* AvrRPS4. *Molecular Plant-Microbe Interactions, 26*, 758–767.

Schmid, M., Davison, T. S., Henz, S. R., Pape, U. J., Demar, M., Vingron, M., ... Lohmann, J. U. (2005). A gene expression map of Arabidopsis thaliana development. *Nature genetics, 37*, 501–506.
Schweizer, F., Fernández-Calvo, P., Zander, M., Diez-Diaz, M., Fonseca, S., Glauser, G., ... Reymond, P. (2013). *Arabidopsis* basic helix-loop-helix transcription factors MYC2, MYC3, and MYC4 regulate glucosinolate biosynthesis, insect performance, and feeding behavior. *Plant Cell, 25*, 3117–3132.
Segarra, S., Mir, R., Martínez, C., & León, J. (2010). Genome-wide analyses of the transcriptomes of salicylic acid-deficient versus wild type plants uncover Pathogen and Circadian Controlled 1 (PCC1) as a regulator of flowering time in *Arabidopsis*. *Plant, Cell & Environment, 33*, 11–22.
Sheard, L. B., Tan, X., Mao, H., Withers, J., Ben-Nissan, G., Hinds, T. R., ... He, S. Y. (2010). Jasmonate perception by inositol-phosphate-potentiated COI1-JAZ co-receptor. *Nature, 468*, 400–405.
Shimizu, T., Toumoto, A., Ihara, K., Shimizu, M., Kyogoku, Y., Ogawa, N., ... Hakoshima, T. (1997). Crystal structure of PHO4 bHLH domain–DNA complex: flanking base recognition. *EMBO Journal, 16*, 4689–4697.
Sievers, F., Wilm, A., Dineen, D., Gibson, T. J., Karplus, K., Li, W., ... Söding, J. (2011). Fast, scalable generation of high-quality protein multiple sequence alignments using Clustal Omega. *Molecular systems biology*, 7.
Skirycz, A., Reichelt, M., Burow, M., Birkemeyer, C., Rolcik, J., Kopka, J., ... Szopa, J. (2006). DOF transcription factor AtDof1. 1 (OBP2) is part of a regulatory network controlling glucosinolate biosynthesis in *Arabidopsis*. *Plant Journal, 47*, 10–24.
Smith, B. J., & Kirkegaard, J. A. (2002). In vitro inhibition of soil microorganisms by 2-phenylethyl isothiocyanate. *Plant Pathology, 51*, 585–593.
Smolen, G., & Bender, J. (2002). *Arabidopsis* cytochrome P450 *cyp83B1* mutations activate the tryptophan biosynthetic pathway. *Genetics, 160*, 323–332.
Smolen, G. A., Pawlowski, L., Wilensky, S. E., & Bender, J. (2002). Dominant alleles of the basic helix-loop-helix transcription factor ATR2 activate stress-responsive genes in *Arabidopsis*. *Genetics, 161*, 1235–1246.
Sønderby, I. E., Burow, M., Rowe, H. C., Kliebenstein, D. J., & Halkier, B. A. (2010). A complex interplay of three R2R3 MYB transcription factors determines the profile of aliphatic glucosinolates in *Arabidopsis*. *Plant Physiology, 153*, 348–363.
Sønderby, I. E., Hansen, B. G., Bjarnholt, N., Ticconi, C., Halkier, B. A., & Kliebenstein, D. J. (2007). A systems biology approach identifies a R2R3 MYB gene subfamily with distinct and overlapping functions in regulation of aliphatic glucosinolates. *PLoS One, 2*.
Song, S., Qi, T., Fan, M., Zhang, X., Gao, H., Huang, H., ... Xie, D. (2013). The bHLH subgroup IIId factors negatively regulate jasmonate-mediated plant defense and development. *PLoS Genetics, 9*, e1003653.
Stracke, R., Werber, M., & Weisshaar, B. (2001). The R2R3-MYB gene family in *Arabidopsis thaliana*. *Current Opinion in Plant Biology, 4*, 447–456.
Toufighi, K., Brady, S. M., Austin, R., Ly, E., & Provart, N. J. (2005). The botany array resource: e-Northerns, expression angling, and promoter analyses. *Plant Journal, 43*, 153.
Vorwerk, S., Biernacki, S., Hillebrand, H., Janzik, I., Müller, A., Weiler, E. W., & Piotrowski, M. (2001). Enzymatic characterization of the recombinant *Arabidopsis thaliana* nitrilase subfamily encoded by the NIT 2/NIT 1/NIT 3-gene cluster. *Planta, 212*, 508–516.
Wang, X., Niu, Q.-W., Teng, C., Li, C., Mu, J., Chua, N.-H., & Zuo, J. (2009). Overexpression of PGA37/MYB118 and MYB115 promotes vegetative-to-embryonic transition in *Arabidopsis*. *Cell Research, 19*, 224–235.

Winter, D., Vinegar, B., Nahal, H., Ammar, R., Wilson, G. V., & Provart, N. J. (2007). An "electronic fluorescent pictograph" browser for exploring and analyzing large-scale biological data sets. *PLoS One, 2*.

Xu, J., Meng, J., Meng, X., Zhao, Y., Liu, J., Sun, T., ... Zhang, S. (2016). Pathogen-responsive MPK3 and MPK6 reprogram the biosynthesis of indole glucosinolates and their derivatives in *Arabidopsis* immunity. *Plant Cell*. tpc. 00871.02015.

Yatusevich, R., Mugford, S. G., Matthewman, C., Gigolashvili, T., Frerigmann, H., Delaney, S., ... Kopriva, S. (2010). Genes of primary sulfate assimilation are part of the glucosinolate biosynthetic network in *Arabidopsis thaliana*. *Plant Journal, 62*, 1–11.

Zang, Y. X., Kim, D. H., Park, B. S., & Hong, S. B. (2009). Metabolic engineering of indole glucosinolates in Chinese cabbage hairy roots expressing *Arabidopsis* CYP79B2, CYP79B3, and CYP83B1. *Biotechnology and Bioprocess Engineering, 14*, 467–473.

Zhang, Y., Cao, G., Qu, L.-J., & Gu, H. (2009). Involvement of an R2R3-MYB transcription factor gene AtMYB118 in embryogenesis in *Arabidopsis*. *Plant Cell Reports, 28*, 337–346.

Zhang, Y., Li, B., Huai, D., Zhou, Y., & Kliebenstein, D. J. (2015). The conserved transcription factors, MYB115 and MYB118, control expression of the newly evolved benzoyloxy glucosinolate pathway in *Arabidopsis thaliana*. *Frontiers of Plant Science, 6*.

Zhao, Y., Hull, A. K., Gupta, N. R., Goss, K. A., Alonso, J., Ecker, J. R., ... Celenza, J. L. (2002). Trp-dependent auxin biosynthesis in *Arabidopsis*: involvement of cytochrome P450s CYP79B2 and CYP79B3. *Genes & Development, 16*, 3010–3012.

Zhao, Z., Zhang, W., Stanley, B. A., & Assmann, S. M. (2008). Functional proteomics of *Arabidopsis thaliana* guard cells uncovers new stomatal signaling pathways. *Plant Cell, 20*, 3210–3226.

Zimmermann, I. M., Heim, M. A., Weisshaar, B., & Uhrig, J. F. (2004). Comprehensive identification of *Arabidopsis thaliana* MYB transcription factors interacting with R/B-like bHLH proteins. *Plant Journal, 40*, 22–34.

CHAPTER FIVE

Glucosinolate Synthesis in the Context of Plant Metabolism

S. Kopriva[1], T. Gigolashvili
University of Cologne, Cologne, Germany
[1]Corresponding author: E-mail: skopriva@uni-koeln.de

Contents

Abstract

The glucosinolate biosynthesis pathway is an important component of plant secondary metabolism. The transcriptional regulation of this pathway is well described, including the nature of corresponding transcription factors. However, glucosinolate synthesis requires also cofactors and intermediates derived from primary metabolism. In particular the pathway has to be well coordinated with sulphur metabolism, through the donor of activated sulphate, 3′-phosphoadenosine-5′-phosphosulphate (PAPS). The availability of PAPS and other intermediates thus affects the rate of glucosinolate synthesis. Interestingly, very recently three enzymes connected to glucosinolate synthesis have been shown to undergo redox regulation, pointing to the complexity of the control of the pathway. In addition, glucosinolate metabolism is divided across multiple cellular compartments including plastids, cytosol, vacuole, endomembrane structures, as well as mitochondria. Therefore, a suite of transporters are necessary to ensure the exchange

Advances in Botanical Research, Volume 80
ISSN 0065-2296
http://dx.doi.org/10.1016/bs.abr.2016.07.002

of intermediates between the compartments but also to enable the long-distance transport between plant organs and tissues. Here we review our current knowledge on how glucosinolate synthesis is connected to general plant metabolism, on the newly discovered redox regulation of the pathway and on the role of various transporters in glucosinolate biosynthesis and their inter- and intracellular distribution. We believe that these processes hold a key for successful engineering of glucosinolate pathways and improved yield and nutritional properties of crops.

1. INTRODUCTION

Given the interest in glucosinolates for improving plant defence, increasing value of crops, improving health and giving flavour to our cooking, it is not surprising the pathway of their synthesis is well understood. Most of the genes encoding the biosynthetic enzymes have been identified, as well as transporters necessary for moving the metabolic intermediates in the cell and the final products between the cells and organs (Sonderby, Geu-Flores, & Halkier, 2010). The biosynthesis is rather complex with more than 40 genes participating, creating the variety of glucosinolates products, depending on amino acid used for the synthesis and its modifications. The biosynthesis can be divided roughly into three modules: (1) chain elongation of methionine to produce aliphatic glucosinolates of differing chain length, (2) the core glucosinolates structure synthesis and (3) side chain modification (Sonderby et al., 2010). The good understanding of glucosinolates biosynthesis enabled synthetic approaches in yeast and tobacco, which normally do not synthesize glucosinolates (Geu-Flores et al., 2009; Mikkelsen et al., 2012; Moldrup et al., 2012). However, the initial efforts revealed that glucosinolate synthesis is more complex than just the interplay of the biosynthetic enzymes and that other factors have to be taken into account to be able to produce glucosinolates according to demand (Moldrup, Geu-Flores, Olsen, & Halkier, 2011).

To be able to manipulate glucosinolate synthesis, it is necessary to understand the regulation of the synthesis as well as translocation, storage and turnover. The pathway is coordinately regulated on a transcriptional level by two small groups of MYB factors specific for aliphatic and indolic glucosinolates and a group of bHLH factors (see, chapter 4 in this volume) (Frerigmann, Berger, & Gigolashvili, 2014; Gigolashvili, Berger, et al., 2007; Gigolashvili, Yatusevich, Berger, Muller, & Flugge, 2007; Malitsky et al., 2008; Schweizer et al., 2013; Sonderby et al., 2007). However, apart of the transcriptional regulation, other processes contribute to the control of glucosinolate synthesis and accumulation, as the glucosinolates are an integral and important part of plant metabolism. Indeed, whole metabolic reconstruction and mathematical modelling revealed that glucosinolate synthesis uses up to 15% of total photosynthetic energy (Bekaert, Edger, Hudson, Pires, & Conant, 2012). In this chapter, the metabolic processes connected with glucosinolate synthesis, mainly the availability of precursors and transport processes, which now form an integral part of the pathway, will be reviewed.

2. CONNECTION OF GLUCOSINOLATES WITH SULPHUR METABOLISM

Glucosinolates are primarily seen as amino acid-derived secondary compounds, but they are also an important component of sulphur metabolism. Each glucosinolate contains two sulphur atoms, in the thioglucosidic bond and the sulphate modifying the core structure, the aliphatic ones possess mostly, but not always, a third atom derived from methionine, unless it is lost during the side chains modifications, e.g., in synthesis of allyl glucosinolates. With concentration in the range of 0.5—5 μmol/g FW in mature Arabidopsis leaves, glucosinolates form a large pool of organic sulphur, similar to size of glutathione pool. This is corroborated by several flux analyses quantifying incorporation of sulphur from sulphate into glucosinolates and primary reduced sulphur compounds cysteine, glutathione and proteins. Under standard conditions the flux of sulphur into glucosinolates is in the same range (4—6% of sulphate taken up) as the flux through primary assimilation (Huseby et al., 2013; Mugford, Lee, Koprivova, Matthewman, & Kopriva, 2011). Interestingly, however, the partitioning of sulphate is influenced by diurnal rhythms, as the maximum incorporation into primary reduced compounds precedes the maximum incorporation into glucosinolates, which take place in the end of the light period (Huseby et al., 2013). It is thus not surprising that glucosinolates synthesis and accumulation is well connected with sulphur metabolism.

Indeed, sulphur limitation leads to reduction of glucosinolate accumulation (Blake-Kalff, Harrison, Hawkesford, Zhao, & Mcgrath, 1998). This is an active process; genes involved in glucosinolate degradation, such as myrosinases, are induced by sulphate deficiency (Hirai et al., 2005). On the other hand, genes for glucosinolate synthesis are strongly downregulated by sulphate starvation (Hirai et al., 2003, 2005; Maruyama-Nakashita, Nakamura, Tohge, Saito, & Takahashi, 2006; Nikiforova et al., 2003).

Glucosinolates have been long considered as a storage for sulphur which can buffer the reduced availability during limitation periods; however, new experiments with *Brassica* species came to different conclusions, so this function of glucosinolates needs to be confirmed (Aghajanzadeh, Hawkesford, & De Kok, 2014). Nevertheless, the downregulation of glucosinolates synthesis genes is a part of the sulphate starvation response in Arabidopsis, under direct or indirect control of the central regulator of sulphate limitation response SLIM1 (Frerigmann & Gigolashvili, 2014; Maruyama-Nakashita et al., 2006). The coordinated regulation of the pathway actually enabled to identify the genes for sulphotransferases catalysing the last step of glucosinolate core synthesis from transcriptomics and metabolomics time series of sulphate limitation (Hirai et al., 2005). The linkage is strengthened by a spatial colocalization of sulphate assimilation and glucosinolate synthesis. Expression of genes for both pathways is enriched in bundle sheath cells of Arabidopsis, again in a coordinated manner (Aubry, Smith-Unna, Boursnell, Kopriva, & Hibberd, 2014).

However, glucosinolates are regulated not only by sulphate nutrition, but also other nutrients affect glucosinolate synthesis and accumulation. Effects of nitrogen limitation depend strongly on sulphur supply; at adequate sulphate availability, plants accumulate aliphatic glucosinolates, possibly as a storage of the surplus sulphur (Gerendas et al., 2009; Hirai et al., 2004; Li et al., 2007). Interestingly, nitrogen source used for plant nutrition affects glucosinolate accumulation; the synthesis is higher in plants grown on ammonium than on nitrate (Marino et al., 2016). Similarly, phosphate deficiency leads to increased glucosinolate synthesis and accumulation (Pant et al., 2015). Again, this is a transcriptionally regulated process, as the pathway genes are upregulated, and it is dependent on the central regulator of phosphate deficiency response, PHR1 (Pant et al., 2015). Also deficiency of another essential macronutrient, potassium, leads to accumulation of glucosinolates (Troufflard et al., 2010). Here the primary driver for increased glucosinolate synthesis is induction of synthesis of jasmonate and other oxylipins, which in turn triggers increased expression of glucosinolate synthesis genes. Indeed, the potassium deficiency-dependent induction of indolic glucosinolates has been abolished in jasmonate insensitive *coi1-16* mutants (Troufflard et al., 2010). Since the aliphatic glucosinolates were increased by K deficiency in both genotypes, another mechanism linking potassium and glucosinolates, independent from jasmonate must be postulated.

The metabolic links between sulphur metabolism and glucosinolates synthesis are on three levels. First is the availability of methionine for the aliphatic glucosinolates, second the provision of sulphur for thioglucoside linkage of glucose by glutathione, and third the donor of active sulphate, 3′-phosphoadenosine 5′-phosphosulphate (PAPS). Surprisingly little is known about the links of methionine and glucosinolates. Methionine is an essential amino acid, and mutants unable to produce it are lethal. On the other hand, several mutants overaccumulating methionine have been isolated, but to our knowledge, no data on glucosinolate levels in these plants have been reported (Kim et al., 2002). Similarly, reduced glutathione levels did not affect glucosinolate accumulation in a *pad2* (phytoalexine-deficient) mutant of γ-glutamylcysteine synthetase (γECS), the first enzyme in glutathione synthesis (Schlaeppi, Bodenhausen, Buchala, Mauch, & Reymond, 2008). However, the induced synthesis of glucosinolates observed in plants exposed to herbivory was disrupted in the mutant. The induction of glucosinolate synthesis was indeed compromised by low glutathione, as only its addition and not changes in redox potential restored their levels (Schlaeppi et al., 2008). Thus under normal conditions even low glutathione levels are sufficient to provide sulphur for the glucosinolates, but when the rate of synthesis increases, glutathione might limit glucosinolate biosynthesis. However, glutathione synthesis is dependent on the provision of cysteine and sulphate assimilation. Increasing sulphate reduction rate by expression of APR-B isoform of adenosine 5′-phosphosulphate (APS) reductase did not affect glucosinolate synthesis and accumulation (Mugford et al., 2011). Also reduction in the flux through primary sulphate assimilation in *apr2* mutants, deficient in the major isoform of APS reductase, did not affect glucosinolate levels (Mugford et al., 2011). Thus, it seems that at least under normal conditions sulphate reduction and glutathione are not limiting for glucosinolate synthesis.

2.1 3′-Phosphoadenosine-5′-Phosphosulphate Is Important for Glucosinolate Synthesis

On the other hand, availability of PAPS has been shown to have a great impact on glucosinolate synthesis even at normal conditions. Analysis of Arabidopsis mutants with diminished production of PAPS revealed strong reduction in glucosinolate contents, approximately by 80% (Mugford et al., 2009). PAPS is synthesized from an intermediate of primary sulphate assimilation, APS by APS kinase. APS is thus a branching point in sulphur metabolism, as it can either be reduced by APS reductase to sulphite, further reduced to sulphide and incorporated into cysteine, which is the first product of sulphate assimilation and donor of reduced sulphur for most metabolic reactions (Fig. 1) (Takahashi, Kopriva, Giordano, Saito, & Hell, 2011). The reduction in PAPS synthesis has been caused by disruption of two major isoforms of APS kinase (out of four genes) in *apk1 apk2* mutants (Mugford et al., 2009). Interestingly, the mutants accumulated large concentrations of the desulpho-glucosinolate precursors, which normally are almost not detectable in wild-type plants. Indeed, the whole pathway of glucosinolate synthesis has been induced in *apk1 apk2* mutants, including the transcripts of the two groups of MYB factors, which in turn caused the increased mRNA levels of the biosynthetic genes (Mugford et al., 2009). Analysis of further APS kinase mutants revealed that PAPS is an essential compound, as loss of *APK1*, *APK3* and *APK4* was lethal (Mugford, Matthewman, Hill, & Kopriva, 2010). This is, however, not caused by lack of glucosinolates, as other mutants without these compounds are viable. Further it can be concluded that the compartmentation of APS synthesis in plastids and cytosol does not affect glucosinolate synthesis per se, as these plants were

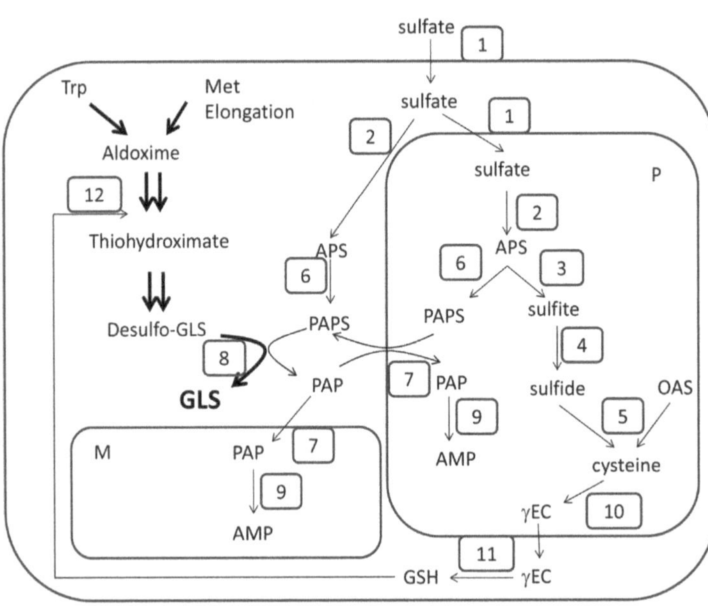

Figure 1 ***Integration of glucosinolate synthesis and sulphur metabolism.*** Shown are reactions of sulphur metabolism connected with glucosinolate synthesis in chloroplast (P), mitochondria (M), and cytosol. The enzymatic and transport steps are indicated by numbers: 1, sulphate transporter; 2, ATP sulphurylase; 3, APS reductase; 4, sulphite reductase; 5, O-acetylserine thiollyase; 6, APS kinase; 7, PAPS transporter; 8, sulphotransferase; 9, 3′(2′),5′-bisphosphate nucleotidase (FRY1/SAL1); 10, γ-glutamylcysteine synthetase; 11, glutathione synthetase; 12, GSH transferase and γ-glutamyl peptidase.

viable, but the phenotype of mutants in which APS was produced only in the plastids (*apk3*) or only in cytosol (*apk1 apk2 apk4*) could be explained only when a PAPS transport across the plastid envelope takes place (see later) (Mugford et al., 2010).

APS kinase, and its interplay with APS reductase, is responsible not only for delivery of PAPS for glucosinolate synthesis, but also for the partitioning of sulphur between primary and secondary metabolism (Mugford et al., 2011). The reduced synthesis of glucosinolates in *apk1 apk2* mutants resulted in increased flux through primary reductive assimilation and to accumulation of reduced sulphur compounds, glutathione and cysteine (Mugford et al., 2011, 2009). This increase in metabolite accumulation and flux is higher than would be caused only by redirecting of APS from the PAPS branch to the reductive branch of the pathway. The increased sulphate reduction rate thus must be a regulated consequence of the diminished glucosinolate synthesis. Indeed, analysis of the MYB factors controlling glucosinolate synthesis revealed that genes for ATP sulphurylase and APS kinase, which are necessary to synthesize PAPS, are part of the same regulatory circuit (Yatusevich et al., 2010). This is not surprising, as the *apk1 apk2* mutants proved that PAPS is clearly important for glucosinolate synthesis. However, also genes of the reductive pathway APS reductase and sulphite reductase are regulated by the MYB factors, showing a more fundamental role of the transcription factors and glucosinolate synthesis in control of sulphur metabolism. Upregulation of the transcript levels and activity of APS reductase in plants overexpressing MYB28 and MYB51 thus seems to be the driver of the flux increase as measured in *apk1 apk2* and MYB51-expressing plants (Mugford et al., 2011; Yatusevich et al., 2010).

While the reduction in activity of APS kinase strongly reduced glucosinolate synthesis, increasing PAPS production by overexpressing the enzyme did not show any effects on its concentration (Mugford et al., 2011). Irrespective of targeting the APS kinase to plastids or cytosol, glucosinolate levels were not affected, even though transcript levels of several genes, e.g., MAM-L and SOT17, were induced in the transgenic lines. Unexpectedly, overexpression of APS kinase increased activity of APS reductase, but this also did not lead to changes in accumulation of cysteine and glutathione. The mechanism for this alteration of sulphate assimilation is unknown, but may point to a so far unknown function of PAPS in sulphur regulatory networks (Mugford et al., 2011). The importance of APS kinase for glucosinolate synthesis has been corroborated in a completely different manner. In the synthetic biology approach to engineer glucosinolate synthesis into a noncrucifer species, the yield of the metabolites was unexpectedly low (Moldrup et al., 2011). The transgenic tobacco plants transiently expressing six genes for synthesis of benzylglucosinolate, however, accumulated large concentrations of the desulpho-precursor, similar to the *apk1 apk2* plants. Indeed, overexpression of *APK2* isoform of APS kinase resulted in a large (16-fold) increase in concentration of benzylglucosinolate (Moldrup et al., 2011). This shows that, although APS kinase is ubiquitous enzyme, its activity in Arabidopsis and presumably in other glucosinolate forming species is adjusted to higher levels to provide sufficient PAPS for the synthesis of glucosinolates and that this activity may form a bottleneck in engineering the pathway in nonhost species.

2.2 Recycling of Sulphation By-Products Affects Glucosinolate Accumulation

The transfer of sulphate from PAPS to the desulpho-glucosinolates results in formation of a second product, 5′-phosphoadenosine 3′-phosphate (PAP). PAP is normally rapidly metabolized to AMP and phosphate by an enzyme named Fiery1 (FRY1) after the genetic screen in which it was identified (Xiong et al., 2001). *FRY1* is an enigmatic gene that has been found in numerous genetic screens for phenotypes as diverse as leaf venation density, jasmonate synthesis and response to ABA (Robles et al., 2010; Rodriguez, Chetelat, Majcherczyk, & Farmer, 2010; Xiong et al., 2001). These pleiotropic phenotypes have rarely been mechanistically understood, because the enzyme dephosphorylates two kinds of substrates, dinucleotide phosphates or inositol phosphates, that are involved in regulation of large number of developmental processes (Xiong et al., 2001). Whereas initially most phenotypes were ascribed to interference with inositol signalling, once the inhibition of RNA processing by PAP was discovered, PAP has been considered as a very important cellular signal (Chan, Wirtz, Phua, Estavillo, & Pogson, 2013; Estavillo et al., 2011; Gy et al., 2007). In particular, PAP is considered as a retrograde signal connecting oxidative stress in plastids with nuclear responses (Estavillo et al., 2011).

But where is the link of PAP and glucosinolates? Since the synthesis of glucosinolates requires high activity of APS kinase and PAPS production, it also results in a high rate of PAP production, which needs to be removed by FRY1. Thus, loss of this enzyme should lead to accumulation of PAP and interference with glucosinolate synthesis. Indeed, analysis of a *fou8* allele of *fry1* mutant revealed reduced accumulation of glucosinolates and accumulation of desulpho-precursors (Lee et al., 2012). The metabolic phenotype thus resembled that of *apk1 apk2* mutant, but with a lower degree, i.e., the glucosinolates were reduced by 50% and not 90%, the levels of desulphoglucosinolates reached only ca. 20% of those in *apk1 apk2*, and not all genes of the glucosinolate pathway were induced in *fou8*. PAP has accumulated, as expected, but surprisingly also PAPS levels were higher in *fou8* than in wild-type plants. The effects of the loss of *FRY1* may be caused by accumulation of PAP, which can either directly inhibit the sulphotransferases and so shift the reaction equilibrium on the site of the substrates PAPS and desulphoglucosinolates, or inhibit the transport of PAPS from the plastids to the cytosol (see later). The analysis of this mutant, however, showed that also *FRY1* has to be considered part of the glucosinolate biosynthetic network, which is corroborated by the upregulation of *FRY1* expression in the *apk1 apk2* mutants (Mugford et al., 2009). Interestingly, the lower glucosinolate synthesis in *fou8* did not redirect the flux of sulphur into the primary metabolites as in *apk1 apk2* (Lee et al., 2012). In contrast, the concentration of glutathione and cysteine were significantly reduced in *fou8* plants. This has been connected with a coordinated reduction of all sulphur compounds and total S (Lee et al., 2012). The mechanisms for this effect on sulphur homeostasis are, however, not known.

3. REDOX REGULATION

A typical mechanism of posttranscriptional regulation is regulation by reduction or oxidation. Many enzymes in plant primary and secondary metabolism are redox regulated, including enzymes involved in glucosinolate synthesis, particularly the methionine elongation cycle. Isopropylmalate dehydrogenase (IPMDH) catalyses oxidation and decarboxylation of the products of methylthioalkylmalate synthase (MAM) reaction and produces methionine (or elongated methionine) elongated by a single methylene unit (He et al., 2009). This enzyme is involved in primary metabolism as well, in the biosynthesis of leucine. It has been found in a screen for proteins interacting with thioredoxin (Lemaire et al., 2004) pointing to a possibility of redox regulation. Indeed, the activity of Arabidopsis IPMDH increased with increasing concentration of thioredoxin (He et al., 2009). The redox regulation affected both V_{max} and K_M of the substrates, with the reduced form being more active. The protein possesses two redox active cysteines, C232 and C390, which form a disulphide bond. When these cysteine residues were exchanged for serine, the enzyme was no more responsive to inhibition by oxidation (He et al., 2009). While the regulation of IPMDH activity was redox regulated also in plant extracts, the physiological role of the regulation is unknown.

A second enzyme in the pathway that is redox regulated is the APS kinase. Resolving the crystal structure of the plant enzyme revealed presence of a disulphide bond between two monomers (Ravilious, Nguyen, Francois, & Jez, 2012). Indeed, mutation of either of the cysteine residues, C86 or C119, abolished formation of dimers. The structural difference is accompanied by changes in enzyme activity, the reduction of the disulphide bond results in higher activity of the enzyme, so that the two mutated proteins show higher activity than the native enzyme, insensitive to redox regulation (Ravilious et al., 2012). In addition to the 17-fold increased catalytic efficiency, the reduced protein also shows much lower level of substrate inhibition. APS has an inhibitory effect, because to form efficient ternary complex, ATP has to bind first and create optimal active site configuration (Ravilious & Jez, 2012). Binding of APS as first does not trigger these necessary conformational changes and has inhibitory effect on the reaction. The redox regulation of APS kinase is even more interesting, when taken in context of regulation of APS reductase, which is also present in plastids and competes with APS kinase for the common substrate. APS reductase activity is also redox regulated; the activity is increased by oxidation and diminished by reduction, opposite to the regulation of APS kinase (Bick et al., 2001; Kopriva & Koprivova, 2004). This allows to speculate that changes in redox environment in plastids can direct the flux of sulphur to primary or secondary metabolism. Thus, in oxidizing conditions, reduced compounds, such as glutathione, would be preferentially made, as glutathione is important for detoxification of reactive oxygen species. On contrary, in reduced environment, when the cell is in a balanced reduced state, the sulphur will flow towards PAPS and synthesis of glucosinolates and other secondary compounds. Again, this hypothesis needs to be tested in vivo. Remarkably, however, APS kinase is not redox regulated in organisms that use PAPS for sulphate reduction, such as yeast, fungi or some proteobacteria and cyanobacteria (Ravilious et al., 2012).

The next redox-regulated enzyme connected to glucosinolate synthesis is FRY1 (Chan et al., 2016). Similar to APS kinase, FRY1 (SAL1) is also more active in reduced state, although the difference in catalytic efficiency (determined as k_{cat}/K_M) is with approximately threefold not as high as in case of APS kinase. Oxidation results in dimerization of FRY1, formation of intramolecular disulphide bond, and glutathionylation of the protein, each contributing to reduction of the enzyme activity (Chan et al., 2016). Such coordinated redox regulation of the two proteins makes a very good sense, because increased PAPS production by APS kinase leads to higher sulphation rate, and thus formation of PAP, that needs to be detoxified by FRY1. It would be interesting to test whether also the transport of these metabolites across plastid envelope is affected by redox changes. However, for FRY1 there might be another function of the oxidative inactivation. Since PAP is considered to be a retrograde signal between plastids and the nucleus, e.g., during drought stress (Estavillo et al., 2011), the inactivation of FRY1 by oxidative stress is the most suitable mechanism for accumulation of the signal. The importance of the redox regulation of the PAPS and PAP metabolism for glucosinolate synthesis, however, remains to be demonstrated.

4. TRANSPORTERS IN GLUCOSINOLATE SYNTHESIS

The plant cells possess highly compartmentalized metabolic networks, which is also the case for glucosinolate biosynthesis. Plastids, cytosol, vacuole, endomembrane structures, peroxisomes and also mitochondria are embedded in the cytosol, cooperating altogether with their specific set of enzymes, to produce, activate or catabolize glucosinolates. Nevertheless, until recently, the focus in glucosinolate research has been mainly on the identification of biosynthetic genes and regulators, and less on transporters. In this section of our review, we will focus on specific intra- and intercellular transporters, which are known to be involved in synthesis, storage or utilization of glucosinolates.

4.1 Transport of Aliphatic Glucosinolate Intermediates: Chloroplastic Transporter of Keto Acids

Methionine (Met) is an essential constituent for the production of aliphatic glucosinolates, and it is synthesized both in cytosolic and chloroplastic compartments in *Arabidopsis thaliana*. To enter the glucosinolate biosynthetic pathway, Met needs to be chain elongated in chloroplasts. This chain elongation process comprises biosynthetic reactions reminiscent of those during branched-chain amino acids biosynthesis. A very first reaction in biosynthesis of Met-derived glucosinolates is a deamination of Met by a cytosolic branched-chain amino acid aminotransferase 4 (BCAT4) (Schuster, Knill, Reichelt, Gershenzon, & Binder, 2006). Three following reactions are known to take place in chloroplasts as all enzymes required for these processes were shown to be localized in chloroplasts (Gigolashvili, Yatusevich, et al., 2009; Sawada, Toyooka, et al., 2009). They include several cycles of (1) condensation with acetyl-CoA catalysed by a MAM, (2) isomerization by an isopropylmalate isomerase (IPMI) and (3) oxidative decarboxylation by an IPMDH (Gigolashvili, Yatusevich, et al., 2009; Knill, Reichelt, Paetz, Gershenzon, & Binder, 2009; Sawada, Kuwahara, et al., 2009, Sawada, Toyooka, et al., 2009). Elongated keto acid can be aminated by BCATs (or exported from plastids and aminated in cytosol) to yield the side-chain elongated methionine (homoMet), which is further channelled into glucosinolate biosynthesis or it can proceed through another round of chain elongations to yield dihomoMet, trihomoMet etc.

Thus, to make the biosynthesis of aliphatic glucosinolates and chain-elongation possible, Met needs to enter chloroplasts as 4-methylthio-2-oxobutanoate (MTOB). This transporter of 2-keto acids was independently identified by Gigolashvili, Yatusevich, et al. (2009) and Sawada, Toyooka, et al. (2009) as bile acid transporter 5 (BAT5 or BASS5) based on coexpression and activation by MYB28. BAT5 had been annotated as a bile acid transporter based on sequence homology to mammalian transporters, which has been reported as misleading (Jensen, Halkier, & Burow, 2014), as plants do not metabolize or transport bile acids. BAT5 or BASS5 transport not bile but keto acids generated in deamination reaction by BCAT4 and BCAT3 (Knill, Schuster, Reichelt, Gershenzon, & Binder, 2008; Schuster et al., 2006). In support of the in vivo function of BAT5, knockout mutants contain strongly reduced levels of aliphatic glucosinolates; however, feeding these plants with elongated methionine, some aliphatic glucosinolates with the respective chain length were synthesized (Gigolashvili, Yatusevich, et al., 2009). Conversely, metabolic complementation experiments with 2-keto acids such as MTOB, 5-methylthio-2-oxopentanoate (MTOP) or 6-methylthio-2-oxohexanoate (MTOH) caused increased accumulation of aliphatic glucosinolates with respective chain length only in wild-type and not *bat5* mutant plants supporting the role of BAT5 as a facilitator of keto acids transport across inner chloroplast membrane (Gigolashvili, Yatusevich, et al., 2009; Sawada, Toyooka, et al., 2009).

To further process the glucosinolate core biosynthetic pathway, chain-elongated Met needs to leave a chloroplast as all subsequent enzymes are associated with the cytosol. The second transport protein linking chloroplastic and cytosolic compartments, however, remains unknown. It is also still open whether elongated 2-keto acids or amino acids are exported from the chloroplast into the cytoplasm. Theoretically, the export of both keto and amino acids is possible as the transamination of keto acids could be catalysed by both chloroplastic BCAT3, BCAT5 or even BCAT2 (Knill et al., 2008) and cytoplasmic BCAT4 (Schuster et al., 2006). The proteins homologous to BAT5, e.g., BAT3 and BAT4, in *Arabidopsis* are obviously not important to transport elongated 2-keto acids or amino acids, as the triple *bat3 bat4 bat5* mutant was not more affected in glucosinolate biosynthesis than the single *bat5* mutant. Indeed, the BAT2 or BASS2 is associated with the biosynthesis of pyruvate (Furumoto et al., 2011), which is an important metabolic precursor for terpenoids, fatty acids and branched-chain amino acids in plants. Thus, more chloroplastic transporters linking chloroplast and cytosol and involved in chain-elongation process of Met are expected to be identified in future. Alternatively, it is possible that the BAT5 or BASS5 is able of exporting the chain-elongated 2-keto acids back into the cytoplasm. However, the direct evidence for this is missing, as attempts to reveal this ability in vitro in artificial membranes failed because of higher hydrophobicity of substrates leading to unspecific membrane binding and permeation, which override the specific measurements (Gigolashvili & Kopriva, 2014).

4.2 Transport of Cosubstrate PAPS: Chloroplastic PAPS/PAP Antiporter in Sulphation of Glucosinolates

Sulphation of desulphoglucosinolates requires PAPS, which functions as a sulphate donor. Sulphation is essential, since unless sulphated, glucosinolates are not transported within the plants, stored or used as a defence compounds. All known sulphotransferases, which sulphate glucosinolates, are localized in cytosol (Hirschmann, Krause, & Papenbrock, 2014). Conversely, the PAPS-synthesizing APS kinases are localized in both the chloroplast and the cytosol. However, the main pool of PAPS required for the sulphation of glucosinolates is produced in the plastids and needs therefore to be transported into the cytosol where the sulphation reaction takes place (Hirschmann et al., 2014; Klein & Papenbrock, 2004; Mugford et al., 2010; Piotrowski et al., 2004). Indeed, specific PAPS transporter 1 (PAPST1) has been identified in Arabidopsis (Gigolashvili et al., 2012). This transporter was formerly known as thylakoid ADP/ATP carrier or TAAC (Thuswaldner et al., 2007), and its function as antiporter which exchanges PAPS against PAP was lately demonstrated (Gigolashvili et al., 2012).

In line with previous observations this work showed that despite existence of two parallel plastidic and cytoplasm-localized pathways for the production of PAPS (Mugford et al., 2010, 2009), the main pool of PAPS required for the biosynthesis of glucosinolates is produced in chloroplasts and transported into the cytosol via PAPST1. Remarkably, Arabidopsis plants, which are not able to produce PAPS are lethal (Mugford et al., 2010, 2009). The PAPS transporter explains why the single loss-of-function mutant of APK3, isoform missing the cytoplasmic APS kinase contains normal levels of glucosinolates, and, also why the triple mutants of APS kinase isoforms with sole activity localized in chloroplasts are viable.

Analysis of glucosinolates in *papst1/taac* mutant revealed about 50–70% reduction of methionine-derived glucosinolates and no changes in levels of indolic glucosinolates compared to corresponding wild-type plants (Gigolashvili et al., 2012; Gigolashvili & Kopriva, 2014). Furthermore, *papst1/taac* mutant is slightly retarded in growth compared to corresponding wild-type plants (Col-*0* or WS-*4*) (Thuswaldner et al., 2007; Yin et al., 2010) resembling the phenotype of plants lacking the plastidic APS kinase isoforms *apk1 apk2* and *apk1 apk2 apk4* (Mugford et al., 2010, 2009). However, this phenotype was associated with the absence of other sulphated compounds such as disulphated pentapeptide phytosulphokine (PSK), peptide containing sulphated tyrosine 1 (PSY1), root growth factors but not glucosinolates (Gigolashvili et al., 2012; Gigolashvili & Kopriva, 2014; Koprivova & Kopriva, 2016).

As a by-product of sulphation reaction, the toxic PAP is produced in the cytosol and needs to be detoxified in chloroplasts by FRY1 (Estavillo et al., 2011). This is enabled again by the function of PAPST1. PAPST1 thus coordinates PAPS production in chloroplasts with its use in the cytosol and subsequent degradation of PAP again in chloroplasts (Chan et al., 2013). However, interestingly, FRY1 is also present in the mitochondria (Chan et al., 2013). Correspondingly, at least in rice, PAPST1 is dually targeted to the plastid and mitochondrial outer membranes (Xu et al., 2013). The role of mitochondria in PAP and PAPS metabolism, however, needs to be established. Remarkably, in contrast to PAPS transporters known from mammals, which are members of the nucleotide-sugar transporter family, the PAPST1 transporter belongs to the mitochondrial carrier family (Haferkamp, Fernie, & Neuhaus, 2011). Finally, it needs to be mentioned that because the reduction of glucosinolate levels in *papst1/taac* mutant is not as dramatic as in *apk1 apk2* mutants (Mugford et al., 2009), an additional exporter of PAPS in plastid envelopes has to be postulated and is expected to be identified in future.

4.3 Intracellular or Short-Distance Transport of Glucosinolates in the Storage and Defence

The classical 'mustard oil bomb' model (Grubb & Abel, 2006; Halkier & Gershenzon, 2006; Sonderby et al., 2010; Wittstock & Burow, 2010; Wittstock, Kliebenstein, Lambrix, Reichelt, & Gershenzon, 2003) proposes spatial compartmentalization of myrosinases and glucosinolates in different compartments and is triggered upon tissue disruption by, e.g., insect feeding. In accordance with this model, after completion of biosynthesis, which is signified by sulphation of desulpho-glucosinolates, native glucosinolates need to be imported into the vacuole, stored there or further transported for long-term storage into the S-cells (Gigolashvili & Kopriva, 2014; Gomez et al., 2009; Jørgensen, Nour-Eldin, & Halkier, 2015; Koroleva, Gibson, Cramer, & Stain, 2010).

In *A. thaliana*, glucosinolates have been reported to be stored to high levels (>130 mM) in specific sulphur-rich cells called also S-cells (Koroleva et al., 2010). This concentration is about 20 times higher in comparison to surrounding tissue. Upon damage by herbivores, the S-cell works as a high-pressure mustard bomb as the large glucosinolate content is released producing the deterring hydrolysis products (Koroleva et al., 2010). Proteomics on cytoplasm from S-cells detected no glucosinolate biosynthetic enzymes, suggesting that transport processes are necessary for accumulation of glucosinolates in S-cells (Koroleva & Cramer, 2011). It can be suggested that glucosinolates synthesized in the neighbouring phloem parenchyma cells can be transported into the S-cells. This conclusion is supported by the vasculature-specific expression of branched-chain aminotransferase 4 (BCAT4) (Schuster et al., 2006), chloroplastic transporter of keto acids BAT5 (Gigolashvili, Yatusevich, et al., 2009), PAPST1 providing activated sulphate for the final biosynthetic step of glucosinolate biosynthesis (Gigolashvili et al., 2012), as well as R2R3 MYB transcription factors controlling production of indolic and aliphatic glucosinolates (Gigolashvili, Berger, & Fluegge, 2009). It is currently not known how glucosinolates are transported into the vacuoles of the S-cells or also in other cells.

Vacuolar localization of glucosinolates has been suggested long time ago for mesophyll cells, but it was experimentally demonstrated using water-free environment gradient centrifugation of *Arabidopsis* leaf tissue only in a recent past (Krueger et al., 2011). Shuffling of sulphated secondary metabolites into vacuole requires intracellular transporter. However, no specific transporter of native glucosinolates localized in the tonoplast is known at the moment. Nevertheless, the vacuolar transporters of other secondary metabolites such as flavonoids are well known, and it is possible that these transporters do also import glucosinolates into the vacuole.

One of the driving forces for the transport of secondary metabolites into vacuole has been inferred to be a proton gradient between the cytoplasm and the vacuole, which is maintained by H^+-ATPases (and H^+-PPases in the tonoplast) (Gigolashvili & Kopriva, 2014; Gomez et al., 2009; Petrussa et al., 2013). Once these compounds are moved into vacuoles, the acidic pH inside the vacuolar compartment and the acylation of flavonoids are both necessary for the induction of a conformational modification, responsible for the appropriate trapping and retention of the metabolites (Kitamura, 2006; Petrussa et al., 2013). Along with this, the ATP-binding cassette (ABC) transporters have also been reported in sequestration of secondary metabolites into the vacuole (Petrussa et al., 2013; Yazaki, 2005; Zhao & Dixon, 2010). Due to broad substrate specificity of this type of transporters, the ABC transporters are good candidates for the transport of native glucosinolates into the vacuole, which still await identification.

An ability to facilitate the transport of glucosinolate catabolism products has been reported for the PDR8 (Bednarek et al., 2009; Fuchs et al., 2016; Lu et al., 2015; Stein et al., 2006) subfamily of ABC transporter family in *A. thaliana*. PDR8 or PEN3 (mutant of which shows a 'penetration' phenotype upon exposure to nonhost pathogens of Arabidopsis) is localized in plasma membrane of leaves where it can export biologically active products of glucosinolate catabolism to the site of invasion of pathogens (Bednarek et al., 2009; Fuchs et al., 2016; Lu et al., 2015; Stein et al., 2006). PDR8/PEN3 has been suggested to be mainly involved in the transport of substrate derived from 4-methoxy-indole-3-ylmethyl-glucosinolate (4MOI3M).

Genetic analysis of *pdr8* mutant revealed that PDR8 or PEN3 acts in the same pathway as PEN2 (a myrosinase-like protein, absence of which has the same 'penetration' phenotype), because, in contrast to *pen1 pen3* double mutants, *pen2 pen3* did not show enhanced *Blumeria graminis f. sp. hordei* penetration rates in comparison to the single mutants (Lipka et al., 2005; Stein et al., 2006). Interestingly, *pdr8* mutant shows enhanced resistance against the adapted powdery mildew *Golovinomyces cichoracearum*, which might be the consequence of an intracellular overaccumulation of potentially toxic indolic glucosinolate hydrolysis products (Stein et al., 2006).

Furthermore, a series of genetic, biochemical and physiological experiments indicate that PEN2 – tail-anchored protein with dual-membrane targeting to peroxisomes and mitochondria (Fuchs et al., 2016; Lu et al., 2015) – is able of hydrolysing indol-3-ylmethylglucosinolate (I3M) and 4MOI3M (Bednarek et al., 2009; Lipka et al., 2005) and is required for production of 4-O-b-D-glucosylindol-3-yl formamide (Lu et al., 2015). Deglucosylated precursor molecules of the latter have recently been proposed to represent the critical PEN3 transport substrate in powdery mildew entry control (Fuchs et al., 2016; Lu et al., 2015). Importantly, only unidentified hydrolysis products of 4MOI3M derivatives, which are synthesized by the cytochrome P450 monooxygenase CYP81F2, seem to be relevant for PEN3-dependent transport activities. Nevertheless, our knowledge about cellular mechanisms that orchestrate PEN2-dependent hydrolysis of indolic glucosinolate derivatives and subsequent energy-dependent discharge of toxic products by PEN3 clearly lags behind (Fuchs et al., 2016).

Altogether, existing data indicate the role of PEN3 in the ATP-dependent transport of enzymatically activated glucosinolates that are toxic for invading pathogens. In contrast to the classical mustard oil bomb, which involves cellular compartmentalization of myrosinases and glucosinolate glucosides in distinct cells (Grubb & Abel, 2006), glucosinolate-based PEN pathway is based on cell-autonomous subcellular compartmentalization and concerted metabolon activity (Fuchs et al., 2016).

To fully understand the way plant activates its chemical defences and to be able to engineer resistant plants, we need to better understand the transport networks followed by glucosinolate biosynthesis and the route from site of synthesis to site of storage and activation upon demand.

4.4 Long-Distance Transport of Glucosinolates

It has long been known that glucosinolates undergo also a long distance transport, which is, e.g., needed for loading of glucosinolates into seeds. More than 25 years ago, before the first glucosinolate transporter could be identified, it was postulated that such glucosinolate-specific transporters should exist (Gijzen, Mcgregor, & Séguin-Swartz, 1989). Experiments with radiolabeled glucosinolates had demonstrated that glucosinolates applied to leaves can move through the phloem to seeds, where they were detected (Brudenell, Griffiths, Rossiter, & Baker, 1999). Additionally, it was also shown that leaf protoplasts as well as rapeseed embryos, when incubated in glucosinolate-containing media, were able to accumulate glucosinolates (Chen & Halkier, 2000; Gijzen et al., 1989). Recently, the long anticipated transporters of glucosinolates named as glucosinolate transporters 1 and 2 (GTR1 and GTR2) were identified (Nour-Eldin et al., 2012) and characterized (Andersen & Halkier, 2014; Andersen et al., 2013; Jørgensen, Olsen, et al., 2015; Madsen, Kunert, Reichelt, Gershenzon, & Halkier, 2015; Madsen, Olsen, Nour-Eldin, & Halkier, 2014). Physiological experiments in *Xenopus oocytes* have demonstrated that GTR1 and GTR2 transporter mature glucosinolates (Nour-Eldin et al., 2012). The double mutant *gtr1 gtr2* is not able to import glucosinolates into the seeds and as a consequence, they accumulated up to 10-fold levels in leaves. The disruption of transport to seeds can potentially be used to engineer a novel mechanism to reduce glucosinolate content in rapeseed.

Tissue-specific expression of GTRs in above-ground tissues along with the analysis of glucosinolate levels in *gtr1* and *gtr2* mutants suggested their additional role in phloem loading as well as in distribution of glucosinolates within the leaf (Andersen et al., 2013; Nour-Eldin & Halkier, 2013). Micrografting experiments with *gtr* mutants revealed the role of rosettes as the major place for the biosynthesis and storage of short-chained aliphatic glucosinolates. It was shown that the activity of the glucosinolate importers GTR1 and GTR2 is required for intraleaf distribution of glucosinolates both along the vasculature and to the leaf margins (Andersen & Halkier, 2014; Andersen et al., 2013; Madsen et al., 2014). According to the proposed model for glucosinolate distribution patterns in leaves, glucosinolates are moved via connecting plasmodesmata to storage cells in the outer defensive perimeters along gradients created by vacuolar importers. Thus, the proposed role of the two plasma membrane-localized GTR importers in leaves includes not only import into phloem for long-distance transport to other organs, but also symplasmic domains for intraleaf transport (Jørgensen, Nour-Eldin, et al., 2015).

Belowground, GTRs were mainly found in cortex cells and root vasculature (Andersen et al., 2013), altogether suggesting important role of GTRs both below- and above-ground and in bidirectional transport between roots and shoots. Furthermore, GTRs seem to be important for loading the seeds with aliphatic and indole glucosinolates. Nonetheless, the distribution of indolic glucosinolates is hardly affected by loss of these transporters (Andersen et al., 2013). It can be also shown that indolic glucosinolates, which are synthesized in rosette and roots can be transported within the plants in the absence of GTRs, suggesting the existence of additional, yet unidentified, indolic glucosinolate-specific transporter. Moreover, not only indolic glucosinolates behave differently from aliphatic ones, but also among the aliphatic glucosinolates the short- and long-chained glucosinolates are transported differently. For example, grafting experiments with double GTR mutants showed that plants with *gtr1 gtr2* rootstock overaccumulated long-chained aliphatic glucosinolates in the rosette (Andersen et al., 2013). This experiment revealed that roots could be important sink place for accumulation as well as for the export of long-chained aliphatic glucosinolates into the soil. In line with this suggestion, Arabidopsis roots were shown to exude various metabolites, including glucosinolates and their breakdown products (Mönchgesang et al., 2016; Strehmel, Böttcher, Schmidt, & Scheel, 2014), which are important for the microbial community in the soil (Bressan et al., 2009). A large-scale metabolomics study of five root cell types reported that aliphatic glucosinolates predominantly accumulate in the cortex layer, whereas indolic glucosinolates predominantly localize to the columella and less in cortex cells (Moussaieff et al., 2013). Nevertheless, the identity of aliphatic or indolic glucosinolate transporters in roots, which can be involved in the exudation of glucosinolates, remains unknown.

5. CONCLUSIONS

Clearly, although glucosinolate biosynthesis is considered a secondary pathway, it is firmly embedded in plant metabolism. Disruptions of glucosinolate synthesis trigger far-reaching consequences for primary sulphur metabolism and intermediates and precursors connected with sulphate assimilation. On the other hand, availability of the sulphur-containing precursors has a large impact on glucosinolate accumulation. Glucosinolate synthesis is a major producer of PAP, an important cellular signal, which connects the pathway with a large range of metabolic and developmental processes. The pathway is dependent on precise coordination of processes in multiple compartments and so on various transport processes. The new findings on metabolic interactions of glucosinolate synthesis are important to better understand the complex control of the pathway and improve the success of genetic engineering of these metabolites. Although the core pathway is well understood, several transporters necessary for optimal distribution of glucosinolates and their precursors are still unknown. Given that the redox regulation of APS kinase and FRY1 has been discovered only very recently, there might be many more new regulatory mechanisms that await discovery. Therefore, this review of the metabolic context of glucosinolate synthesis pathway is far from being ultimate.

ACKNOWLEDGEMENT

The authors are supported by the Deutsche Forschungsgemeinschaft (EXC 1028).

REFERENCES

Aghajanzadeh, T., Hawkesford, M. J., & De Kok, L. J. (2014). The significance of glucosinolates for sulfur storage in *Brassicaceae* seedlings. *Frontiers in Plant Science, 5*, 704.

Andersen, T. G., & Halkier, B. A. (2014). Upon bolting the GTR1 and GTR2 transporters mediate transport of glucosinolates to the inflorescence rather than roots. *Plant Signaling & Behavior, 9*, e27740.

Andersen, T. G., Nour-Eldin, H. H., Fuller, V. L., Olsen, C. E., Burow, M., & Halkier, B. A. (2013). Integration of biosynthesis and long-distance transport establish organ-specific glucosinolate profiles in vegetative Arabidopsis. *The Plant Cell (Online), 25*, 3133–3145.

Aubry, S., Smith-Unna, R. D., Boursnell, C. M., Kopriva, S., & Hibberd, J. M. (2014). Transcript residency on ribosomes reveals a key role for the *Arabidopsis thaliana* bundle sheath in sulfur and glucosinolate metabolism. *The Plant Journal, 78*, 659–673.

Bednarek, P., Pislewska-Bednarek, M., Svatos, A., Schneider, B., Doubsky, J., Mansurova, M., … Schulze-Lefert, P. (2009). A glucosinolate metabolism pathway in living plant cells mediates broad-spectrum antifungal defense. *Science, 323*, 101–106.

Bekaert, M., Edger, P. P., Hudson, C. M., Pires, J. C., & Conant, G. C. (2012). Metabolic and evolutionary costs of herbivory defense: systems biology of glucosinolate synthesis. *The New Phytologist, 196*, 596–605.

Bick, J. A., Setterdahl, A. T., Knaff, D. B., Chen, Y., Pitcher, L. H., Zilinskas, B. A., & Leustek, T. (2001). Regulation of the plant-type 5′-adenylyl sulfate reductase by oxidative stress. *Biochemistry, 40*, 9040–9048.

Blake-Kalff, M. M., Harrison, K. R., Hawkesford, M. J., Zhao, F. J., & Mcgrath, S. P. (1998). Distribution of sulfur within oilseed rape leaves in response to sulfur deficiency during vegetative growth. *Plant Physiology, 118*, 1337–1344.

Bressan, M., Roncato, M.-A., Bellvert, F., Comte, G., El Zahar Haichar, F., Achouak, W., & Berge, O. (2009). Exogenous glucosinolate produced by *Arabidopsis thaliana* has an impact on microbes in the rhizosphere and plant roots. *The ISME Journal, 3*, 1243–1257.

Brudenell, A., Griffiths, H., Rossiter, J., & Baker, D. (1999). The phloem mobility of glucosinolates. *Journal of Experimental Botany, 50*, 745–756.

Chan, K. X., Mabbitt, P. D., Phua, S. Y., Mueller, J. W., Nisar, N., Gigolashvili, T., … Pogson, B. J. (2016 Aug 2). Sensing and signaling of oxidative stress in chloroplasts by inactivation of the SAL1 phosphoadenosine phosphatase. *Proc Natl Acad Sci U S A, 113*(31), E4567–E4576.

Chan, K. X., Wirtz, M., Phua, S. Y., Estavillo, G. M., & Pogson, B. J. (2013). Balancing metabolites in drought: the sulfur assimilation conundrum. *Trends in Plant Science, 18*, 18–29.

Chen, S., & Halkier, B. A. (2000). Characterization of glucosinolate uptake by leaf protoplasts of *Brassica napus*. *Journal of Biological Chemistry, 275*, 22955–22960.

Estavillo, G. M., Crisp, P. A., Pornsiriwong, W., Wirtz, M., Collinge, D., Carrie, C., … Pogson, B. J. (2011). Evidence for a SAL1-PAP chloroplast retrograde pathway that functions in drought and high light signaling in Arabidopsis. *The Plant Cell, 23*, 3992–4012.

Frerigmann, H. (2016). Glucosinolate Regulation in a Complex Relationship – MYC and MYB – No One Can Act Without Each Other. In S. Kopriva (Ed.). *Glucosinolates* (Vol. 80). in this volume.
Frerigmann, H., Berger, B., & Gigolashvili, T. (2014). bHLH05 is an interaction partner of MYB51 and a novel regulator of glucosinolate biosynthesis in Arabidopsis. *Plant Physiology, 166*, 349–369.
Frerigmann, H., & Gigolashvili, T. (2014). Update on the role of R2R3-MYBs in the regulation of glucosinolates upon sulfur deficiency. *Frontiers in Plant Science, 5*, 626.
Fuchs, R., Kopischke, M., Klapprodt, C., Hause, G., Meyer, A. J., Schwarzländer, M., … Lipka, V. (2016). Immobilized subpopulations of leaf epidermal mitochondria mediate PEN2-dependent pathogen entry control in Arabidopsis. *The Plant Cell, 28*. TPC2015-00887-RA.
Furumoto, T., Yamaguchi, T., Ohshima-Ichie, Y., Nakamura, M., Tsuchida-Iwata, Y., Shimamura, M., … Westhoff, P. (2011). A plastidial sodium-dependent pyruvate transporter. *Nature, 476*, 472–475.
Gerendas, J., Podestat, J., Stahl, T., Kubler, K., Bruckner, H., Mersch-Sundermann, V., & Muhling, K. H. (2009). Interactive effects of sulfur and nitrogen supply on the concentration of sinigrin and allyl isothiocyanate in Indian mustard (*Brassica juncea* L.). *Journal of Agricultural and Food Chemistry, 57*, 3837–3844.
Geu-Flores, F., Nielsen, M. T., Nafisi, M., Moldrup, M. E., Olsen, C. E., Motawia, M. S., & Halkier, B. A. (2009). Glucosinolate engineering identifies a gamma-glutamyl peptidase. *Nature Chemical Biology, 5*, 575–577.
Gigolashvili, T., Berger, B., & Fluegge, U.-I. (2009). Specific and coordinated control of indolic and aliphatic glucosinolate biosynthesis by R2R3-MYB transcription factors in *Arabidopsis thaliana*. *Phytochemistry Reviews, 8*, 3–13.
Gigolashvili, T., Berger, B., Mock, H. P., Muller, C., Weisshaar, B., & Flugge, U. I. (2007). The transcription factor HIG1/MYB51 regulates indolic glucosinolate biosynthesis in *Arabidopsis thaliana*. *The Plant Journal, 50*, 886–901.
Gigolashvili, T., Geier, M., Ashykhmina, N., Frerigmann, H., Wulfert, S., Krueger, S., … Flügge, U.-I. (2012). The Arabidopsis thylakoid ADP/ATP carrier TAAC has an additional role in supplying plastidic phosphoadenosine 5′-phosphosulfate to the cytosol. *The Plant Cell (Online), 24*, 4187–4204.
Gigolashvili, T., & Kopriva, S. (2014). Transporters in plant sulfur metabolism. *Frontiers in Plant Science, 5*, 442.
Gigolashvili, T., Yatusevich, R., Berger, B., Muller, C., & Flugge, U. I. (2007). The R2R3-MYB transcription factor HAG1/MYB28 is a regulator of methionine-derived glucosinolate biosynthesis in *Arabidopsis thaliana*. *The Plant Journal, 51*, 247–261.
Gigolashvili, T., Yatusevich, R., Rollwitz, I., Humphry, M., Gershenzon, J., & Fluegge, U.-I. (2009). The plastidic bile acid transporter 5 is required for the biosynthesis of methionine-derived glucosinolates in *Arabidopsis thaliana*. *The Plant Cell, 21*, 1813–1829.
Gijzen, M., Mcgregor, I., & Séguin-Swartz, G. (1989). Glucosinolate uptake by developing rapeseed embryos. *Plant Physiology, 89*, 260–263.
Gomez, C., Terrier, N., Torregrosa, L., Vialet, S., Fournier-Level, A., Verriès, C., … Cheynier, V. (2009). Grapevine MATE-type proteins act as vacuolar H^+-dependent acylated anthocyanin transporters. *Plant Physiology, 150*, 402–415.
Grubb, C. D., & Abel, S. (2006). Glucosinolate metabolism and its control. *Trends in Plant Science, 11*, 89–100.
Gy, I., Gasciolli, V., Lauressergues, D., Morel, J. B., Gombert, J., Proux, F., … Mallory, A. C. (2007). Arabidopsis FIERY1, XRN2, and XRN3 are endogenous RNA silencing suppressors. *The Plant Cell, 19*, 3451–3461.

Haferkamp, I., Fernie, A. R., & Neuhaus, H. E. (2011). Adenine nucleotide transport in plants: much more than a mitochondrial issue. *Trends in Plant Science, 16*, 507–515.
Halkier, B. A., & Gershenzon, J. (2006). Biology and biochemistry of glucosinolates. *Annual Review of Plant Biology, 57*, 303–333.
He, Y., Mawhinney, T. P., Preuss, M. L., Schroeder, A. C., Chen, B., Abraham, L., … Chen, S. (2009). A redox-active isopropylmalate dehydrogenase functions in the biosynthesis of glucosinolates and leucine in Arabidopsis. *The Plant Journal, 60*, 679–690.
Hirai, M. Y., Fujiwara, T., Awazuhara, M., Kimura, T., Noji, M., & Saito, K. (2003). Global expression profiling of sulfur-starved Arabidopsis by DNA macroarray reveals the role of O-acetyl-l-serine as a general regulator of gene expression in response to sulfur nutrition. *The Plant Journal, 33*, 651–663.
Hirai, M. Y., Klein, M., Fujikawa, Y., Yano, M., Goodenowe, D. B., Yamazaki, Y., … Saito, K. (2005). Elucidation of gene-to-gene and metabolite-to-gene networks in arabidopsis by integration of metabolomics and transcriptomics. *The Journal of Biological Chemistry, 280*, 25590–25595.
Hirai, M. Y., Yano, M., Goodenowe, D. B., Kanaya, S., Kimura, T., Awazuhara, M., … Saito, K. (2004). Integration of transcriptomics and metabolomics for understanding of global responses to nutritional stresses in *Arabidopsis thaliana*. *Proceedings of the National Academy of Sciences of the United States of America, 101*, 10205–10210.
Hirschmann, F., Krause, F., & Papenbrock, J. (2014). The multi-protein family of sulfotransferases in plants: composition, occurrence, substrate specificity, and functions. *Frontiers in Plant Science, 5*, 556.
Huseby, S., Koprivova, A., Lee, B. R., Saha, S., Mithen, R., Wold, A. B., … Kopriva, S. (2013). Diurnal and light regulation of sulphur assimilation and glucosinolate biosynthesis in Arabidopsis. *Journal of Experimental Botany, 64*, 1039–1048.
Jensen, L. M., Halkier, B. A., & Burow, M. (2014). How to discover a metabolic pathway? An update on gene identification in aliphatic glucosinolate biosynthesis, regulation and transport. *Biological Chemistry, 395*, 529–543.
Jørgensen, M. E., Nour-Eldin, H. H., & Halkier, B. A. (2015). Transport of defense compounds from source to sink: lessons learned from glucosinolates. *Trends in Plant Science, 20*, 508–514.
Jørgensen, M. E., Olsen, C. E., Geiger, D., Mirza, O., Halkier, B. A., & Nour-Eldin, H. H. (2015). A functional EXXEK motif is essential for proton coupling and active glucosinolate transport by NPF2.11. *Plant and Cell Physiology, 56*, 2340–2350.
Kim, J., Lee, M., Chalam, R., Martin, M. N., Leustek, T., & Boerjan, W. (2002). Constitutive overexpression of cystathionine gamma-synthase in Arabidopsis leads to accumulation of soluble methionine and S-methylmethionine. *Plant Physiology, 128*, 95–107.
Kitamura, S. (2006). Transport of flavonoids: from cytosolic synthesis to vacuolar accumulation. In *The science of flavonoids* (pp. 123–146). Springer.
Klein, M., & Papenbrock, J. (2004). The multi-protein family of Arabidopsis sulphotransferases and their relatives in other plant species. *Journal of Experimental Botany, 55*, 1809–1820.
Knill, T., Reichelt, M., Paetz, C., Gershenzon, J., & Binder, S. (2009). *Arabidopsis thaliana* encodes a bacterial-type heterodimeric isopropylmalate isomerase involved in both Leu biosynthesis and the Met chain elongation pathway of glucosinolate formation. *Plant Molecular Biology, 71*, 227–239.
Knill, T., Schuster, J., Reichelt, M., Gershenzon, J., & Binder, S. (2008). Arabidopsis branched-chain aminotransferase 3 functions in both amino acid and glucosinolate biosynthesis. *Plant Physiology, 146*, 1028–1039.
Kopriva, S., & Koprivova, A. (2004). Plant adenosine 5′-phosphosulphate reductase: the past, the present, and the future. *Journal of Experimental Botany, 55*, 1775–1783.

Koprivova, A., & Kopriva, S. (2016). Sulfation pathways in plants. *Chemico-Biological Interactions*.

Koroleva, O. A., & Cramer, R. (2011). Single-cell proteomic analysis of glucosinolate-rich S-cells in *Arabidopsis thaliana*. *Methods, 54*, 413–423.

Koroleva, O. A., Gibson, T. M., Cramer, R., & Stain, C. (2010). Glucosinolate-accumulating S-cells in Arabidopsis leaves and flower stalks undergo programmed cell death at early stages of differentiation. *Plant Journal, 64*, 456–469.

Krueger, S., Giavalisco, P., Krall, L., Steinhauser, M.-C., Bussis, D., Usadel, B., … Steinhauser, D. (2011). A topological map of the compartmentalized *Arabidopsis thaliana* leaf metabolome. *PLoS One, 6*, e17806.

Lee, B. R., Huseby, S., Koprivova, A., Chetelat, A., Wirtz, M., Mugford, S. T., … Kopriva, S. (2012). Effects of fou8/fry1 mutation on sulfur metabolism: is decreased internal sulfate the trigger of sulfate starvation response? *PLoS One, 7*, e39425.

Lemaire, S. D., Guillon, B., Le Marechal, P., Keryer, E., Miginiac-Maslow, M., & Decottignies, P. (2004). New thioredoxin targets in the unicellular photosynthetic eukaryote *Chlamydomonas reinhardtii*. *Proceedings of the National Academy of Sciences of the United States of America, 101*, 7475–7480.

Li, S., Schonhof, I., Krumbein, A., Li, L., Stutzel, H., & Schreiner, M. (2007). Glucosinolate concentration in turnip (*Brassica rapa* ssp. *rapifera* L.) roots as affected by nitrogen and sulfur supply. *Journal of Agricultural and Food Chemistry, 55*, 8452–8457.

Lipka, V., Dittgen, J., Bednarek, P., Bhat, R., Wiermer, M., Stein, M., … Scheel, D. (2005). Pre-and postinvasion defenses both contribute to nonhost resistance in Arabidopsis. *Science, 310*, 1180–1183.

Lu, X., Dittgen, J., Piślewska-Bednarek, M., Molina, A., Schneider, B., Svatoš, A., … Bednarek, P. (2015). Mutant allele-specific uncoupling of PENETRATION3 functions reveals engagement of the ATP-binding cassette transporter in distinct tryptophan metabolic pathways. *Plant Physiology, 168*, 814–827.

Madsen, S. R., Kunert, G., Reichelt, M., Gershenzon, J., & Halkier, B. A. (2015). Feeding on leaves of the glucosinolate transporter mutant gtr1gtr2 reduces fitness of *Myzus persicae*. *Journal of Chemical Ecology, 41*, 975–984.

Madsen, S. R., Olsen, C. E., Nour-Eldin, H. H., & Halkier, B. A. (2014). Elucidating the role of transport processes in leaf glucosinolate distribution. *Plant Physiology, 166*, 1450–1462.

Malitsky, S., Blum, E., Less, H., Venger, I., Elbaz, M., Morin, S., … Aharoni, A. (2008). The transcript and metabolite networks affected by the two clades of Arabidopsis glucosinolate biosynthesis regulators. *Plant Physiology, 148*, 2021–2049.

Marino, D., Ariz, I., Lasa, B., Santamaria, E., Fernandez-Irigoyen, J., Gonzalez-Murua, C., & Aparicio Tejo, P. M. (2016). Quantitative proteomics reveals the importance of nitrogen source to control glucosinolate metabolism in *Arabidopsis thaliana* and *Brassica oleracea*. *Journal of Experimental Botany, 67*, 3313–3323.

Maruyama-Nakashita, A., Nakamura, Y., Tohge, T., Saito, K., & Takahashi, H. (2006). Arabidopsis SLIM1 is a central transcriptional regulator of plant sulfur response and metabolism. *The Plant Cell, 18*, 3235–3251.

Mikkelsen, M. D., Buron, L. D., Salomonsen, B., Olsen, C. E., Hansen, B. G., Mortensen, U. H., & Halkier, B. A. (2012). Microbial production of indolylglucosinolate through engineering of a multi-gene pathway in a versatile yeast expression platform. *Metabolic Engineering, 14*, 104–111.

Moldrup, M. E., Geu-Flores, F., De Vos, M., Olsen, C. E., Sun, J., Jander, G., & Halkier, B. A. (2012). Engineering of benzylglucosinolate in tobacco provides proof-of-concept for dead-end trap crops genetically modified to attract *Plutella xylostella* (diamondback moth). *Plant Biotechnology Journal, 10*, 435–442.

Moldrup, M. E., Geu-Flores, F., Olsen, C. E., & Halkier, B. A. (2011). Modulation of sulfur metabolism enables efficient glucosinolate engineering. *BMC Biotechnology, 11*, 12.
Mönchgesang, S., Strehmel, N., Schmidt, S., Westphal, L., Taruttis, F., Müller, E., ... Scheel, D. (2016). Natural variation of root exudates in *Arabidopsis thaliana*-linking metabolomic and genomic data. *Scientific Reports, 6*, 29033.
Moussaieff, A., Rogachev, I., Brodsky, L., Malitsky, S., Toal, T. W., Belcher, H., ... Aharoni, A. (2013). High-resolution metabolic mapping of cell types in plant roots. *Proceedings of the National Academy of Sciences, 110*, E1232–E1241.
Mugford, S. G., Lee, B. R., Koprivova, A., Matthewman, C., & Kopriva, S. (2011). Control of sulfur partitioning between primary and secondary metabolism. *The Plant Journal, 65*, 96–105.
Mugford, S. G., Matthewman, C. A., Hill, L., & Kopriva, S. (2010). Adenosine-5′-phosphosulfate kinase is essential for Arabidopsis viability. *FEBS Letters, 584*, 119–123.
Mugford, S. G., Yoshimoto, N., Reichelt, M., Wirtz, M., Hill, L., Mugford, S. T., ... Kopriva, S. (2009). Disruption of adenosine-5′-phosphosulfate kinase in arabidopsis reduces levels of sulfated secondary metabolites. *The Plant Cell, 21*, 910–927.
Nikiforova, V., Freitag, J., Kempa, S., Adamik, M., Hesse, H., & Hoefgen, R. (2003). Transcriptome analysis of sulfur depletion in *Arabidopsis thaliana*: interlacing of biosynthetic pathways provides response specificity. *The Plant Journal, 33*, 633–650.
Nour-Eldin, H. H., Andersen, T. G., Burow, M., Madsen, S. R., Jørgensen, M. E., Olsen, C. E., ... Halkier, B. A. (2012). NRT/PTR transporters are essential for translocation of glucosinolate defence compounds to seeds. *Nature, 488*.
Nour-Eldin, H. H., & Halkier, B. A. (2013). The emerging field of transport engineering of plant specialized metabolites. *Current Opinion in Biotechnology, 24*, 263–270.
Pant, B. D., Pant, P., Erban, A., Huhman, D., Kopka, J., & Scheible, W. R. (2015). Identification of primary and secondary metabolites with phosphorus status-dependent abundance in Arabidopsis, and of the transcription factor PHR1 as a major regulator of metabolic changes during phosphorus limitation. *Plant, Cell & Environment, 38*, 172–187.
Petrussa, E., Braidot, E., Zancani, M., Peresson, C., Bertolini, A., Patui, S., & Vianello, A. (2013). Plant flavonoids—biosynthesis, transport and involvement in stress responses. *International Journal of Molecular Sciences, 14*, 14950–14973.
Piotrowski, M., Schemenewitz, A., Lopukhina, A., Müller, A., Janowitz, T., Weiler, E. W., & Oecking, C. (2004). Desulfoglucosinolate sulfotransferases from *Arabidopsis thaliana* catalyze the final step in the biosynthesis of the glucosinolate core structure. *Journal of Biological Chemistry, 279*, 50717–50725.
Ravilious, G. E., & Jez, J. M. (2012). Nucleotide binding site communication in *Arabidopsis thaliana* adenosine 5′-phosphosulfate kinase. *Journal of Biological Chemistry, 287*, 30385–30394.
Ravilious, G. E., Nguyen, A., Francois, J. A., & Jez, J. M. (2012). Structural basis and evolution of redox regulation in plant adenosine-5′-phosphosulfate kinase. *Proceedings of the National Academy of Sciences of the United States of America, 109*, 309–314.
Robles, P., Fleury, D., Candela, H., Cnops, G., Alonso-Peral, M. M., Anami, S., ... Micol, J. L. (2010). The RON1/FRY1/SAL1 gene is required for leaf morphogenesis and venation patterning in Arabidopsis. *Plant Physiology, 152*, 1357–1372.
Rodriguez, V. M., Chetelat, A., Majcherczyk, P., & Farmer, E. E. (2010). Chloroplastic phosphoadenosine phosphosulfate metabolism regulates basal levels of the prohormone jasmonic acid in Arabidopsis leaves. *Plant Physiology, 152*, 1335–1345.
Sawada, Y., Kuwahara, A., Nagano, M., Narisawa, T., Sakata, A., Saito, K., & Hirai, M. Y. (2009). Omics-based approaches to methionine side chain elongation in Arabidopsis:

characterization of the genes encoding methylthioalkylmalate isomerase and methylthioalkylmalate dehydrogenase. *Plant and Cell Physiology, 50*, 1181–1190.

Sawada, Y., Toyooka, K., Kuwahara, A., Sakata, A., Nagano, M., Saito, K., & Hirai, M. Y. (2009). Arabidopsis bile acid: sodium symporter family protein 5 is involved in methionine-derived glucosinolate biosynthesis. *Plant and Cell Physiology, 50*, 1579–1586.

Schlaeppi, K., Bodenhausen, N., Buchala, A., Mauch, F., & Reymond, P. (2008). The glutathione-deficient mutant pad2-1 accumulates lower amounts of glucosinolates and is more susceptible to the insect herbivore *Spodoptera littoralis. The Plant Journal, 55*, 774–786.

Schuster, J., Knill, T., Reichelt, M., Gershenzon, J., & Binder, S. (2006). Branched-chain aminotransferase4 is part of the chain elongation pathway in the biosynthesis of methionine-derived glucosinolates in Arabidopsis. *The Plant Cell, 18*, 2664–2679.

Schweizer, F., Fernandez-Calvo, P., Zander, M., Diez-Diaz, M., Fonseca, S., Glauser, G., … Reymond, P. (2013). Arabidopsis basic helix-loop-helix transcription factors MYC2, MYC3, and MYC4 regulate glucosinolate biosynthesis, insect performance, and feeding behavior. *The Plant Cell, 25*, 3117–3132.

Sonderby, I. E., Geu-Flores, F., & Halkier, B. A. (2010). Biosynthesis of glucosinolates—gene discovery and beyond. *Trends in Plant Science, 15*, 283–290.

Sonderby, I. E., Hansen, B. G., Bjarnholt, N., Ticconi, C., Halkier, B. A., & Kliebenstein, D. J. (2007). A systems biology approach identifies a R2R3 MYB gene subfamily with distinct and overlapping functions in regulation of aliphatic glucosinolates. *PLoS One, 2*, e1322.

Stein, M., Dittgen, J., Sánchez-Rodríguez, C., Hou, B.-H., Molina, A., Schulze-Lefert, P., … Somerville, S. (2006). Arabidopsis PEN3/PDR8, an ATP binding cassette transporter, contributes to nonhost resistance to inappropriate pathogens that enter by direct penetration. *The Plant Cell (Online), 18*, 731–746.

Strehmel, N., Böttcher, C., Schmidt, S., & Scheel, D. (2014). Profiling of secondary metabolites in root exudates of *Arabidopsis thaliana. Phytochemistry, 108*, 35–46.

Takahashi, H., Kopriva, S., Giordano, M., Saito, K., & Hell, R. (2011). Sulfur assimilation in photosynthetic organisms: molecular functions and regulations of transporters and assimilatory enzymes. *Annual Review of Plant Biology, 62*, 157–184.

Thuswaldner, S., Lagerstedt, J. O., Rojas-Stuetz, M., Bouhidel, K., Der, C., Leborgne-Castel, N., … Spetea, C. (2007). Identification, expression, and functional analyses of a thylakoid ATP/ADP carrier from Arabidopsis. *Journal of Biological Chemistry, 282*, 8848–8859.

Troufflard, S., Mullen, W., Larson, T. R., Graham, I. A., Crozier, A., Amtmann, A., & Armengaud, P. (2010). Potassium deficiency induces the biosynthesis of oxylipins and glucosinolates in *Arabidopsis thaliana. BMC Plant Biology, 10*, 172.

Wittstock, U., & Burow, M. (2010). Glucosinolate breakdown in Arabidopsis: mechanism, regulation and biological significance. *The Arabidopsis Book/American Society of Plant Biologists, 8*, e0134.

Wittstock, U., Kliebenstein, D. J., Lambrix, V., Reichelt, M., & Gershenzon, J. (2003). Glucosinolate hydrolysis and its impact on generalist and specialist insect herbivores. *Recent Advances in Phytochemistry, 37*, 101–125.

Xiong, L., Lee, B., Ishitani, M., Lee, H., Zhang, C., & Zhu, J. K. (2001). FIERY1 encoding an inositol polyphosphate 1-phosphatase is a negative regulator of abscisic acid and stress signaling in Arabidopsis. *Genes & Development, 15*, 1971–1984.

Xu, J., Yang, J., Wu, Z., Liu, H., Huang, F., Wu, Y., … Wu, P. (2013). Identification of a dual-targeted protein belonging to the mitochondrial carrier family that is required for early leaf development in rice. *Plant Physiology, 161*, 2036–2048.

Yatusevich, R., Mugford, S. G., Matthewman, C., Gigolashvili, T., Frerigmann, H., Delaney, S., ... Kopriva, S. (2010). Genes of primary sulfate assimilation are part of the glucosinolate biosynthetic network in *Arabidopsis thaliana*. *The Plant Journal, 62*, 1–11.
Yazaki, K. (2005). Transporters of secondary metabolites. *Current Opinion in Plant Biology, 8*, 301–307.
Yin, L., Lundin, B., Bertrand, M., Nurmi, M., Solymosi, K., Kangasjarvi, S., ... Spetea, C. (2010). Role of thylakoid ATP/ADP carrier in photoinhibition and photoprotection of photosystem II in arabidopsis. *Plant Physiology, 153*, 666–677.
Zhao, J., & Dixon, R. A. (2010). The 'ins' and 'outs' of flavonoid transport. *Trends in Plant Science, 15*, 72–80.

CHAPTER SIX

Glucosinolate Breakdown

U. Wittstock[1], E. Kurzbach, A.-M. Herfurth, E.J. Stauber
Technische Universität Braunschweig, Braunschweig, Germany
[1]Corresponding author: E-mail: u.wittstock@tu-bs.de

Contents

Abstract

Most known biological roles of the glucosinolate–myrosinase system depend on its functioning as a 'chemical bomb'. This term describes chemical defences which are based on the accumulation of inert components in separate compartments and rapid

Advances in Botanical Research, Volume 80
ISSN 0065-2296
http://dx.doi.org/10.1016/bs.abr.2016.06.006

release of large amounts of defensive chemicals through mixing of these components and their reaction with each other upon tissue disruption. The essential components of the 'chemical bomb' formed by the glucosinolate–myrosinase system are the water-soluble glucosinolates and their hydrolytic enzymes, myrosinases. The defensive chemicals released upon mixing of glucosinolates and myrosinases are the rather lipophilic, irritant and toxic isothiocyanates (mustard oils; reflected in the term 'mustard oil bomb') and their derivatives. In addition, so-called specifier proteins may promote the formation of alternative products such as nitriles and epithionitriles. As a consequence, tissue disruption typically leads to the release of complex mixtures of glucosinolate breakdown products. Apart from directly defensive roles (e.g., as toxins), the breakdown products may also be involved in indirect defences (e.g., as signals in tritrophic interactions). Besides glucosinolate breakdown upon tissue disruption, their turnover in intact tissue has been discussed to contribute sulphur and nitrogen for plant growth. In addition, breakdown of glucosinolates in nondisrupted tissue has been observed upon pathogen attack. This chapter summarizes the current knowledge on the protein components of the glucosinolate breakdown machinery and its compartmentation, highlights the relevance of different breakdown product types and their regulated formation, and discusses pathways and roles of glucosinolate breakdown in nondisrupted tissue.

1. INTRODUCTION

The occurrence of glucosinolates throughout the order Brassicales coincides with the occurrence of specific thioglucosidases, the myrosinases, which are able to hydrolyse these compounds (Rodman, Karol, Price, & Sytsma, 1996). While this might already be taken as an indication that glucosinolate hydrolysis must be important for the plant, numerous experimental studies have demonstrated relevant biological effects of glucosinolate breakdown products (reviewed in Wittstock, Kliebenstein, Lambrix, Reichelt, & Gershenzon, 2003). Moreover, the use of genetic tools has provided convincing evidence for a joint role of glucosinolates and their breakdown machinery in plant defence (Barth & Jander, 2006; Bednarek et al., 2009; Lambrix, Reichelt, Mitchell-Olds, Kliebenstein, & Gershenzon, 2001). The structure of glucosinolates is unique among natural products with respect to the presence of a sulphate group and a glucose molecule linked by a thioglucoside bond within one molecule. Both moieties are bound to an aldoxime core carrying a variable side chain which can contain an additional sulphur atom (Fig. 1). This structure, the diversity of reactions that may occur upon its hydrolysis and the properties of the breakdown products have fascinated scientists for more than a century. Insights

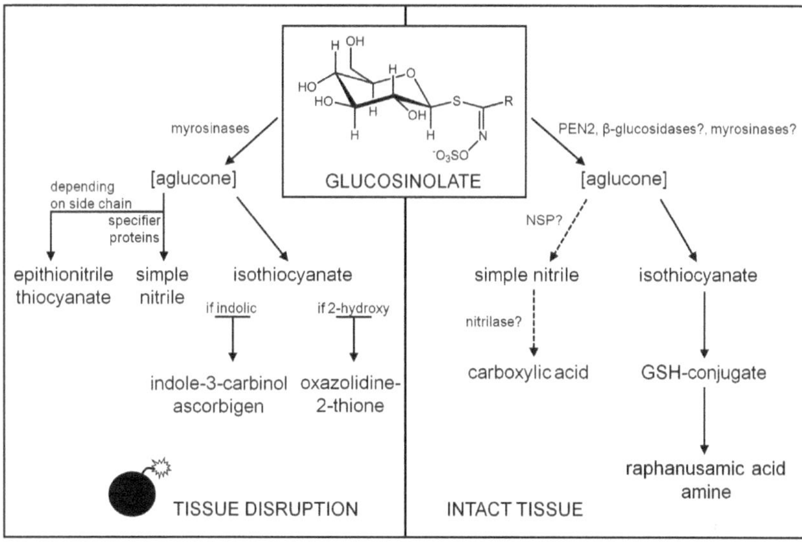

Figure 1 ***Glucosinolate structure and scheme of glucosinolate breakdown.*** Glucosinolates are thioglucosides of (*Z*)-*N*-hydroximinosulphate esters with variable side chains (R). Upon tissue disruption (left), glucosinolates are hydrolysed by myrosinases yielding an unstable aglucone. Spontaneous rearrangement of the aglucone leads to formation of a toxic isothiocyanate ('mustard oil bomb'; Matile, 1980) which can undergo further reactions. In the presence of specifier proteins, alternative products are formed instead of isothiocyanates. Glucosinolate breakdown is also observed in intact tissue and upon pathogen attack (right). See text for references and details. See Figs 3–5 for examples of chemical structures.

from chemical, biochemical, genetic and ecological studies give a multifaceted picture of the glucosinolate–myrosinase system and its role in plant–environment interactions. This chapter provides an overview on the present knowledge of the protein components of the glucosinolate breakdown machinery, their spatial organization and joint action at different levels of plant defence as well as in processes not directly related to biotic interactions.

2. MYROSINASES

2.1 Definition, General Properties and Products

Myrosinases are thioglucosidases (thioglucoside glucohydrolases, EC 3.2.1.147) that catalyse the initial step of bioactivation of glucosinolates, the hydrolysis of their thioglucosidic bond (reviewed in Rask et al., 2000). They are members of glycoside hydrolase family I along with other β-glycosidases that are involved in plant defence (Morant et al., 2008; Xu et al., 2004; see also CAZy (http://www.cazy.org, Lombard, Golaconda Ramulu, Drula, Coutinho, & Henrissat, 2014)). Myrosinases are usually composed of two identical 55–65 kDa polypeptides which are heavily glycosylated resulting in a native molecular weight of 120–150 kDa of the dimeric proteins (Björkman & Janson, 1972; Loebers, Müller-Uri, & Kreis, 2014; Lönnerdal & Janson, 1973; Wang et al., 2009; Zhou, Tokuhisa, Bevan, & Esen, 2012). The dimeric structure is maintained through tetrahedral coordination of a Zn^{2+} by two conserved residues of each monomer (His 56 and Asp 70 in the *Sinapis alba* myrosinase) (Burmeister et al., 1997).

Common properties of myrosinases are their (heat) stability (with temperature optima of up to 70°C) and their activation by ascorbic acid at low millimolar concentrations (typically 1–2 mM) (Andersson et al., 2009; Björkman & Janson, 1972; Björkman & Lönnerdal, 1973; Ettlinger, Dateo, Harrison, Mabry, & Thompson, 1961; Loebers et al., 2014; Tsuruo, Yoshida, & Hata, 1967; Wang et al., 2009; Zhou et al., 2012). The pH optima can be broad (e.g., *Arabidopsis thaliana* TGG1, TGG2 and TGG4, *Armoracia rusticana* ArMY1, *Carica papaya* TGG2) or rather narrow (e.g., *Brassica napus* myrosinases, *A. rusticana* ArMY2; Andersson et al., 2009; Björkman & Lönnerdal, 1973; Härtel & Brandt, 2002; Li & Kushad, 2005; Loebers et al., 2014; Wang et al., 2009; Zhou et al., 2012). Myrosinases accept glucosinolates of different structural types as substrates, but differ in their affinity to individual glucosinolates and the efficiency of their conversion (Bernardi, Finiguerra, Rossi, & Palmieri, 2003; Björkman & Lönnerdal, 1973; Loebers et al., 2014; Zhou et al., 2012). Among myrosinases with high substrate specificity, a myrosinase from *Crambe abyssinica* seeds has been shown to hydrolyse the (2*S*)-epimer of 2-hydroxy-3-butenylglucosinolate much more efficiently than the (2*R*)-epimer (Bernardi et al., 2003). Low activity of myrosinases with O-glucosides has also been described (Durham & Poulton, 1990; Li & Kushad, 2005; Loebers et al., 2014; Zhou et al., 2012). Glucosinolate-containing species typically possess several myrosinase isoforms which may differ in their enzymatic properties and/or their spatial or temporal expression in different organs or developmental stages (e.g., Eriksson, Ek, Xue, Rask, & Meijer, 2001; James & Rossiter, 1991; Lenman et al., 1993; Loebers et al., 2014; Thangstad, Winge, Husebye, & Bones, 1993; Ueda et al., 2006; Wang et al., 2009; reviewed in Rask et al., 2000; Wittstock & Burow, 2010). In some cases, substrate specificity of myrosinases roughly matches the glucosinolate profile of the plant organ in which the enzyme is expressed indicating some degree of specialization. For example, ArMY2 from

A. rusticana has a high affinity for 2-phenylethylglucosinolate and is expressed exclusively in young roots with high content of 2-phenylethylglucosinolate while storage roots with high allylglucosinolate content express a different isoenzyme (Loebers et al., 2014).

An array of methods has been applied to determine myrosinase activity. Most of these methods rely on photometric detection of glucose release, detection of breakdown products by gas chromatography (GC) or photometric determination of substrate decay (reviewed in Kleinwächter & Selmar, 2004). Activity measurements in plant extracts are complicated by formation of insoluble complexes between certain myrosinase isoforms and lectins termed myrosinase-binding proteins (Eriksson et al., 2002, 2001; Taipalensuu, Eriksson, & Rask, 1997; Travers-Martin, Kuhlmann, & Müller, 2008; see Section 2.4.). The affinity of myrosinases to lectins has also been exploited to purify myrosinases by Concanavalin A affinity chromatography (Palmieri, Iori, & Leoni, 1986; Pessina, Thomas, Palmieri, & Luisi, 1990). Besides studies of myrosinases purified from plant material, the characterization of purified recombinant myrosinases has become increasingly important. Although myrosinase activity has been obtained upon expression in *Escherichia coli*, eukaryotic hosts such as the yeast *Pichia pastoris* have proved most suitable for high level expression of active myrosinases (e.g., Andersson et al., 2009; Härtel & Brandt, 2002; Loebers et al., 2014; Wang et al., 2009). *Arabidopsis thaliana* TGG1 purified from rosettes of *tgg2* mutant plants retained more than 70% of its activity upon deglycosylation by treatment with glycoprotein glycosidases for 18 h (Zhou et al., 2012). However, after 72 h of incubation with deglycosylating enzymes, myrosinase activity was basically lost indicating that glycosylation may not be required for activity but for stability in various environments, for example, the herbivore gut (Zhou et al., 2012).

As described in Section 2.2, products of myrosinase-catalysed glucosinolate hydrolysis are an isothiocyanate (mustard oil), β-D-glucose, and sulphate. Depending on its side chain, an isothiocyanate can undergo further reactions. The first isothiocyanate-derived compound identified was goitrin (5-ethenyl-1,3-oxazolidine-2-thione) which is a cyclization product of the isothiocyanate generated upon hydrolysis of (*S*)-2-hydroxy-3-butenylglucosinolate (Astwood, Greer, & Ettlinger, 1949). Its name refers to the observation that intake of high amounts of this compound through Brassica vegetables may cause an enlargement of the thyroid gland (goiter) (reviewed in Fenwick, Heaney, & Mullin, 1983). As in case of (*S*)-2-hydroxy-3-butenylglucosinolate, isothiocyanates of other glucosinolates with a hydroxy group at C2 of the side chain also undergo spontaneous cyclization to produce goitrin derivatives with corresponding residues at the C5 of the 1,3-oxazolidine-2-thione ring (reviewed in Agerbirk & Olsen, 2012). Further rearrangement to a 1,3-thiazolidine-2-one was described upon breakdown of (*R*)-2-hydroxy-2-(4-hydroxy)phenylethylglucosinolate (Agerbirk & Olsen, 2015). Isothiocyanates derived from indole glucosinolates are unstable and react with various nucleophiles under release of a thiocyanate ion (SCN^-) and structurally diverse products (reviewed in Agerbirk, De Vos, Kim, & Jander, 2009). Among these products, indole-3-carbinols are major compounds obtained in vitro upon reaction of indolic isothiocyanates with water as nucleophile. In the presence of ascorbic acid as a nucleophile, for example, in plant homogenates, indolic isothiocyanates and indole-3-carbinols form adducts with ascorbate known as ascorbigens (Agerbirk, Olsen, & Sørensen, 1998). The isothiocyanate derived from 4-hydroxybenzylglucosinolate undergoes similar reactions yielding a thiocyanate ion, 4-hydroxybenzyl alcohol and/or ascorbigens (Buskov et al., 2000). A broad spectrum of other products that are not isothiocyanate-derived can be formed in the presence of specifier proteins (see Section 3). But also in the absence of specifier proteins, alternative routes of glucosinolate breakdown may be favoured, that is, formation of simple nitriles at $pH < 5$ or in the presence of Fe^{2+} (usually >0.1 mM) (Vaughn & Berhow, 2005; reviewed in Fenwick et al., 1983).

2.2 Mechanism

First ideas about the reaction mechanism of myrosinase arose after the stereochemistry of the glucosinolate structure had been clarified using allylglucosinolate as a prototype (Benn & Ettlinger, 1965; Ettlinger & Lundeen, 1956; Waser & Watson, 1963). Glucosinolates were identified as 1-β-D-thioglucosides in which the thioglucosyl moiety is linked to the sulphate group via a C=N bond in *Z*-configuration. This allowed for postulation of a reaction mechanism in which enzymatic removal of the glucosyl group would be sufficient to initiate a Lossen-type rearrangement for isothiocyanate formation (Ettlinger & Lundeen, 1957). Sulphate would be released without the need for an additional sulphatase activity. The postulate was experimentally confirmed upon elucidation of the crystal structure of a myrosinase isolated from *S. alba* seeds and subsequent analysis of structures of the enzyme with bound ascorbate and/or transition state analogues (Fig. 2) (Burmeister et al., 1997; Burmeister, Cottaz, Rollin, Vasella, & Henrissat, 2000). This

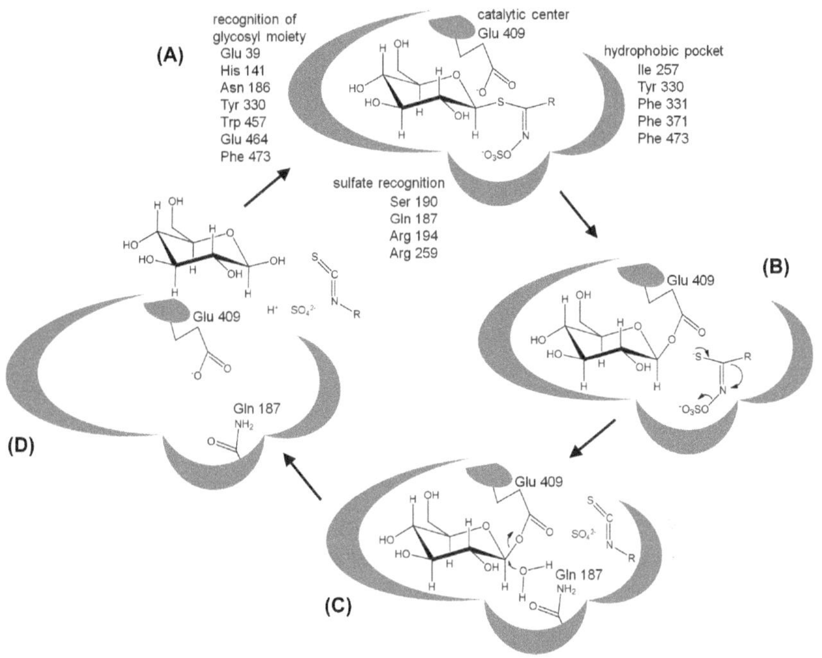

Figure 2 ***Scheme of glucosinolate binding and catalysis in the active site of myrosinase.*** (A) Each myrosinase monomer has a substrate-binding pocket with subsites for recognition of the glucosyl moiety, the sulphate group and the side chain (R). Residues important for substrate recognition in each subsite are listed. The catalytic Glu residue is highlighted. To fit the active site, the substrate has to be in a distorted conformation (not shown). (B) Nucleophilic attack of Glu 409 at the anomeric carbon initiates departure of the aglucone and results in formation of a glucosyl-enzyme intermediate. The aglucone undergoes a Lossen-type rearrangement. (C) While the resulting isothiocyanate diffuses away, Gln 187 (which replaces the acid/base catalyst Glu found in O-glycosidases) positions a water molecule close to the anomeric carbon to enable hydrolysis of the glucosyl-enzyme. (D) Isothiocyanate, sulphate and β-D-glucose are released as products of the enzymatic reaction. For references see Burmeister et al. 2000.

identified a striking difference to O-glycosidases, the replacement of the acid/base catalyst Glu in the conserved TFNEP motif by a Gln.

In O-glycosidases, the Glu residue in the I/VTENG motif acts as nucleophile attacking the anomeric carbon while a second Glu residue in the TFNEP motif serves as acid catalyst protonating the glycosidic oxygen to facilitate departure of the aglycone (Davies & Henrissat, 1995; Morant et al., 2008). As result of this first step of catalysis, a glycosyl-enzyme intermediate is formed. In a second step, a water molecule is activated by the Glu residue of the TFNEP motif now acting as general base catalyst to hydrolyse the glycosyl-enzyme intermediate (Davies & Henrissat, 1995; Morant et al., 2008). As the glucosinolate aglucone is an excellent leaving group, glucosinolate hydrolysis appears to not require protonation in the first catalytic step (Burmeister et al., 2000). Moreover, the Glu residue of the TFNEP motif would interfere with recognition of the sulphate group (Fig. 2) (Burmeister et al., 2000). This may explain, why this amino acid is replaced by a conserved Gln in myrosinases (Gln 187 in *S. alba* myrosinase) (Burmeister et al., 2000). This Gln residue is also sufficient for the second step of catalysis as the glycosyl-myrosinase intermediate seems to be more reactive than that of O-glycosidases and is hydrolysed without deprotonation by a general base catalysis (Burmeister et al., 2000). Gln 187 apparently serves in positioning of a water molecule for subsequent hydrolysis (Fig. 2) (Burmeister et al., 2000). Structural analysis also delivered an explanation for the activating effect of ascorbic acid. Ascorbate occupies the subsite for sulphate recognition in the substrate-binding pocket upon departure of the aglucone and supports the hydrolysis of the glucosyl-enzyme intermediate by serving as general base catalyst (Burmeister et al., 2000).

2.3 Atypical Plant Myrosinases and Myrosinases From Insects and Microbes

In contrast to the classical myrosinases described in the preceding section, *A. thaliana* PEN2 was identified as a glucosinolate-hydrolysing thioglucosidase with two catalytic Glu residues (Bednarek et al., 2009; Clay, Adio, Denoux, Jander, & Ausubel, 2009). It has therefore been termed an 'atypical' myrosinase. PEN2 is required for induced defence against certain fungal pathogens (see Section 7; see, chapter 7 in this volume). Myrosinases with two catalytic Glu residues have also been cloned and characterized from insects, namely *Brevicoryne brassicae* (Hemiptera: Aphididae) and *Phyllotreta striolata* (Coleoptera: Chrysomelidae) (Beran et al., 2014; Husebye et al., 2005; Jones, Bridges, Bones, Cole, & Rossiter, 2001; Pontoppidan, Ekbom, Eriksson, & Meijer, 2001; see, chapter 8 in this volume). The insect myrosinases as well as PEN2 also accept O-glucosides as substrates, but conversion rates are much lower than those for glucosinolates (Bednarek et al., 2009; Beran et al., 2014; Pontoppidan et al., 2001). As one may expect based on the role of ascorbate in plant myrosinases, the aphid myrosinase has been shown to be inhibited rather than activated by ascorbic acid (Pontoppidan et al., 2001). Besides plants and insects, certain bacteria, including species of the human gut flora, and fungi also produce myrosinases (e.g., Meulenbeld & Hartmans, 2001; Mullaney, Kelly, McGhie, Ansell, & Heyes, 2013; Reese, Clapp, & Mandels, 1958; Tani, Ohtsuru, & Hata, 1974). The presence of myrosinase-producing bacteria in the human gut affects the bioavailability of breakdown products and may therefore influence health effects of glucosinolate-containing foods in humans (reviewed in Johnson, 2002; see, chapter 9 in this volume). Identification and characterization of such myrosinases would certainly contribute to a better understanding of the fate of glucosinolates consumed with Brassica vegetables and help to explore glucosinolates as plant-based pharmaceuticals or nutraceuticals.

2.4 Myrosinase-Binding Proteins and Myrosinase-Associated Proteins

Studies on myrosinases in plant extracts, especially of *B. napus* and *S. alba* seeds, have often been complicated by the formation of complexes with molecular weights of up to 800 kDa (Eriksson et al., 2001; Lenman, Rodin, Josefsson, & Rask, 1990). Certain antibodies which recognized the complexes did not bind to myrosinases indicating the presence of additional proteins which were termed myrosinase-binding proteins (MBPs) (Lenman et al., 1990). More detailed studies with *B. napus* identified several different MBPs with molecular weights of up to 110 kDa and differential expression upon wounding, methyljasmonate treatment and aphid infestation (Falk, Taipalensuu, Ek, Lenman, & Rask, 1995; Geshi & Brandt, 1998; Pontoppidan, Hopkins, Rask, & Meijer, 2003; Taipalensuu, Falk, & Rask, 1996). The cellular localization of MBPs may change during development. For example, MBPs have been localized in myrosin cells in 7-day-old *B. napus* seedlings, but were present in most cells except myrosin cells in *B. napus* seeds and two-day-old seedlings (Andréasson, Jørgensen, Höglund, Rask, & Meijer, 2001; Eriksson et al., 2002; Geshi & Brandt, 1998). *B. napus* MBPs possess lectin activity, that is, they are able to bind to carbohydrates such as oligosaccharides at the surface of glycoproteins (Geshi & Brandt, 1998; Taipalensuu, Eriksson, et al., 1997). Sequence analysis predicted the presence of several lectin domains (Taipalensuu, Falk, Ek, & Rask, 1997). Although myrosinases of at least the aboveground parts of *A. thaliana* seem to be soluble, a large number of homologs to *B. napus* MBPs are encoded in the *A. thaliana* genome (Barth & Jander, 2006; Burow et al., 2009; Capella, Menossi, Arruda, & Benedetti, 2001; Zhou et al., 2012). The three-dimensional structure of one of these proteins (At3g16450) has been elucidated by NMR spectroscopy (Takeda et al., 2008). This confirmed its composition of two lectin domains. However, the *A. thaliana* protein only possessed moderate lectin activity (Takeda et al., 2008). Although MBPs have been demonstrated to be required for myrosinase complex formation in *B. napus* seed homogenates, their biological roles are still unknown (Eriksson et al., 2002). *B. napus* plants without functional MBPs due to expression of an antisense constructs did not differ from wild-type plants with respect to myrosinase localization, glucosinolate breakdown, performance of beetles and flea beetles on seeds or cotyledons, respectively, and pathogen growth (Eriksson et al., 2002).

Another group of proteins which form complexes with myrosinases and have been suspected to play a role in glucosinolate metabolism are the myrosinase-associated proteins (MyAPs) first identified in seeds of *B. napus* (Taipalensuu et al., 1996). MyAPs are encoded by a gene family in *B. napus* with some wound-, methyljasmonate- or aphid-inducible representatives (Pontoppidan et al., 2003; Taipalensuu, Andréasson, Eriksson, & Rask, 1997; Taipalensuu et al., 1996). Their biological roles in *B. napus* are still unknown. Two genes which encode MyAP-like proteins in *A. thaliana* have been studied with some detail, the *EPITHIOSPECIFIER-MODIFIER1* (*ESM1*) gene and the *modified vacuolar phenotype 1* (*MVP1*) gene (Agee et al., 2010; Zhang, Ober, & Kliebenstein, 2006). While MVP1 appears to play a role in endomembrane trafficking, forms complexes with the myrosinase TGG2 and affects localization of TGG2:GFP fusion proteins, *ESM1* impacts the outcome of glucosinolate breakdown by an unknown mechanism (Agee et al., 2010; Zhang et al., 2006). If *ESM1* is functional, alternative product formation, namely nitrile formation, is suppressed in favour of isothiocyanate formation upon glucosinolate hydrolysis in leaf homogenates (Zhang et al., 2006; see also Section 5).

3. SPECIFIER PROTEINS

3.1 Specifier Protein Types

Based on the observation that breakdown of certain structural classes of glucosinolates may yield products other than isothiocyanates, the first specifier protein was isolated from *C. abyssinica* in 1973 and named *e*pithio*s*pecifier *p*rotein (ESP) (Gmelin & Virtanen, 1959; Tookey, 1973; Virtanen, 1965). This name refers to the ability of ESP to promote formation of epithionitriles, that is, nitriles with a terminal thiirane moiety (Fig. 3G), upon

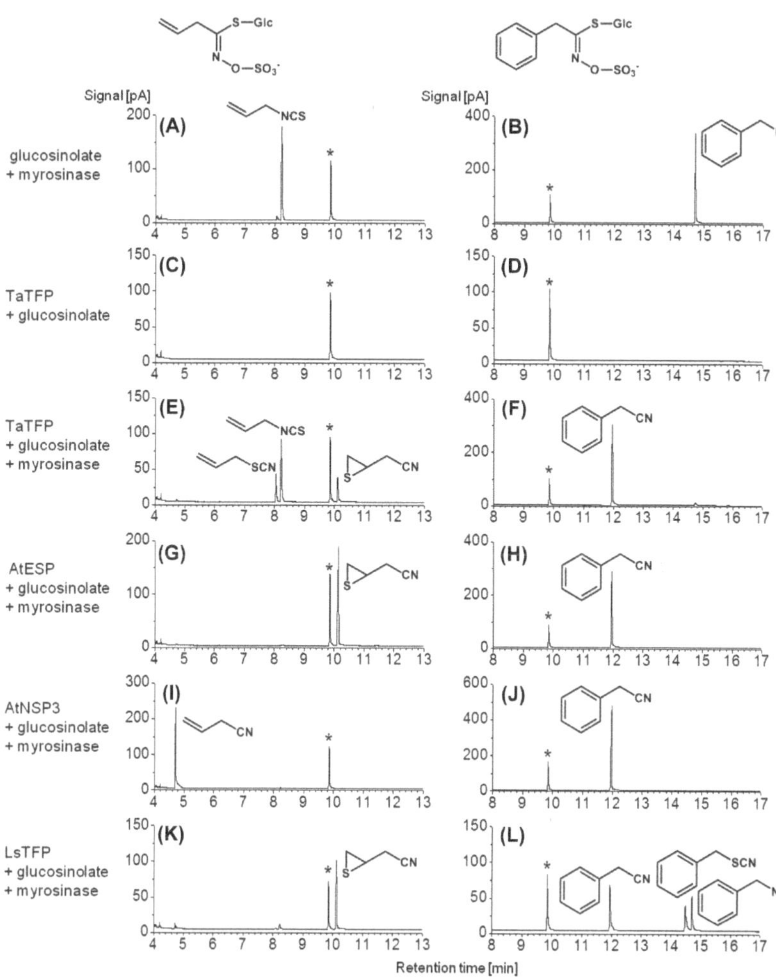

Figure 3 ***In vitro activity of specifier proteins expressed in* E. coli.** Crude extracts of *E. coli* expressing TaTFP (C-F), AtESP (G, H), AtNSP3 (I, J) or LsTFP (K, L) (or extraction buffer only: A, B) were incubated with 2 mM allylglucosinolate (left) or 1 mM benzylglucosinolate (right) and 0.005 units purified myrosinase from *Sinapis alba* seeds (or buffer only: C, D) for 40 min. After addition of an internal standard (*), reaction mixtures were extracted with dichloromethane and organic phases analysed by GC. Shown are chromatograms obtained with flame ionization detection. For the analysis of TaTFP, AtESP and AtNSP3, *E. coli* extract corresponding to 750 μg total protein was used. LsTFP was analysed with 5 mg total protein (in assays with allylglucosinolate) and 1 mg (in assays with benzylglucosinolate). Products of allylglucosinolate breakdown were: allylcyanide (3-butenonitrile, 4.71 min), allylthiocyanate (8.04 min), allylisothiocyanate (8.21 min), 3,4-epithiobutanenitrile (10.13 min). Products of benzylglucosinolate breakdown were: benzylcyanide (phenylacetonitrile, 11.96 min), benzylthiocyanate (14.47 min), benzylisothiocyanate (14.70 min).

glucosinolate breakdown without possessing myrosinase activity itself. Epithionitrile formation is confined to alkenylglucosinolates with a terminal double bond. Labelling studies demonstrated that the sulphur of the thiirane ring originates from the thioglucosidic bond of the glucosinolate, that is, it is captured by the side-chain double bond upon deglycosylation (Brocker & Benn, 1983). It has been discussed since its discovery if ESP acts as a separate enzyme catalysing this reaction of the aglucone or if it serves as a cofactor of myrosinase (reviewed in Wittstock & Burow, 2007). As the aglucone is only a transient intermediate that has never been isolated, it has been difficult to test the role of ESP in epithionitrile formation biochemically. Cloning and characterization of the first specifier protein, ESP from *A. thaliana* (accession Landsberg *erecta* (*Ler*); AtESP), in 2001 was followed by identification of different types of specifier proteins from *A. thaliana* and other species (Burow, Bergner, Gershenzon, & Wittstock, 2007; Burow et al., 2009; Kissen & Bones, 2009; Kuchernig, Backenköhler, Lübbecke, Burow, & Wittstock, 2011; Kuchernig, Burow, & Wittstock, 2012; Lambrix et al., 2001; Matusheski et al., 2006). Besides giving first insights into the evolution of this group of proteins, this has also opened new possibilities for studying their molecular mechanisms.

As of mid-2016, 19 specifier proteins have been cloned and characterized after heterologous expression of their cDNAs in *Escherichia coli*. In agreement with the original definition (Tookey, 1973), they do not themselves convert glucosinolates, but affect the outcome of myrosinase-catalysed glucosinolate hydrolysis (Fig. 3A–F). In the presence of specifier proteins, simple nitriles, epithionitriles and organic thiocyanates are formed instead of the corresponding isothiocyanates (Fig. 3E–L). As specifier proteins do not hydrolyse glucosinolates, a typical assay for specifier protein activity is composed of a glucosinolate, specifier protein of interest and myrosinase (added at last) in an appropriate buffer. After an incubation time, breakdown products can be quantified, typically by GC of dichloromethane extracts. The presence of specifier proteins in the reaction mixture affects primarily product proportions, not total amount of breakdown products (although an impact on total amount of breakdown products has also been observed (Burow, Markert, Gershenzon, & Wittstock, 2006; Kuchernig et al., 2011)). With increasing amounts of specifier protein, the amount of non-isothiocyanate products increases and the amount of isothiocyanate decreases. Isothiocyanate formation can be completely abolished when high amounts of specifier proteins are used (Fig. 3F–K). However, in some cases isothiocyanate formation does not decrease below a certain proportion (Fig. 3E).

According to their product profiles, specifier proteins are grouped into three different types: *n*itrile-*s*pecifier *p*roteins (NSPs), ESPs and *t*hiocyanate-*f*orming *p*roteins (TFPs). NSPs promote the formation of simple nitriles regardless of the glucosinolate side-chain (Fig. 3I and J) (Burow et al., 2009; Kissen & Bones, 2009). Simple nitriles are also formed in the presence of ESP unless the glucosinolate side chain possesses a terminal double bond to enable epithionitrile formation (Fig. 3G and H) (Burow, Matkert, et al., 2006; Matusheski et al., 2006). This is very similar to the reactivity of TFPs, but TFPs have the additional ability to promote the formation of organic thiocyanates (Fig. 3E and F). Organic thiocyanate formation has only been observed for three glucosinolates, namely allyl-, benzyl- and 4-methylthiobutylglucosinolate. It has been proposed that the ability to form stable carbocations is a common feature of the aglucones of these three glucosinolates and a prerequisite for organic thiocyanate formation (Lüthy & Benn, 1977). Interestingly, each of the TFPs identified until mid-2016 promotes thiocyanate formation only for one single glucosinolate (either allyl- or benzylglucosinolate; Fig. 3E, F, K and L) (Burow, Bergner, et al., 2007; Kuchernig et al., 2011). Thus, besides formation of distinct product types, specifier proteins appear to act specifically with respect to the glucosinolate side chain. While simple nitriles may also be formed under certain conditions in the absence of specifier proteins (pH < 5, presence of Fe^{2+} at ≥ 0.1 mM), formation of epithionitriles and organic thiocyanates is strictly dependent on specifier proteins (Burow, Bergner, et al., 2007; Burow, Markert, et al., 2006; Kuchernig et al., 2011).

3.2 Fe^{2+} Dependency and Mechanism

Alternative product formation in the presence of purified specifier proteins is generally increased at the expense of isothiocyanate formation when the reactions are supplemented with Fe^{2+} or, to a lower extent, when other metal ions are added (Burow, Bergner, et al., 2007; Burow et al., 2009; Burow, Markert, et al., 2006; Kissen & Bones, 2009; Kuchernig et al., 2011; Matusheski et al., 2006). A molecular modelling study identified a potential Fe^{2+}-binding site (EXXXDXXXH) that is conserved in all known specifier proteins with the exception of AtNSP5 (EXXXHXXXH) (Brandt et al., 2014). Individual substitution of the conserved amino acids by the corresponding amides or Ala, respectively, led to a complete loss of specifier protein activity of AtESP and TFPs from *Lepidium sativum* (LsTFP) and *Thlaspi arvense* (TaTFP) regardless of Fe^{2+} supplementation (Brandt et al., 2014).

Thus, the ability to coordinate Fe^{2+} appears to be essential for specifier protein activity.

Addition of the myrosinase activator ascorbic acid (see Section 2.1) to reaction mixtures composed of AtESP, allylglucosinolate and myrosinase in appropriate buffer with Fe^{2+} led to an increase of the total amount of breakdown products formed (Burow, Markert, et al., 2006). However, this increase was entirely due to enhanced isothiocyanate formation while the total amount of the corresponding epithionitrile remained constant arguing against a role of AtESP as a classical cofactor of myrosinase. Furthermore, specifier proteins promote alternative product formation specifically with respect to glucosinolate side chain and product type and possess an Fe^{2+}-binding site. Thus, it seems reasonable to hypothesize that specifier proteins are enzymes which catalyse the conversion of the glucosinolate aglucone to non-isothiocyanate products. The mechanisms of product formation are, however, poorly understood (Foo et al., 2000; Rossiter et al., 2007; reviewed in Wittstock & Burow, 2007). The recent elucidation of the three-dimensional structure of TaTFP will greatly benefit further investigations of possible reaction mechanisms. As previously proposed based on structural motif searches in combination with homology modelling, TaTFP, which crystallized as a dimer, adopts a six-blade kelch-type β-propeller fold (Gumz et al., 2015). Molecular docking studies with Fe^{2+} and the aglucones of allyl- and benzylglucosinolate supported the presence of an active site composed of amino acid residues of loops protruding at the bottom side of the kelch structure, including the Fe^{2+}-binding triad EXXXDXXXH. Based on previous results obtained by molecular modelling with other kelch proteins as templates, these active site elements seem to be functionally conserved in specifier proteins of different types, but the aglucone side-chain is arranged differently due to structural variation of its binding pocket (Brandt et al., 2014). With this information in hand, it has now become a realistic goal to elucidate the mechanisms of alternative product formation, the role of specifier proteins in these reactions and the structural determinants of their specificity. It will also be interesting to clarify in which way specifier proteins and myrosinases interact to allow the glucosinolate aglucone to pass from the active site of myrosinase to that of the specifier protein without rearranging to an isothiocyanate.

3.3 Evolution

The ability to form simple nitriles, epithionitriles and organic thiocyanates is relatively widespread within the Brassicaceae, but rare outside this family. In a screening of 22 species of the Brassicaceae and seven species from five other families of the Brassicales, 14 species of the Brassicaceae produced non-isothiocyanate products when plant parts such as leaves and seedlings were homogenized (Kuchernig et al., 2012). Among plants from the other families, only *Tropaeolum majus* produced minor amounts of simple nitrile (Kuchernig et al., 2012). Attempts to isolate specifier protein cDNA from species outside the Brassicaceae failed (Kuchernig et al., 2012). Thus, it is well possible that the diversification of glucosinolate breakdown products by action of specifier proteins is a relatively recent development that has contributed to the success of the Brassicaceae. Amino acid sequence identity between ESPs and TFPs is higher (59—98%) than between ESPs or TFPs and NSPs (about 45—65%) (Kuchernig et al., 2012). Phylogenetic analyses identified NSPs as the evolutionary oldest specifier proteins and suggested that ESPs evolved in the course of the *Atα* whole genome duplication experienced by the core Brassicaceae 31—36 million years ago or as a consequence of local gene duplications and neofunctionalization in this time window or before (Kuchernig et al., 2012). The ability to promote organic thiocyanate formation most likely arose from ESP-forming activity twice independently in different lineages of the Brassicaceae (Kuchernig et al., 2012). Thus, natural selection seems to have favoured structural diversification at the level of glucosinolate breakdown through the evolution of specifier proteins with extended capabilities.

Most NSPs identified up to now belong to the NSP1 cluster based on phylogenetic analysis while only one representative of the more ancestral NSP5 cluster has been characterized (Kuchernig et al., 2012). In contrast to all other known specifier proteins, NSPs of the NSP1 cluster possess one or two jacalin-like lectin domains of unknown function which are linked to the N-terminus of the kelch domain region (Kuchernig et al., 2012). Although the lectin domains might possibly affect NSP activity, they are not generally required for nitrile-forming activity as NSP5 does not possess such a domain (Burow et al., 2009). However, presence of these domains delineates a link to MBPs which are composed entirely of jacalin-like lectin domains (see Section 2.4). Future research should address if presence of the jacalin-like lectin domain in specifier proteins of the NSP1 cluster affects their interaction with myrosinase and which consequences a changed interaction might have.

In *A. thaliana*, the closest relative of specifier proteins is encoded by a gene (At3g07720) whose transcription is highly induced in roots upon iron deficiency (Buckhout, Yang, & Schmidt, 2009; Kuchernig et al.,

2012; Stein & Waters, 2012; Urzica et al., 2012). The Fe^{2+}-binding triad is only partially conserved in this protein (EXXXSXXXH). As of mid-2016, the biological role of the ancestral protein has not been elucidated, but a role in Fe^{2+} homeostasis seems likely. Thus, it is tempting to speculate that the ability of the ancestral proteins to bind Fe^{2+} allowed specifier protein activity to evolve under environmental selection pressures. It would be interesting to compare the Fe^{2+} affinity of the protein encoded by At3g07720, AtNSP5 (EXXXHXXXH) and specifier proteins with the conserved Fe^{2+}-binding triad (EXXXDXXXH).

4. COMPARTMENTATION

4.1 The Principle of a 'Chemical Bomb'

If a plant relies on reactive metabolites (such as isothiocyanates) or universally toxic compounds (such as cyanide) as chemical defences, it has to take measures to prevent self-intoxication (Wittstock & Gershenzon, 2002). One possibility is to store these chemicals in an inactive or inert form, for example as a glycoside, separately from an enzyme that catalyses the conversion to the active form, and only upon cellular disintegration, for example by a chewing herbivore, substrate and enzyme are mixed and defensive chemicals be generated. Although this happens only locally at the site of attack, the 'chemical bomb' provides an efficient defence: the reactive or toxic metabolites are typically released rapidly and in high amounts as both the precursors and the activating enzymes can be stored safely at high concentrations. Among the various possibilities of separate storage of glucosinolates as the 'explosive' and myrosinases as the 'fuse' within plant tissue, most of the available data point at a distribution in different, specialized cells (see Sections 4.2 and 4.3). At least in *A. thaliana* flower stalks, specialized cells containing either glucosinolates or myrosinases are located in close proximity or next to each other between the phloem and the endodermis (Andréasson et al., 2001; Husebye, Chadchawan, Winge, Thangstad, & Bones, 2002; Koroleva et al., 2000). This increases the likelihood of simultaneous disruption of both cell types leading to a burst of the 'mustard oil bomb'. However, there are also reports on the cooccurrence of glucosinolates and myrosinases within the same cells raising the question how detonation of the 'bomb' is prevented in these cells (Koroleva & Cramer, 2011). Moreover, the distribution of glucosinolate-containing and myrosinase-containing cells in a tissue varies with developmental stage, organ and species (see the following).

4.2 Storage of Glucosinolates

Due to their water-solubility, glucosinolates are thought to be stored in the vacuole (reviewed in Andréasson & Jørgensen, 2003). This has been confirmed experimentally by the analysis of isolated vacuoles of roots of *A. rusticana*, immunohistochemical detection in *Brassica juncea* embryos using an allylglucosinolate antibody and cell sap analysis of so-called S-cells of *A. thaliana* flower stalks (see the following) (Grob & Matile, 1979; Helmlinger, Rausch, & Hilgenberg, 1983; Kelly, Bones, & Rossiter, 1998; Koroleva et al., 2000; reviewed in Andréasson & Jørgensen, 2003).

Although glucosinolates probably occur throughout the tissue, they accumulate to high concentrations in specialized cells termed S-cells. These cells have first been described in flower stalks of *A. thaliana* based on energy-dispersive X-ray microanalysis in combination with electron scanning microscopy of freeze-dried sections and single cell sampling and analysis (Koroleva et al., 2000). Elemental analysis of different cell types by this technique revealed high sulphur content in epidermal cells and in groups of large cells located between the phloem and the endodermis (starch sheath). The phloem-associated sulphur-rich cells were termed S-cells because sulphur concentrations of up to 350 mM were determined in their cell sap (Koroleva et al., 2000). Glucosinolates were estimated to account for 84% of the S-cell sulphur content (corresponding to a glucosinolate concentration of 146 mM) based on glucose release upon incubation of cell sap with myrosinase (Koroleva et al., 2000). High concentrations of aliphatic and indolic glucosinolates (e.g., 96 mM 4-methylsulphinylbutylglucosinolate) have later been confirmed by LC-MS analysis of cell sap taken from flower stalks with a glass microcapillary (Koroleva, Gibson, Cramer, & Stain, 2010). S-cells of up to 1 mm length have been observed in longitudinal sections of the *A. thaliana* flower stalk (Koroleva et al., 2000). In cross sections, the S-cells appear as large, irregularly shaped cells with a diameter of about 30 μm (Koroleva et al., 2000). They have thin cell walls, large vacuoles and do not seem to contain plastids (Andréasson et al., 2001; Husebye et al., 2002; Koroleva et al., 2010). Ultrastructural analysis and enzymatic DNA nick detection in S-cell nuclei showed that S-cells of the *A. thaliana* flower stalk display features of programmed cell death underlining their high degree of metabolic specialization (Koroleva et al., 2010). Because of their similarity to nonarticulated laticifers, an evolutionary link between S-cells and nonarticulated laticifers has been proposed (Koroleva et al., 2010). These branched, tubular storage compartments for latex and low molecular weight compounds in, for example, Euphorbiaceae and Apocynaceae undergo apoptotic processes during development. Interestingly, S-cells have been demonstrated to contain myrosinases, likely localized in a different subcellular compartment (Koroleva & Cramer, 2011).

Sulphur-rich cells which fulfil the description of the S-cells mentioned in the preceding paragraph have also been identified along the midvein of rosette leaves, the midvein and lateral veins of cauline leaves and all main veins of pedicle and siliques of *A. thaliana* (Koroleva et al., 2010). In addition, sulphur-rich cells were found at the outer margin of mature rosette leaves of *A. thaliana* (Koroleva et al., 2010). In contrast to the earlier mentioned S-cells, the latter cells did not accumulate high sulphur levels in younger developmental stages (Koroleva et al., 2010). However, they could be a storage site of glucosinolates in agreement with a report on uneven distribution of glucosinolates in *A. thaliana* rosette leaves and high glucosinolate content at the leaf margin (Koroleva et al., 2010; Shroff, Vergara, Muck, Svatoš, & Gershenzon, 2008). For sepals it has been demonstrated by MALDI mass spectrometry imaging that cells at the leaf margin contain glucosinolates, in particular high levels of indol-3-ylmethylglucosinolate, while S-cells near the phloem in the middle of the basal part seem to be high in 4-methylsulphinylbutylglucosinolate (Sarsby, Towers, Stain, Cramer, & Koroleva, 2012). Epidermal cells of the sepals also contain high sulphur levels indicating storage of glucosinolates as a defensive barrier (Sarsby et al., 2012). Although glucosinolate accumulation is likely to happen through similar structures in other plants than *A. thaliana*, we are not aware of any reports on the presence of S-cells in other species.

4.3 Storage of Myrosinases

As myrosinases are glycoproteins, they are presumably synthesized at the rough endoplasmic reticulum and targeted to their final destination after glycosylation in the Golgi-apparatus. Immunogold-labelling studies conducted with embryos and seedlings of different species of the Brassicaceae and rosette leaves of *A. thaliana* demonstrated that myrosinases are stored inside vacuoles referred to as myrosin grains (Andréasson et al., 2001; Höglund, Lenman, & Rask, 1992; Thangstad, Evjen, & Bones, 1991; Ueda et al., 2006). Using transmission electron microscopy, these were described to be filled with electron dense, finely granular material with a homogenous distribution of the gold label indicating presence of myrosinase, but without globoid inclusions (Andréasson et al., 2001; Höglund et al., 1992; Thangstad et al., 1991; Ueda et al., 2006). In *B. napus* embryos,

Table 1 History of the Terms 'Myrosinase' and 'Myrosin Cell'

Year	
1839	Bussy described 'myrosine' as a protein substance required for release of essential oil from seeds of *Brassica nigra*.
1884	Heinricher identified a protein-rich cell-type in Brassicaceae which differed from surrounding cells in morphology, light refraction and staining properties.
1890	Guignard introduced the term 'myrosin cell' for the protein-rich cells described by Heinricher that stain strongly with Millon's reagent and Anilin Blue Black.

According to Andréasson, E., & Jørgensen, L. B. (2003). Localization of plant myrosinases and glucosinolates. In J. T. Romeo (Ed.), *Recent advances in phytochemistry: Vol. 37. Integrative phytochemistry: From ethnobotany to molecular ecology* (pp. 79–99). Amsterdam, Elsevier.

myrosin grains within one cell were associated with each other to form a continuous vacuolar system termed myrosin body (Andréasson et al., 2001). In agreement with the immunogold-labelling studies, myrosinases were identified in the rosette leaf vacuolar proteome of *A. thaliana* (Carter et al., 2004). When the *A. thaliana* myrosinases TGG1 or TGG2 were transiently overexpressed in *A. thaliana* as fusion proteins with a C-terminal GFP, epidermal cells of cotyledons showed fluorescence in the ER, in ER bodies and in transvacuolar strands (Agee et al., 2010). In the light of the earlier studies, it remains uncertain if this result mirrors a physiological situation in a certain developmental stage, stunted trafficking or improper targeting due to overexpression and GFP fusion.

Myrosinase-containing cells are single isodiametric or prosenchymatic cells that are scattered within different tissues throughout various organs of the species studied until mid-2016 (reviewed in Andréasson & Jørgensen, 2003; Bones & Rossiter, 1996; Kissen, Rossiter, & Bones, 2009; Wittstock & Burow, 2010). Besides stomatal guard cells, myrosinase-containing cells are typically found among phloem parenchyma cells at the abaxial side of the phloem and in the mesophyll or cortex (reviewed in Andréasson & Jørgensen, 2003; Bones & Rossiter, 1996; Kissen, Rossiter, et al., 2009; Wittstock & Burow, 2010). The discovery of these cells and their designation as myrosin cells dates back to the 19th century when their morphological characteristics and staining properties attracted the attention of scientists (Table 1) (reviewed in Andréasson & Jørgensen, 2003; Bones & Rossiter, 1996). As myrosin cells were clearly distinguishable from surrounding parenchyma cells they were classified as idioblasts, that is, specialized cells embedded in the parenchyma (Heinricher, 1884). Due to their high vacuolar protein content, myrosin cells stain strongly with protein dyes such as Millon's reagent (Heinricher, 1884). Using protein dyes, myrosin cells were detected in various species of a range of families of the Brassicales (Jørgensen, 1981). With the availability of specific antibodies directed against one or several myrosinase isoforms, immunocytochemical studies in several species of the Brassicaceae provided evidence that myrosin cells actually accumulate myrosinases (e.g., Andréasson et al., 2001; Bones, Thangstad, Haugen, & Espevik, 1991; Höglund et al., 1992; Husebye et al., 2002; Thangstad, Evjen, & Bones, 1991; Thangstad, Iversen, Slupphaug, & Bones, 1990). In parallel, in situ hybridization experiments with *S. alba* embryos and seedlings and *B. napus* embryos revealed the presence of myrosinase transcripts in scattered cells (Lenman et al., 1993; Xue, Pihlgren, & Rask, 1993). Myrosinase promoter-reporter gene fusions were used in *B. napus* seeds, embryos and seedlings and *A. thaliana* seedlings and mature plants to demonstrate myrosinase promoter activity in scattered cells whose identity as myrosin cells was proven by immunolocalization (Husebye et al., 2002; Thangstad et al., 2004). Thus, myrosin cells appear to be the site of both transcription and accumulation of myrosinases. Besides developmental and organ-specific regulation of expression of myrosinase isoforms, different isoforms may also accumulate in certain subtypes of myrosin cells (Barth & Jander, 2006; Lenman et al., 1993; Zhao, Zhang, Stanley, & Assmann, 2008).

Studies in *A. thaliana* showed that myrosin cells of the phloem parenchyma develop from ground tissue meristem (rather than from the procambium) under control of the basic helix-loop-helix transcription factor FAMA (Li & Sack, 2014; Shirakawa, Ueda, Nagano, et al., 2014). T-DNA insertion lines with lacking FAMA expression failed to produce myrosin cells, guard cells and the aboveground myrosinases TGG1 and TGG2 (Li & Sack, 2014; Shirakawa, Ueda, Nagano, et al., 2014). A study of endocytosis and vacuolar trafficking-related proteins indicated that expression of FAMA in ground meristem cells at the abaxial side of the phloem is directed by auxin concentrations which are modulated by the auxin efflux carrier PIN1 (Shirakawa, Ueda, Shimada, Kohchi, & Hara-Nishimura, 2014).

PEN2 has been designated as an atypical myrosinase involved in defence against fungal pathogens (see Sections 2.3 and 7.2 and chapter 7 in this volume). It is targeted to peroxisomes and the outer mitochondrial membrane (Fuchs et al., 2016). Thus, PEN2 does not only differ from 'typical' myrosinases in its active site amino acid composition, but also in its subcellular localization.

4.4 Storage of Specifier Proteins

As specifier proteins depend on myrosinases and glucosinolates for activity, their spatial separation from glucosinolates or myrosinase is not necessary to prevent a premature burst of the 'mustard oil bomb'. However, based on signal peptide prediction servers, all known specifier proteins are likely to be cytosolic. No specifier protein was detected in the vacuolar proteome of leaves from *A. thaliana* Col-0 (Carter et al., 2004). Immunocytochemical studies have only been conducted in *A. thaliana* with focus on ESP using an antibody directed against an ESP-specific peptide. This confirmed a cytosolic localization, but also gave indications for a possible localization in the nucleus (Burow, Rice, Hause, Gershenzon, & Wittstock, 2007; Miao & Zentgraf, 2007). ESP was localized almost exclusively in the epidermis of all aboveground organs of *A. thaliana* L*er* except the anthers (Burow, Rice, et al., 2007). Thus, it will almost certainly get mixed with glucosinolates and myrosinases when a chewing herbivore attacks the plant. ESP was also present in cells tentatively identified as S-cells in the stem (Burow, Rice, et al., 2007). In the Cape Verde Island (Cvi) accession, but not in L*er*, ESP also accumulated in putative S-cells of the leaves (Burow, Rice, et al., 2007). Based on immunolocalization, stomatal guard cells of L*er* plants do not contain ESP, while it seemed to be present in guard cells of the Wassilewskija (WS) accession, together with NSP1 and NSP5, based on mass-spectrometric analysis (Burow, Rice, et al., 2007; Zhao et al., 2008). Thus, there seems to be some variation in ESP expression patterns among *A. thaliana* accessions. Studies of other specifier protein types and other species are needed to evaluate if the observations made so far are representative for specifier protein localization in general.

5. STRUCTURAL DIVERSITY OF BREAKDOWN PRODUCTS FORMED UPON TISSUE DISRUPTION AND THEIR IMPACT ON PLANT DEFENCE

5.1 Variation of Glucosinolate Breakdown Within a Plant

Glucosinolate breakdown is initiated when the spatial separation of myrosinases and glucosinolates is abolished. Depending on glucosinolate profile and presence of specifier proteins and other factors, a mixture of few or several structurally diverse breakdown products will be generated. The number of possible breakdown products that can principally be formed from a single glucosinolate depends on the side chain structure. For example,

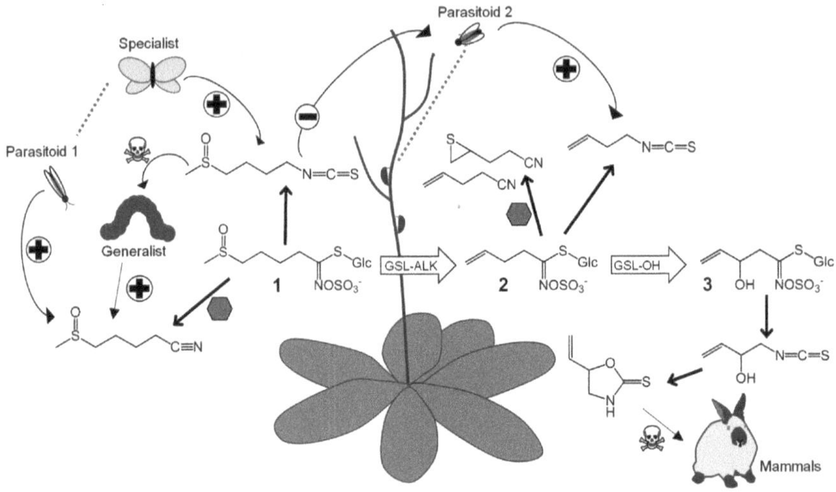

Figure 4 ***Impact of structural diversity on biotic interactions.*** Schematic representation of the interplay between biosynthesis and breakdown of glucosinolates. Minor differences in the glucosinolate side chain may have dramatic effects on the potential to generate various breakdown products. These, in turn may have differential effects on generalist and specialist herbivores of different feeding guilds as well as on their parasitoids. *Block arrows* indicate biosynthetic pathways, *bold arrows with open heads* stand for myrosinase-catalysed breakdown (with grey hexagons for specifier proteins) and *arrows with filled heads* highlight plant-herbivore and plant–insect interactions. The *dotted line* indicates herbivore-parasitoid associations. **1**: 4-methylsulphinylbutylglucosinolate, **2**: 3-butenylglucosinolate, **3**: 2-hydroxy-3-butenylglucosinolate. For details see text.

allylglucosinolate may be converted to four different product types (isothiocyanate, simple nitrile, epithionitrile and organic thiocyanate; Fig. 3E and I) while most methylsulphinylalkylglucosinolates give rise to only two different product types (isothiocyanate and simple nitrile). Hydroxylation of the alkenylglucosinolates at C2 of the side-chain allows the isothiocyanate to form yet another type of product, cyclic 1,3-oxazolidine-2-thione derivatives such as goitrin (Fig. 4). Thus, variation in glucosinolate biosynthesis has the potential to dramatically affect the outcome of glucosinolate breakdown. Besides glucosinolate structure, the type and quantity of specifier protein present in the tissue determines product profiles. In other words, a plant may control the outcome of glucosinolate breakdown in a certain tissue or stage by regulating accumulation of glucosinolates or specifier protein activity or both. *ESM1* and *MVP1* have been identified as additional regulators of the outcome of glucosinolate breakdown in *A. thaliana* (Agee et al., 2010; Zhang et al., 2006; see Section 2.4).

Organ-specific and developmental regulation of the outcome of glucosinolate breakdown is evident from several studies. For example, organic thiocyanate and simple nitrile formation from benzylglucosinolate, the major glucosinolate of *L. sativum*, is restricted to seedlings, flowers and seeds of this plant (Burow, Bergner, et al., 2007). In *A. thaliana*, roots and seedlings of several accessions, including Columbia-0 (Col-0), generate mainly simple nitrile from added allylglucosinolate although leaf homogenates produce mainly the corresponding isothiocyanate indicating organ-specific NSP activity (Wentzell & Kliebenstein, 2008; Witzel et al., 2015). ESP activity (measured as epithionitrile-forming activity of protein extracts supplemented with allylglucosinolate) and ESP protein (detected by immunoblotting and immunolocalization) is only found in the aboveground organs in the accession L*er* (Burow, Rice, et al., 2007). In rosette and cauline leaves, ESP occurs throughout the epidermis except the stomata, but not in the mesophyll (Burow, Rice, et al., 2007). The type of breakdown product formed is also affected by plant age, growing density, and infection with a fungal pathogen in several *A. thaliana* accessions (Wentzell & Kliebenstein, 2008; Witzel et al., 2015). Insect herbivory induces simple nitrile formation in *A. thaliana* Col-0 rosettes which otherwise produce mainly isothiocyanates upon disruption (Burow et al., 2009). These and other studies conducted in *A. thaliana* also demonstrate that the ability to generate a range of different breakdown product types is subject to natural variation and regulation (Lambrix et al., 2001; see, chapter 3 in this volume). The biological relevance of variation in breakdown product types is not well understood, but they certainly differ in their physicochemical properties (e.g., volatility) and/or biological activities (e.g., toxicity) (reviewed in Wittstock et al., 2003). Structural diversification may therefore provide a selective advantage to the plant under certain environmental conditions, for example, high load of specialized herbivores (see, chapter 2 in this volume). In such a case, release of a volatile breakdown product that attracts predators might be more advantageous to the plant than production of the 'standard toxin'. Examples of studies demonstrating the effectiveness of different glucosinolate breakdown product types in direct and indirect plant defence are discussed in Section 5.2.

Variation in glucosinolate breakdown product types and differences in their biological acitivities are also relevant for health effects of glucosinolate-containing vegetables in humans (see, chapter 9 in this volume). For example, sulphoraphane, the isothiocyanate derived from 4-methylsulphinylbutylglucosinolate (Fig. 4), induces phase II detoxification enzymes in cell-based assays much more strongly than the corresponding simple nitrile. However, cultivars of broccoli, a major source of sulphoraphane in the human diet, vary largely with respect to the proportion of isothiocyanate:nitrile formed upon homogenization (Matusheski et al., 2006).

5.2 Effects on Direct and Indirect Defence

Experiments with isolated isothiocyanates in artificial diets or media demonstrated their toxicity to, for example, insect herbivores, nematodes, fungi and oomycetes (e.g., Agrawal & Kurashige, 2003; Donkin, Eiteman, & Williams, 1995; El Sayed, Louveaux, Mavratzotis, Rollin, & Quinsac, 1996; Greenhalgh & Mitchell, 1976; Li, Eigenbrode, Stringam, & Thiagarajah, 2000; Stotz et al., 2011). Although harmful effects have also been found for other breakdown products, isothiocyanates often seem to possess higher toxicity than other products, especially upon contact or ingestion, and have therefore been regarded as major defensive chemicals of glucosinolate-producing plants (reviewed in Wittstock et al., 2003). However, biological activities and defensive potential of other types of breakdown products, for example, epithionitriles and organic thiocyanates, have rarely been investigated (Wittstock et al., 2003).

A nice demonstration of the defensive potential of isothiocyanates versus simple nitriles has been provided by QTL mapping of *A. thaliana* *Ler* × Col-0 recombinant inbred (RI) lines using glucosinolate breakdown product types, leaf damage by larvae of the generalist herbivore *Trichoplusia ni* (Lepidoptera: Noctuidae) and larval weight gain as experimentally determined variables (Jander, Cui, Nhan, Pierce, & Ausubel, 2001; Lambrix et al., 2001). A genetic locus termed *TASTY* correlated with high levels of plant damage and larval weight gain and coincided with a genetic locus determining most of the variation in isothiocyanate versus simple nitrile formation (Jander et al., 2001; Lambrix et al., 2001). A more detailed analysis of the locus led to identification of the *ESP* gene (Lambrix et al., 2001). As both parental accessions lacked alkenylglucosinolates, presence of ESP in the RI lines led to high levels of simple nitrile versus isothiocyanate formation. *T. ni* larvae preferred nitrile-producing over isothiocyanate-producing lines supporting a defensive role of isothiocyanates (Lambrix et al., 2001). However, differential effects could have been due to different breakdown product types or side chains as *A. thaliana* *Ler* produces hydroxyalkylglucosinolates which are lacking in Col-0. Therefore, another study expressed the *ESP* gene from the *Ler* accession in the Col-0 accession which does not normally express a functional ESP. This allowed biotests to be conducted with plants accumulating the same glucosinolates (mainly methylsulphinylalkylglucosinolates), but producing either mainly simple nitriles or mainly isothiocyanates (Fig. 4) (Burow, Müller, Gershenzon, & Wittstock, 2006). Confirming the results obtained with RI lines, larvae of the generalist herbivore *Spodoptera littoralis* (Lepidoptera: Noctuidae) performed better on nitrile-producing than on isothiocyanate-producing lines (Burow, Müller, et al., 2006). This raises the question why plants invest in specifier proteins if their products provide less protection to the plant than isothiocyanates.

One possible answer to this question may lay in the ability of herbivores specialized on glucosinolate-containing plants to exploit glucosinolates and their breakdown products, especially isothiocyanates, as host finding and identification cues (reviewed in Hopkins, van Dam, & van Loon, 2009; see, chapter 2 in this volume). A potential role of non-isothiocyanate products of glucosinolate breakdown could therefore be to interfere with attraction of ovipositing herbivores. The simple nitrile- or isothiocyanate-producing *A. thaliana* lines obtained by transfer of a functional ESP to the Col-0 accession (see preceding paragraph) have been used to test if oviposition by the specialist *Pieris rapae* (Lepidoptera: Pieridae) is affected by the changed breakdown product profile. It turned out that ovipositing *P. rapae* females laid fewer eggs on nitrile-producing than on isothiocyanate-producing lines when plants were damaged prior to the experiments (Fig. 4) (Mumm et al., 2008). This could be due to reduced isothiocyanate emission (as isothiocyanates are host identification cues for *P. rapae*; reviewed in Hopkins et al., 2009) and/or due to increased indole-3-acetonitrile formation (Fig. 5) (as indole-3-acetonitrile acts as oviposition deterrent for *P. rapae* according to De Vos, Kriksunov, & Jander, 2008). Thus, the presence or absence of certain glucosinolate breakdown products may modulate the behaviour of ovipositing females, and this may lead to plant protection. The effects depend on breakdown product types and are likely to be herbivore-specific. For example, the oviposition deterrent to *P. rapae*, indole-3-acetonitrile, did not influence oviposition by females of another specialist herbivore, *Plutella xylostella* (Lepidoptera: Plutellidae) (Sun, Sønderby, Halkier, Jander, & de Vos, 2009).

As another line of defence, herbivore-damaged plants emit volatiles which may be recognized by enemies of the herbivores, such as parasitic wasps, and guide them to their hosts (reviewed in Heil, 2014). Experimental studies suggest that glucosinolate breakdown products may play a role in

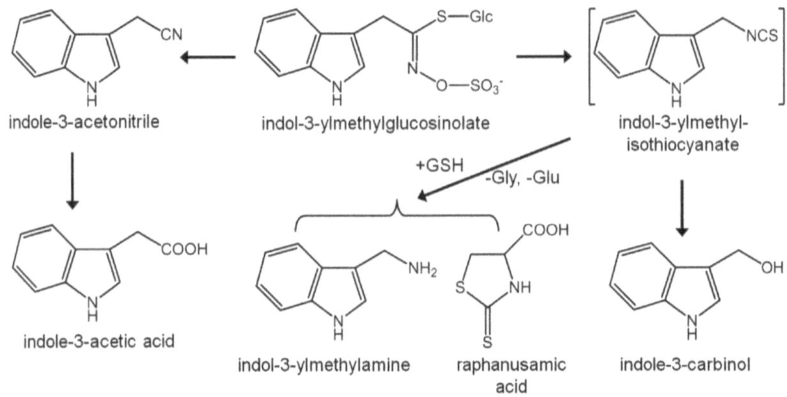

Figure 5 ***Scheme of product formation upon hydrolysis of indol-3-ylmethylglucosinolate.*** Hydrolysis in the presence of specifier proteins (left) yields a nitrile that can be hydrolysed to the carboxylic acid by nitrilases. The isothiocyanate formed in the absence of specifier proteins (right) is unstable and yields an alcohol upon reaction with water. Upon conjugation with glutathione (GSH), further metabolism may yield the corresponding amine and raphanusamic acid.

parasitoid attraction, but also indicate that their effects might be quite specific with respect to the parasitoid and its feeding experience, the glucosinolate breakdown product type and side chain structure. *Diaeretiella rapae* (Hymenoptera: Braconidae) is an endoparasitoid of aphids which feed on glucosinolate-containing plants, for example, the specialist aphid *Brevicoryne brassicae* (Hemiptera: Aphididae). When different breakdown product types of 3-butenyl- and 2-propenylglucosinolate were tested, electroantennogram responses of *D. rapae* females were strongest towards the isothiocyanates. Other product types (simple nitriles and epithionitriles) were only perceived for 3-butenylglucosinolate (Pope et al., 2008). When the isolated compounds were tested for attractiveness in a Y-tube olfactometer, the wasps preferred the isothiocyanates (Fig. 4) (Pope et al., 2008). The wasps' preference for 3-butenyl- or 2-propenylisothiocyanate was affected by the presence of the corresponding glucosinolates in the plant that was used to raise the aphids (*B. brassicae*) which were supplied to the parasitoid cultures (Pope et al., 2008). A study conducted with *A. thaliana* Col-5 confirmed the importance of alkenylisothiocyanates for host finding (Kissen, Pope, et al., 2009). *A. thaliana* Col-5 wildtype plants accumulate 4-methylsulphinylbutylglucosinolate as major glucosinolate, but did not emit breakdown products of this glucosinolate upon infestation with *B. brassicae*, probably due to low volatility of the isothiocyanate (Fig. 4). Transgenic lines engineered to accumulate 3-butenylglucosinolate instead (due to overexpression of a 2-oxoglutarate dependent dioxygenase from *Brassica nigra*) emitted the corresponding 3-butenylisothiocyanate and were more attractive to *D. rapae* than wildtype plants when tested in a Y-tube olfactometer (Fig. 4) (Kissen, Pope, et al., 2009). Another study was setup to test if the breakdown product type affects attractiveness of *P. rapae*—infested plants to the larval endoparasitoid *Cotesia rubecula* (Hymenoptera: Braconidae) using transgenic *A. thaliana* lines with or without ectopic ESP expression in wind tunnel experiments (Mumm et al., 2008). Parasitoids preferred simple nitrile-producing lines over isothiocyanate-producing lines indicating that simple nitriles might also be involved in tritrophic interactions (Fig. 4) (Mumm et al., 2008).

Taken together, individual glucosinolate breakdown products may have specific effects on certain herbivores or pathogens and they may also be involved in indirect defence responses. A tight regulation of biosynthesis and breakdown of glucosinolates may enable a plant to adjust the profile of breakdown products specifically to the needs imposed by the environmental conditions (see, chapter 2 in this volume).

6. BREAKDOWN INSIDE HERBIVORES

The sudden release of glucosinolate breakdown products upon tissue damage ('mustard oil bomb') represents a powerful means of direct plant defence. However, several herbivore species have acquired adaptations that allow them to feed on glucosinolate-containing plants with impunity. Some of the most feared pests on glucosinolate-containing crops, *P. xylostella* and *P. rapae*, possess gut proteins which protect them from toxic isothiocyanates by facilitating other breakdown pathways. In larvae of *P. xylostella*, high amounts of a gut sulphatase 'disarm' the 'mustard oil bomb' by quickly converting ingested glucosinolates to desulphoglucosinolates (Ratzka, Vogel, Kliebenstein, Mitchell-Olds, & Kroymann, 2002). In this way, ingested plant myrosinases are deprived of their substrates. Within Lepidoptera, glucosinolate sulphatase activity has only been reported for *P. xylostella*, but other insect herbivores, including specialists and generalists, chewing insects and phloem feeders, have also been shown to employ a sulphatase to prevent formation of harmful glucosinolate breakdown products upon ingestion of glucosinolate-containing plant material (reviewed in Jeschke, Gershenzon, & Vassão, 2015; Winde & Wittstock, 2011; see, chapter 8 in this volume). Larvae of the specialist *P. rapae* possess a gut NSP which redirects glucosinolate hydrolysis to the formation of simple nitriles at the expense of isothiocyanate formation (Wittstock et al., 2004). Simple nitriles are either excreted or further metabolized (Stauber et al., 2012; Vergara et al., 2006). The larval NSP is unrelated to plant specifier proteins, but has roughly the same ability to promote simple nitrile formation upon myrosinase-catalysed hydrolysis of glucosinolates as plant NSPs, that is, it collaborates with ingested plant myrosinase to enable simple nitrile formation. NSP activity which has only been found in the Pierinae subfamily was recruited from a group of insect-specific proteins termed major allergens (Fischer, Wheat, Heckel, & Vogel, 2008). It has been identified as an evolutionary key innovation that allowed Pierid species to colonize glucosinolate-containing plants only about 10 million years after occurrence of glucosinolates in the Brassicales (about 80 million years ago) (Wheat et al., 2007). As a consequence of their alternative breakdown pathways, herbivores equipped with gut sulphatase or NSP may be sensitive to isothiocyanates (as has been demonstrated for *P. xylostella* and *P. rapae* (Agrawal & Kurashige, 2003; Li et al., 2000)), but are normally not exposed to toxic concentrations of these compounds.

Besides herbivores which enforce special glucosinolate breakdown pathways, there are also specialized feeders which prevent any breakdown of ingested glucosinolates until the compounds have reached a 'safe' compartment within the insect, separate from ingested plant myrosinase, where storage or further metabolism can take place to serve the insect's defence (reviewed in Jeschke et al., 2015; Müller, 2009; Winde & Wittstock, 2011; see, chapter 8 in this volume). The specialist aphid *Brevicoryne brassicae* and the specialist flea beetle *Phyllotreta striolata* (Coleoptera: Chrysomelidae) possess their own myrosinase to generate isothiocyanates from sequestered glucosinolates when attacked by predators (Beran et al., 2014; Kazana et al., 2007; see Section 2.3.). While phloem feeders are likely to avoid detonation of the 'mustard oil bomb' by their feeding mode, it is presently not known how chewing herbivores manage to take up intact glucosinolates from the gut before they can be degraded by ingested myrosinase.

If a herbivore is unable to prevent detonation of the 'mustard oil bomb', it has to deal with the reactive isothiocyanates and the other products generated by the plant's glucosinolate breakdown machinery. A common pathway of isothiocyanate detoxification is their conjugation with glutathione, a classical phase-II detoxification reaction (Schramm, Vassão, Reichelt, Gershenzon, & Wittstock, 2012). However, while it removes reactive isothiocyanates, this pathway may deplete cells of glutathione if the herbivore is exposed to high isothiocyanate levels (Jeschke, Gershenzon, & Vassão, 2016; see, chapter 8 in this volume).

7. BREAKDOWN IN NONDISRUPTED TISSUE

7.1 Signs of Turnover in Intact Tissue

The idea that glucosinolates are also broken down in intact tissue of a living plant has been fuelled by observations that glucosinolate content may decline in certain developmental stages and under sulphur deficiency, and varies diurnally. For example, total glucosinolate content per individuum decreases by about 30% during seed germination and seedling development in *B. napus* and *A. thaliana* (Brown, Tokuhisa, Reichelt, & Gershenzon, 2003; McGregor, 1988; Petersen, Chen, Hansen, Olsen, & Halkier, 2002). Individual glucosinolates are lost to different extent indicating some specificity in this degradation (Brown et al., 2003; McGregor, 1988; Petersen et al., 2002). Similar observations have been made at later developmental stages (Brown et al., 2003; Clossais-Besnard & Larher, 1991; Petersen et al., 2002). Glucosinolate degradation during seed-to-seedling transition was also detected in a radiolabelling experiment. When seeds of *A. thaliana* Col-0 which contained ^{14}C-labelled 4-hydroxybenzylglucosinolate (due to previous application of the tracer to leaves of flowering plants), the content of radiolabelled glucosinolate in the whole plant was reduced to about 70% (relative to seed content) in the cotyledon stage, and further declined until no radiolabelled glucosinolate was detectable in plants of the rosette stage (Petersen et al., 2002). Several studies reported a decline in glucosinolates when plants were deprived of sulphur (reviewed in Falk, Tokuhisa, & Gershenzon, 2007; see, chapter 10 in this volume). For example, leaf and root glucosinolate content decreases upon transfer of two-to-three-week-old *A. thaliana* Col-0 from sulphur-sufficient to sulphur-deficient medium indicating degradation (Hirai et al., 2004). Diurnal variation of glucosinolate levels, which cannot be explained by allocation, has been detected in seedlings and young plants of *A. thaliana* Col-0 and seedlings of *Brassica oleracea* var. *capitata* (Brassicaceae) (Huseby et al., 2013; Rosa & Rodrigues,

1998). Besides these observations, glucosinolate metabolism in intact tissue has been proposed to result in accumulation of alliarinoside, a γ-hydroxynitrile glucoside, in *Alliaria petiolata* (Frisch et al., 2015). Although the validity of the pathway in intact tissue still needs to be demonstrated, feeding experiments with radiolabelled Met showed that both allylglucosinolate and alliarinoside are biosynthesized from Met and supported the conversion of allylglucosinolate to the alliarinoside aglucone rather than branching into two separate biosynthetic pathways at the aldoxime intermediate (Frisch et al., 2015).

The physiological relevance of glucosinolate turnover in uninfected tissue is poorly understood. Most research in this context has been conducted in relation to sulphur nutrition (see, chapter 10 in this volume). As glucosinolate content generally rises with sulphur supply and decreases under sulphur deficiency, glucosinolates have been proposed to serve as sulphur storage compounds in addition to their roles as chemical defences (reviewed in Falk et al., 2007; see, chapter 10 in this volume). In a study reported in 2014, root and shoot glucosinolate content per gram plant material decreased in seedlings of *B. juncea* and *B. rapa* under sulphur-deficient conditions, but glucosinolate content relative to total organic sulphur increased two-to-fourfold (Aghajanzadeh, Hawkesford, & De Kok, 2014). Thus, other organic sulphur compounds were much more affected by sulphur deficiency than glucosinolates, putting a role of glucosinolates as sulphur storage compounds into question (Aghajanzadeh et al., 2014).

7.2 β-Glucosidases Involved in Turnover Pathways

The pathway(s) that may allow metabolically safe turnover of glucosinolates still have to be elucidated (Fig. 1). Based on a study conducted with the *tgg1 tgg2* mutant of *A. thaliana* Col-0, the involvement of known myrosinases is unlikely (Barth & Jander, 2006). In seedlings of the *tgg1 tgg2* mutant, myrosinase activity was undetectable, but glucosinolate levels still declined during seedling development (Barth & Jander, 2006). In agreement with this, *TGG2* was downregulated upon sulphur starvation based on transcriptome analysis (Nikiforova et al., 2003). Transcriptome analyses of *A. thaliana* also pinpointed putative thioglucosidases whose transcripts were upregulated upon sulphur deficiency and which are therefore candidates with a possible role in glucosinolate turnover (reviewed in Falk et al., 2007). Among these, *BGLU28* (At2g44460) is regulated by the transcription factor SLIM1 which globally controls responses to sulphur deficiency (Maruyama-Nakashita, Nakamura, Tohge, Saito, & Takahashi, 2006). Among sulphur deficiency-induced genes affected in *slim1* mutants, *BGLU28* showed the strongest decline in expression. In *slim1* mutants grown under sulphur-deficient conditions, *BGLU28* was no longer expressed and aliphatic glucosinolate content in shoots and roots increased two to threefold compared to the parental line (Maruyama-Nakashita et al., 2006). Although many other genes, including glucosinolate biosynthesis genes, were also affected, this strongly supports involvement of *BGLU28* in glucosinolate breakdown upon sulphur deficiency (Maruyama-Nakashita et al., 2006). Together with other candidate proteins (including *A. thaliana* PEN2 and PYK10, see following paragraphs), BGLU28 forms a sister clade to the classical myrosinases in a phylogenetic tree of *A. thaliana* glycoside hydrolase family 1 (Xu et al., 2004). Only few representatives of this clade have been studied with respect to their ability to hydrolyse glucosinolates (Ahn et al., 2010) and, with the exception of PEN2, glucosinolate hydrolase activity has not been proven in vitro for any of these candidates. Besides likely involvement of BGLU28 in sulphur deficiency—induced breakdown of aliphatic glucosinolates, no β-glucosidase involved in developmental, diurnal or sulphur deficiency—induced glucosinolate turnover has been identified.

PEN2 (At2g44490, BGLU26) is an atypical *A. thaliana* myrosinase which is able to hydrolyse indolic glucosinolates but likely not aliphatic glucosinolates (Bednarek et al., 2009). Based on the present data, PEN2 is involved in defence against pathogens rather than in glucosinolate turnover. Together with 4-methoxyindol-3-ylmethylglucosinolate, *PEN2* is required for induced defence against the grass powdery mildew *Blumeria graminis hordei* and for microbe-associated molecular pattern (MAMP)-induced callose formation in *A. thaliana* (Bednarek et al., 2009; Clay et al., 2009; see, chapter 7 in this volume). Consumption of indol-3-ylmethyl- and 4-methoxyindol-3-ylmethylglucosinolate by purified recombinant PEN2 expressed in *P. pastoris* has been demonstrated (Bednarek et al., 2009). Products of PEN2-catalysed 4-methoxyindol-3-ylmethylglucosinolate breakdown have not been identified, but when indol-3-ylmethylglucosinolate is hydrolysed by PEN2 in planta, indol-3-ylmethylamine and raphanusamic acid (4-carboxy-1,3-thiazolidine-2-thione), a cyclic derivative of the corresponding isothiocyanate, are formed (Fig. 5). PEN2 is localized in the outer membrane of a subpopulation of mitochondria with the globular part facing the cytosol (Fuchs et al., 2016). Upon fungal infection, these mitochondria get arrested at the pathogen entry site in close proximity to ER-bound CYP81F2 required for pathogen-induced biosynthesis of the PEN2 substrate 4-methoxyindol-3-ylmethylglucosinolate (Fuchs et al., 2016). Thus, the biosynthetic and breakdown enzymes colocalize specifically at the site of fungal attack supporting a role of PEN2 in the generation of antifungal glucosinolate breakdown products which might be exported locally by the transporter PEN3 to fight off the invading fungus (Fuchs et al., 2016; see, chapter 7 in this volume). Besides PEN2, the classical myrosinase TGG1 may also play a role in antifungal defence responses in *A. thaliana*. Treatment with the fungal toxin fumonisin B1 led to a decrease in indolic glucosinolates and increased expression of *TGG1* rather than *PEN2*, and the possible indolic glucosinolate breakdown products indole-3-carbinole and indole-3-acetonitrile reduced the programmed cell death response induced by the toxin (Zhao et al., 2015; see, chapter 7 in this volume). Thus, glucosinolate-mediated antifungal defence responses are likely specific for certain groups of pathogens.

Arabidopsis thaliana PYK10 (At3g09260, BGLU23) is expressed in seedlings and roots, but not in rosette leaves, of young *A. thaliana* plants based on Western blot analysis and has been suspected to be involved in glucosinolate metabolism (Ahn et al., 2010; Matsushima, Kondo, Nishimura & Hara-Nishimura, 2003; reviewed in Wittstock & Burow, 2010). However, when recombinantly expressed in insect cells, it hydrolysed several hydroxycumarine glucosides, but not allylglucosinolate (Ahn et al., 2010). As indole glucosinolates are the main glucosinolates in *A. thaliana* Col-0 roots (Brown et al., 2003; Petersen et al., 2002), it would be interesting to test if PYK10 accepts indolic glucosinolates as substrates. Furthermore, it might be necessary to consider in expression experiments that the active form of PYK10 forms a large complex with other putative β-glucosidases and lectins upon tissue disruption (Nagano, Matsushima, & Hara-Nishimura, 2005).

7.3 Nitriles and/or Isothiocyanates as Pathway Intermediates?

Assuming that glucosinolate turnover starts with initial hydrolysis by a β-glucosidase, two principle pathways of further conversion have been proposed: NSP-mediated simple nitrile formation followed by nitrilase-catalysed hydrolysis and isothiocyanate formation with subsequent glutathione conjugation (Fig. 1). In contrast to isothiocyanate formation, the nitrilase pathway would mobilize both sulphur atoms and the nitrogen without exposing the plant to toxic isothiocyanates and without the need to invest glutathione (Janowitz, Trompetter, & Piotrowski, 2009).

In support of the nitrilase pathway, the Brassicaceae possess specifier proteins like NSPs (see Section 3) and a special group of nitrilase isoforms, the NIT1 group (with three members NIT1-NIT3 in *A. thaliana*). This group is confined to the Brassicaceae and probably evolved based on duplication of *NIT4* which is ubiquitous in plants and involved in cyanide detoxification (Janowitz et al., 2009; Piotrowski, 2008). *Arabidopsis thaliana* NIT1 and NIT2 catalyse the conversion of a number of glucosinolate-derived nitriles, including those with aliphatic side chains (Janowitz et al., 2009). They are also able to produce indole-3-acetic acid (auxin) from indole-3-acetonitrile as a product of indol-3-ylmethylglucosinolate breakdown, but with only low activity, likely to avoid interference with plant growth (Fig. 5). Two NIT1 homologs from *Capsella rubella*, a plant that accumulates short-chain and long-chain aliphatic glucosinolates, were shown to prefer either the corresponding long-chain or short-chain nitriles (Janowitz et al., 2009). Matching glucosinolate profiles and nitrilase substrate specificities have also been reported for *B. rapa* and *Sinapis* species (Agerbirk, Warwick, Hansen, & Olsen, 2008; Ishikawa et al., 2007; Zhang et al., 2008). Within the *A. thaliana NIT1* group, only *NIT3* is induced by sulphur deprivation which is also associated with changed root morphology (Kutz et al., 2002; Nikiforova et al., 2003). *NIT3* promotor activity (as monitored using *NIT3*-promoter:*GUS* fusion constructs) in roots correlates with a decrease in the content of indol-3-ylmethylglucosinolate in roots which declines faster than that of all other root glucosinolates (Kutz et al., 2002). Based on this observation, it has been proposed that NIT3 regulates root growth and branching through release of auxin from glucosinolate-derived indole-3-acetonitrile (Kutz et al., 2002). However, as activities of NIT3 isoenzymes have always been much lower in vitro than those of NIT1 and NIT2, it is still uncertain if any of the tested compounds serves as substrate under physiological conditions (Janowitz et al., 2009).

Glucosinolate turnover through isothiocyanate formation has been proposed based on the analysis of *A. thaliana* with ectopic expression of the myrosinase TGG4 (which is root-expressed in Col-0 wildtype plants) (Bednarek et al., 2009). DMSO extracts of these plants contained products with structural similarity to glucosinolate breakdown products, but were different from the products obtained upon tissue disruption. The compounds were identified as primary amines with indolic and aliphatic moieties corresponding to glucosinolate side chain structures and raphanusamic acid (Bednarek et al., 2009). Their formation can be explained by decomposition of glutathione conjugates of isothiocyanates derived from glucosinolate breakdown (Bednarek et al., 2009). Thus, when TGG4 was ectopically expressed, isothiocyanates were generated in intact tissue and were further metabolized. Whether such a pathway also functions in wildtype plants is presently unknown.

8. CONCLUSIONS AND PERSPECTIVES

As we are approaching the year 2020, the glucosinolate-myrosinase system presents itself as a highly specialized, dynamic multicomponent system involved in a large number of plant–environment interactions. With less than 150 different structures, glucosinolates represent a relatively small group of specialized metabolites. However, an enormous diversification takes place upon glucosinolate breakdown providing plants with a much larger array of compounds with different physicochemical properties and biological activities. Since the late 1990s, analytical, molecular-biological and genetic tools as well as visualization techniques have been combined to localize the components of the glucosinolate-myrosinase system, to dissect glucosinolate breakdown pathways in disrupted as well as nondisrupted tissue and to explore biological roles of these pathways in plant–herbivore and plant–pathogen interactions. This has not only provided new insights into the protein machinery involved in structural diversification and its genetic bases, but also into the regulation of glucosinolate breakdown at the spatial and temporal scale. Despite this considerable progress, many aspects of glucosinolate breakdown are still poorly understood, and new questions arise as soon as another piece is added to the big puzzle.

The elucidation of specific roles for certain breakdown products of individual glucosinolates raises the question if and how biosynthesis and breakdown machineries have coevolved. For example, only a plant that produces alkenylglucosinolates would be able to benefit from the extended capabilities of ESPs (relative to NSPs). NSPs, however, are likely required for nitrilase-dependent glucosinolate turnover pathways. Such turnover pathways are still intriguing as they also necessitate access of a thioglucosidase to the glucosinolates which otherwise appear to be protected from breakdown. Based on the present knowledge it seems quite likely that this protection is accomplished by the combination of several mechanisms, including spatial and/or temporal separation, and inhibition by, for example, complex formation, which may or may not involve MBPs and MyAPs. Additionally, translocation, that is, intracellular and long-distance transport, might play an important role in bringing substrate and enzyme together. The identity of the thioglucosidases involved has also remained a big mystery. Although the atypical myrosinase PEN2 does not seem to be a major player in glucosinolate turnover in uninfected tissue, it is the first characterized plant myrosinase that appears to convert its glucosinolate substrate without prior cell disruption (Fuchs et al., 2016). As the active site of myrosinases had been thought to be specifically tailored for hydrolysis of glucosinolates, the discovery of PEN2 (and insect myrosinases) with the conserved active site motifs of O-glycosidases was rather surprising. In phylogenetic trees, PEN2 and other (putative) β-glucosidases encoded in the *A. thaliana* genome cluster separately from the classical myrosinases (Xu et al., 2004). Future research will have to show if this cluster is comprised entirely of atypical myrosinases dedicated to glucosinolate metabolism other than the classical 'mustard oil bomb'.

Taken together, the coming years are likely to witness research efforts aimed at identifying new components of the glucosinolate breakdown machinery, a better understanding of how the components of the glucosinolate-myrosinase system interact to produce structurally diverse breakdown products at the right time and the right place and how these interactions are regulated in response to the biotic and abiotic environment.

ACKNOWLEDGEMENTS

The authors wish to thank Anita Backenköhler for providing the chromatograms shown in Fig. 3. Funding of research in the Wittstock Lab by the German Research Foundation and the State Lower Saxony is gratefully acknowledged. We apologize to all authors whose work is not cited in this chapter due to space limitations.

REFERENCES

Agee, A. E., Surpin, M., Sohn, E. J., Girke, T., Rosado, A., Kram, B. W., ... Raikhel, N. V. (2010). Modified vacuole phenotype1 is an *Arabidopsis* myrosinase-associated protein involved in endomembrane protein trafficking. *Plant Physiology, 152*, 120–132.

Agerbirk, N., De Vos, M., Kim, J. H., & Jander, G. (2009). Indole glucosinolate breakdown and its biological effects. *Phytochemistry Reviews, 8*, 101–120.

Agerbirk, N., & Olsen, C. E. (2012). Glucosinolate structures in evolution. *Phytochemistry, 77*, 16–45.

Agerbirk, N., & Olsen, C. E. (2015). Glucosinolate hydrolysis products in the crucifer *Barbarea vulgaris* include a thiazolidine-2-one from a specific phenolic isomer as well as oxazolidine-2-thiones. *Phytochemistry, 115*, 143–151.

Agerbirk, N., Olsen, C. E., & Sørensen, H. (1998). Initial and final products, nitriles, and ascorbigens produced in myrosinase-catalyzed hydrolysis of indole glucosinolates. *Journal of Agricultural & Food Chemistry, 46*, 1563–1571.

Agerbirk, N., Warwick, S. I., Hansen, P. R., & Olsen, C. E. (2008). Sinapis phylogeny and evolution of glucosinolates and specific nitrile degrading enzymes. *Phytochemistry, 69*, 2937–2949.

Aghajanzadeh, T., Hawkesford, M. J., & De Kok, L. J. (2014). The significance of glucosinolates for sulfur storage in Brassicaceae seedlings. *Frontiers in Plant Science, 5*, 704.

Agrawal, A. A., & Kurashige, N. S. (2003). A role for isothiocyanates in plant resistance against the specialist herbivore *Pieris rapae*. *Journal of Chemical Ecology, 29*, 1403–1415.

Ahn, Y. O., Shimizu, B. I., Sakata, K., Gantulga, D., Zhou, Z., Bevan, D. R., & Esen, A. (2010). Scopolin-hydrolyzing β-glucosidases in roots of *Arabidopsis*. *Plant & Cell Physiology, 51*, 132–143.

Andersson, D., Chakrabarty, R., Bejai, S., Zhang, J., Rask, L., & Meijer, J. (2009). Myrosinases from root and leaves of *Arabidopsis thaliana* have different catalytic properties. *Phytochemistry, 70*, 1345–1354.

Andréasson, E., & Jørgensen, L. B. (2003). Localization of plant myrosinases and glucosinolates. In J. T. Romeo (Ed.), *Integrative phytochemistry: From ethnobotany to molecular ecology: Vol. 37. Recent advances in phytochemistry* (pp. 79–99). Amsterdam: Elsevier.

Andréasson, E., Jørgensen, L. B., Höglund, A. S., Rask, L., & Meijer, J. (2001). Different myrosinase and idioblast distribution in *Arabidopsis* and *Brassica napus*. *Plant Physiology, 127*, 1750–1763.

Astwood, E. B., Greer, M. A., & Ettlinger, M. G. (1949). 1-5-Vinyl-2-thiooxazolidone, an antithyroid compound from yellow turnip and from *Brassica* seeds. *Journal of Biological Chemistry, 181*, 121–130.

Barth, C., & Jander, G. (2006). *Arabidopsis* myrosinases tgg1 and tgg2 have redundant function in glucosinolate breakdown and insect defense. *Plant Journal, 46*, 549–562.

Bednarek, P., Piślewska-Bednarek, M., Svatoš, A., Schneider, B., Doubsky, J., Mansurova, M., ... Schulze-Lefert, P. (2009). A glucosinolate metabolism pathway in living plant cells mediates broad-spectrum antifungal defense. *Science, 323*, 101–106.

Benn, H. H., & Ettlinger, M. G. (1965). The synthesis of sinigrin. *Chemical Communications (London), 19*, 445–447.

Beran, F., Pauchet, Y., Kunert, G., Reichelt, M., Wielsch, N., Vogel, H., ... Heckel, D. G. (2014). *Phyllotreta striolata* flea beetles use host plant defense compounds to create their own glucosinolate-myrosinase system. *Proceedings of the National Academy of Sciences of the United States of America, 111*, 7349–7354.

Bernardi, R., Finiguerra, M. G., Rossi, A. A., & Palmieri, S. (2003). Isolation and biochemical characterization of a basic myrosinase from ripe *Crambe abyssinica* seeds, highly specific for epi-progoitrin. *Journal of Agricultural and Food Chemistry, 51*, 2737–2744.

Björkman, R., & Janson, J. C. (1972). Studies on myrosinases. i. Purification and characterization of a myrosinase from white mustard seed (*Sinapis alba* L.). *Biochimica et Biophysica Acta, 276*, 508–518.

Björkman, R., & Lönnerdal, B. (1973). Studies on myrosinases iii. Enzymatic properties of myrosinases from *Sinapis alba* and *Brassica napus* seeds. *Biochimica et Biophysica Acta (Enzymology), 327*, 121–131.

Bones, A. M., & Rossiter, J. T. (1996). The myrosinase-glucosinolate system, its organisation and biochemistry. *Physiologia Plantarum, 97*, 194–208.

Bones, A. M., Thangstad, O. P., Haugen, O. A., & Espevik, T. (1991). Fate of myrosin cells: characterization of monoclonal antibodies against myrosinase. *Journal of Experimental Botany, 42*, 1541–1549.

Brandt, W., Backenköhler, A., Schulze, E., Plock, A., Herberg, T., Roese, E., & Wittstock, U. (2014). Molecular models and mutational analyses of plant specifier proteins suggest active site residues and reaction mechanism. *Plant Molecular Biology, 84*, 173–188.

Brocker, E. R., & Benn, M. H. (1983). The intramolecular formation of epithioalkanenitriles from alkenylglucosinolates by *Crambe abyssinica* seed flour. *Phytochemistry, 22*, 770–772.

Brown, P. D., Tokuhisa, J. G., Reichelt, M., & Gershenzon, J. (2003). Variation of glucosinolate accumulation among different organs and developmental stages of *Arabidopsis thaliana*. *Phytochemistry, 62*, 471–481.

Buckhout, T. J., Yang, T. J. W., & Schmidt, W. (2009). Early iron-deficiency-induced transcriptional changes in *Arabidopsis* roots as revealed by microarray analyses. *BMC Genomics, 10*, 147.

Burmeister, W. P., Cottaz, S., Driguez, H., Iori, R., Palmieri, S., & Henrissat, B. (1997). The crystal structures of *Sinapis alba* myrosinase and a covalent glycosyl-enzyme intermediate provide insights into the substrate recognition and active-site machinery of an S-glycosidase. *Structure, 5*, 663–675.

Burmeister, W. P., Cottaz, S., Rollin, P., Vasella, A., & Henrissat, B. (2000). High resolution x-ray crystallography shows that ascorbate is a cofactor for myrosinase and substitutes for the function of the catalytic base. *Journal of Biological Chemistry, 275*, 39385–39393.

Burow, M. (2016). Complex environments interact with plant development to shape glucosinolate profiles. In S. Kopriva (Ed.), *Glucosinolates: Vol. 80* (in this volume).

Burow, M., Bergner, A., Gershenzon, J., & Wittstock, U. (2007). Glucosinolate hydrolysis in *Lepidium sativum* – identification of the thiocyanate-forming protein. *Plant Molecular Biology, 63*, 49–61.

Burow, M., Losansky, A., Müller, R., Plock, A., Kliebenstein, D. J., & Wittstock, U. (2009). The genetic basis of constitutive and herbivore-induced *ESP*-independent nitrile formation in *Arabidopsis*. *Plant Physiology, 149*, 561–574.

Burow, M., Markert, J., Gershenzon, J., & Wittstock, U. (2006). Comparative biochemical characterization of nitrile-forming proteins from plants and insects that alter myrosinase-catalysed hydrolysis of glucosinolates. *FEBS Journal, 273*, 2432–2446.

Burow, M., Müller, R., Gershenzon, J., & Wittstock, U. (2006). Altered glucosinolate hydrolysis in genetically engineered *Arabidopsis thaliana* and its influence on the larval development of *Spodoptera littoralis*. *Journal of Chemical Ecology, 32*, 2333–2349.

Burow, M., Rice, M., Hause, B., Gershenzon, J., & Wittstock, U. (2007). Cell- and tissue-specific localization and regulation of the epithiospecifier protein in *Arabidopsis thaliana*. *Plant Molecular Biology, 64*, 173–185.

Buskov, S., Hasselstrøm, J., Olsen, C. E., Sørensen, H., Sørensen, J. C., & Sørensen, S. (2000). Supercritical fluid chromatography as a method of analysis for the determination of 4-hydroxybenzylglucosinolate degradation products. *Journal of Biochemical & Biophysical Methods, 43*, 157–174.

Capella, A., Menossi, M., Arruda, P., & Benedetti, C. (2001). *Coi1* affects myrosinase activity and controls the expression of two flower-specific myrosinase-binding protein homologues in *Arabidopsis*. *Planta, 213*, 691–699.

Carter, C., Pan, S., Zouhar, J., Avila, E. L., Girke, T., & Raikhel, N. V. (2004). The vegetative vacuole proteome of *Arabidopsis thaliana* reveals predicted and unexpected proteins. *Plant Cell, 16*, 3285–3303.

Clay, N. K., Adio, A. M., Denoux, C., Jander, G., & Ausubel, F. M. (2009). Glucosinolate metabolites required for an *Arabidopsis* innate immune response. *Science, 323*, 95–101.

Clossais-Besnard, N., & Larher, F. (1991). Physiological role of glucosinolates in *Brassica napus*. Concentration and distribution pattern of glucosinolates among plant organs during a complete life cycle. *Journal of Agricultural and Food Chemistry, 56*, 25–38.

Davies, G., & Henrissat, B. (1995). Structures and mechanisms of glycosyl hydrolases. *Structure, 3*, 853–859.

De Vos, M., Kriksunov, K. L., & Jander, G. (2008). Indole-3-acetonitrile production from indole glucosinolates deters oviposition by *Pieris rapae* (white cabbage butterfly). *Plant Physiology, 146*, 916–926.

Donkin, S. G., Eiteman, M. A., & Williams, P. L. (1995). Toxicity of glucosinolates and their enzymatic decomposition products to *Caenorhabditis elegans*. *Journal of Nematology, 27*, 258–262.

Durham, P. L., & Poulton, J. E. (1990). Enzymic properties of purified myrosinase from *Lepidium sativum* seedlings. *Zeitschrift für Naturforschung C, 45*, 173–178.

El Sayed, G., Louveaux, A., Mavratzotis, M., Rollin, P., & Quinsac, A. (1996). Effects of glucobrassicin, epiprogoitrin and related breakdown products on locusts feeding: *Schouwia purpurea* and desert locust relationships. *Entomologia Experimentalis et Applicata, 78*, 231–236.

Eriksson, S., Andréasson, E., Ekbom, B., Graner, G., Pontoppidan, B., Taipalensuu, J., … Meijer, J. (2002). Complex formation of myrosinase isoenzymes in oilseed rape seeds are dependent on the presence of myrosinase-binding proteins. *Plant Physiology, 129*, 1592–1599.

Eriksson, S., Ek, B., Xue, J., Rask, L., & Meijer, J. (2001). Identification and characterization of soluble and insoluble myrosinase isoenzymes in different organs of *Sinapis alba*. *Physiologia Plantarum, 111*, 353–364.

Ettlinger, M., Dateo, G. P., Harrison, B., Mabry, T. J., & Thompson, C. P. (1961). Vitamin C as a coenzyme: the hydrolysis of mustard oil glucosides. *Proceedings of the National Academy of Sciences of the United States of America, 47*, 1875–1880.

Ettlinger, M. G., & Lundeen, A. J. (1956). The structures of sinigrin and sinalbin; an enzymatic rearrangement. *Journal of the American Chemical Society*, 4172–4173.

Ettlinger, M. G., & Lundeen, A. J. (1957). First synthesis of a mustard oil glucoside; the enzymatic Lossen rearrangement. *Journal of the American Chemical Society*, 1764–1765.

Falk, A., Taipalensuu, J., Ek, B., Lenman, M., & Rask, L. (1995). Characterization of rapeseed myrosinase-binding protein. *Planta, 195*, 387–395.

Falk, K. L., Tokuhisa, J. G., & Gershenzon, J. (2007). The effect of sulfur nutrition on plant glucosinolate content: physiology and molecular mechanisms. *Plant Biology, 9*, 573–581.

Fenwick, G. R., Heaney, R. K., & Mullin, W. J. (1983). Glucosinolates and their breakdown products in food and food plants. *Critical Reviews in Food Science and Nutrition, 18*, 123–201.

Fischer, H. M., Wheat, C. W., Heckel, D. G., & Vogel, H. (2008). Evolutionary origins of a novel host plant detoxification gene in butterflies. *Molecular Biology and Evolution, 25*, 809–820.

Foo, H. L., Grønning, L. M., Goodenough, L., Bones, A. M., Danielsen, B. E., Whiting, D. A., & Rossiter, J. T. (2000). Purification and characterisation of epithiospecifier protein from *Brassica napus*: enzymic intramolecular sulphur addition within alkenyl thiohydroximates derived from alkenyl glucosinolate hydrolysis. *FEBS Letters, 468*, 243–246.

Frisch, T., Motawia, M. S., Olsen, C. E., Agerbirk, N., Møller, B. L., & Bjarnholt, N. (2015). Diversified glucosinolate metabolism: biosynthesis of hydrogen cyanide and of the hydroxynitrile glucoside alliarinoside in relation to sinigrin metabolism in *Alliaria petiolata*. *Frontiers in Plant Science, 6*, 926.

Fuchs, R., Kopischke, M., Klapprodt, C., Hause, G., Meyer, A. J., Schwarzländer, M., … Lipka, V. (2016). Immobilized subpopulations of leaf epidermal mitochondria mediate *PENETRATION2*-dependent pathogen entry control in *Arabidopsis*. *Plant Cell, 28*, 130–145.

Geshi, N., & Brandt, A. (1998). Two jasmonate-inducible myrosinase-binding proteins from *Brassica napus* L. seedlings with homology to jacalin. *Planta, 204*, 295–304.

Gmelin, R., & Virtanen, A. I. (1959). A new type of enzymatic cleavage of mustard oil glucosides. Formation of allylthiocyanate in *Thlaspi arvense* L. and benzylthiocyanate in *Lepidium ruderale* L. and *Lepidium sativum* L. *Acta Chemica Scandinavica, 13*, 1474–1475.

Greenhalgh, R., & Mitchell, N. D. (1976). The involvement of flavour volatiles in the resistance to downy mildew of wild and cultivated forms of *Brassica oleracea*. *New Phytologist, 77*, 391–398.

Grob, K., & Matile, P. (1979). Vacuolar localization of glucosinolates in horseradish root cells. *Plant Science Letters, 14*, 327–335.

Gumz, F., Krausze, J., Eisenschmidt, D., Backenköhler, A., Barleben, L., Brandt, W., & Wittstock, U. (2015). The crystal structure of the thiocyanate-forming protein from *Thlaspi arvense*, a kelch protein involved in glucosinolate breakdown. *Plant Molecular Biology, 89*, 67–81.

Härtel, F. V., & Brandt, A. (2002). Characterization of a *Brassica napus* myrosinase expressed and secreted by *Pichia pastoris*. *Protein Expression and Purification, 24*, 221–226.

Heil, M. (2014). Herbivore-induced plant volatiles: targets, perception and unanswered questions. *New Phytologist, 204*, 297–306.

Heinricher, E. (1884). Über Eiweißstoffe führende Idioblasten bei einigen Cruciferen. *Berichte der Deutschen Botanischen Gesellschaft, 2*, 463–466.

Helmlinger, J., Rausch, T., & Hilgenberg, W. (1983). Localization of newly synthesized indole-3-methylglucosinolate (= glucobrassicin) in vacuoles from horseradish (*Armoracia rusticana*). *Physiologia Plantarum, 58*, 302–310.

Hirai, M. Y., Yano, M., Goodenowe, D. B., Kanaya, S., Kimura, T., Awazuhara, M., … Saito, K. (2004). Integration of transcriptomics and metabolomics for understanding of global responses to nutritional stresses in *Arabidopsis thaliana*. *Proceedings of the National Academy of Sciences of the United States of America, 101*, 10205–10210.

Höglund, A. S., Lenman, M., & Rask, L. (1992). Myrosinase is localized to the interior of myrosin grains and is not associated to the surrounding tonoplast membrane. *Plant Science, 85*, 165–170.

Hopkins, R. J., van Dam, N. M., & van Loon, J. J. A. (2009). Role of glucosinolates in insect-plant relationships and multitrophic interactions. *Annual Review of Entomology, 54*, 57–83.

Huseby, S., Koprivova, A., Lee, B. R., Saha, S., Mithen, R., Wold, A. B., … Kopriva, S. (2013). Diurnal and light regulation of sulphur assimilation and glucosinolate biosynthesis in *Arabidopsis*. *Journal of Experimental Botany, 64*, 1039–1048.

Husebye, H., Arzt, S., Burmeister, W. P., Härtel, F. V., Brandt, A., Rossiter, J. T., & Bones, A. M. (2005). Crystal structure at 1.1 Å resolution of an insect myrosinase from *Brevicoryne brassicae* shows its close relationship to β-glucosidases. *Insect Biochemistry and Molecular Biology, 35*, 1311–1320.

Husebye, H., Chadchawan, S., Winge, P., Thangstad, O. P., & Bones, A. M. (2002). Guard cell- and phloem idioblast-specific expression of thioglucoside glucohydrolase 1 (myrosinase) in *Arabidopsis*. *Plant Physiology, 128*, 1180–1188.

Ishikawa, T., Okazaki, K., Kuroda, H., Itoh, K., Mitsui, T., & Hori, H. (2007). Molecular cloning of *Brassica rapa* nitrilases and their expression during clubroot development. *Molecular Plant Pathology, 8*, 623–637.

James, D. C., & Rossiter, J. T. (1991). Development and characteristics of myrosinase in *Brassica napus* during early seedling growth. *Physiologia Plantarum, 82*, 163–170.

Jander, G., Cui, J. P., Nhan, B., Pierce, N. E., & Ausubel, F. M. (2001). The *TASTY* locus on chromosome 1 of *Arabidopsis* affects feeding of the insect herbivore *Trichoplusia ni*. *Plant Physiology, 126*, 890–898.

Janowitz, T., Trompetter, I., & Piotrowski, M. (2009). Evolution of nitrilases in glucosinolate-containing plants. *Phytochemistry, 70*, 1680–1686.

Jeschke, V., Gershenzon, J., & Vassão, D. G. (2015). Metabolism of glucosinolates and their hydrolysis products in insect herbivores. In R. Jetter (Ed.), *The formation, structure and activity of phytochemicals: Vol. 45. Recent advances in phytochemistry* (pp. 163–194). Cham: Springer International Publishing.

Jeschke, V., Gershenzon, J., & Vassão, D. G. (2016). A mode of action of glucosinolate-derived isothiocyanates: detoxification depletes glutathione and cysteine levels with ramifications on protein metabolism in *Spodoptera littoralis*. *Insect Biochemistry and Molecular Biology, 71*, 37–48.

Jeschke, V., Gershenzon, J., & Vassão, D. G. (2016). Insect detoxification of glucosinolates and their hydrolysis products. In S. Kopriva (Ed.), *Glucosinolates: Vol. 80* (in this volume).

Johnson, I. T. (2002). Glucosinolates in the human diet. Bioavailability and implications for health. *Phytochemistry Reviews, 1*, 183–188.

Jones, A. M. E., Bridges, M., Bones, A. M., Cole, R., & Rossiter, J. T. (2001). Purification and characterisation of a non-plant myrosinase from the cabbage aphid *Brevicoryne brassicae* (L.). *Insect Biochemistry and Molecular Biology, 31*, 1–5.

Jørgensen, L. B. (1981). Myrosin cells and dilated cisternae of the endoplasmic reticulum in the order Capparales. *Nordic Journal of Botany, 1*, 433–445.

Kazana, E., Pope, T. W., Tibbles, L., Bridges, M., Pickett, J. A., Bones, A. M., … Rossiter, J. T. (2007). The cabbage aphid: a walking mustard oil bomb. *Proceedings of the Royal Society Series B Biological Sciences, 274*, 2271–2277.

Kelly, P. J., Bones, A., & Rossiter, J. T. (1998). Sub-cellular immunolocalization of the glucosinolate sinigrin in seedlings of *Brassica juncea*. *Planta, 206*, 370–377.

Kissen, R., & Bones, A. M. (2009). Nitrile-specifier proteins involved in glucosinolate hydrolysis in *Arabidopsis thaliana*. *Journal of Biological Chemistry, 284*, 12057–12070.

Kissen, R., Pope, T. W., Grant, M., Pickett, J. A., Rossiter, J. T., & Powell, G. (2009). Modifying the alkylglucosinolate profile in *Arabidopsis thaliana* alters the tritrophic interaction with the herbivore *Brevicoryne brassicae* and parasitoid Diaeretiella rapae. *Journal of Chemical Ecology, 35*, 958–969.

Kissen, R., Rossiter, J. T., & Bones, A. M. (2009). The 'mustard oil bomb': not so easy to assemble?! Localization, expression and distribution of the components of the myrosinase enzyme system. *Phytochemistry Reviews, 8*, 69–86.

Kleinwächter, M., & Selmar, D. (2004). A novel approach for reliable activity determination of ascorbic acid depending myrosinases. *Journal of Biochemical and Biophysical Methods, 59*, 253–265.

Kliebenstein, D. J., & Cacho, N. I. (2016). Nonlinear selection and a blend of convergent, divergent and parallel evolution shapes: natural variation in glucosinolates. In S. Kopriva (Ed.), *Glucosinolates: Vol. 80* (in this volume).

Koroleva, O. A., & Cramer, R. (2011). Single-cell proteomic analysis of glucosinolate-rich S-cells in *Arabidopsis thaliana*. *Methods: A Companion to Methods in Enzymology, 54*, 413–423.

Koroleva, O. A., Davies, A., Deeken, R., Thorpe, M. R., Tomos, A. D., & Hedrich, R. (2000). Identification of a new glucosinolate-rich cell type in *Arabidopsis* flower stalk. *Plant Physiology, 124*, 599–608.

Koroleva, O. A., Gibson, T. M., Cramer, R., & Stain, C. (2010). Glucosinolate-accumulating S-cells in *Arabidopsis* leaves and flower stalks undergo programmed cell death at early stages of differentiation. *Plant Journal, 64*, 456–469.

Kuchernig, J. C., Backenköhler, A., Lübbecke, M., Burow, M., & Wittstock, U. (2011). A thiocyanate-forming protein generates multiple products upon allylglucosinolate breakdown in *Thlaspi arvense*. *Phytochemistry, 72*, 1699–1709.

Kuchernig, J. C., Burow, M., & Wittstock, U. (2012). Evolution of specifier proteins in glucosinolate-containing plants. *BMC Evolutionary Biology, 12*, 127.

Kutz, A., Müller, A., Hennig, P., Kaiser, W. M., Piotrowski, M., & Weiler, E. W. (2002). A role for nitrilase 3 in the regulation of root morphology in sulphur-starving *Arabidopsis thaliana*. *Plant Journal, 30*, 95–106.

Lambrix, V., Reichelt, M., Mitchell-Olds, T., Kliebenstein, D. J., & Gershenzon, J. (2001). The *Arabidopsis* epithiospecifier protein promotes the hydrolysis of glucosinolates to nitriles and influences *Trichoplusia ni* herbivory. *Plant Cell, 13*, 2793–2807.

Lenman, M., Falk, A., Rodin, J., Höglund, A. S., Ek, B., & Rask, L. (1993). Differential expression of myrosinase gene families. *Plant Physiology, 103*, 703–711.
Lenman, M., Rodin, J., Josefsson, L. G., & Rask, L. (1990). Immunological characterization of rapeseed myrosinase. *European Journal of Biochemistry, 194*, 747–753.
Li, M., & Sack, F. D. (2014). Myrosin idioblast cell fate and development are regulated by the *Arabidopsis* transcription factor FAMA, the auxin pathway, and vesicular trafficking. *Plant Cell, 26*, 4053–4066.
Li, Q., Eigenbrode, S. D., Stringam, G. R., & Thiagarajah, M. R. (2000). Feeding and growth of *Plutella xylostella* and *Spodoptera eridania* on *Brassica juncea* with varying glucosinolate concentrations and myrosinase activities. *Journal of Chemical Ecology, 26*, 2401–2419.
Li, X., & Kushad, M. M. (2005). Purification and characterization of myrosinase from horseradish (*Armoracia rusticana*) roots. *Plant Physiology and Biochemistry, 43*, 503–511.
Loebers, A., Müller-Uri, F., & Kreis, W. (2014). A young root-specific gene (*ArMY2*) from horseradish encoding a MYR II myrosinase with kinetic preference for the root-specific glucosinolate gluconasturtiin. *Phytochemistry, 99*, 26–35.
Lombard, V., Golaconda Ramulu, H., Drula, E., Coutinho, P. M., & Henrissat, B. (2014). The carbohydrate-active enzymes database (CAZy) in 2013. *Nucleic Acids Research, 42*, D490–D495.
Lönnerdal, B., & Janson, J. C. (1973). Studies on myrosinases: ii. Purification and characterization of a myrosinase from rapeseed (*Brassica napus* L.). *Biochimica et Biophysica Acta, 315*, 421–429.
Lüthy, J., & Benn, M. H. (1977). Thiocyanate formation from glucosinolates: a study of the autolysis of allylglucosinolate in *Thlaspi arvense* L. seed flour extracts. *Canadian Journal of Biochemistry, 55*, 1028–1031.
Maruyama-Nakashita, A., Nakamura, Y., Tohge, T., Saito, K., & Takahashi, H. (2006). *Arabidopsis SLIM1* is a central transcriptional regulator of plant sulfur response and metabolism. *Plant Cell, 18*, 3235–3251.
Matile, P. (1980). The mustard oil bomb. Compartmentation of the myrosinase system. *Biochemie und Physiologie der Pflanzen, 175*, 722–731.
Matsushima, R., Kondo, M., Nishimura, M., & Hara-Nishimura, I. (2003). A novel ER-derived compartment, the ER body, selectively accumulates a β-glucosidase with an ER-retention signal in *Arabidopsis*. *Plant Journal, 33*, 493–502.
Matusheski, N. V., Swarup, R., Juvik, J. A., Mithen, R., Bennett, M., & Jeffery, E. H. (2006). Epithiospecifier protein from broccoli (*Brassica oleracea* L. ssp. *italica*) inhibits formation of the anticancer agent sulforaphane. *Journal of Agricultural and Food Chemistry, 54*, 2069–2076.
McGregor, D. I. (1988). Glucosinolate content of developing rapeseed (*Brassica napus* L. 'Midas') seedlings. *Canadian Journal of Plant Science, 68*, 367–380.
Meulenbeld, G. H., & Hartmans, S. (2001). Thioglucosidase activity from Sphingobacterium sp strain OTG1. *Applied Microbiology and Biotechnology, 56*, 700–706.
Miao, Y., & Zentgraf, U. (2007). The antagonist function of *Arabidopsis WRKY53* and *ESR/ESP* in leaf senescence is modulated by the jasmonic and salicylic acid equilibrium. *Plant Cell, 19*, 819–830.
Morant, A. V., Jørgensen, K., Jørgensen, C., Paquette, S. M., Sanchez-Perez, R., Møller, B. L., & Bak, S. (2008). β-Glucosidases as detonators of plant chemical defense. *Phytochemistry, 69*, 1795–1813.
Mullaney, J. A., Kelly, W. J., McGhie, T. K., Ansell, J., & Heyes, J. A. (2013). Lactic acid bacteria convert glucosinolates to nitriles efficiently yet differently from Enterobacteriaceae. *Journal of Agricultural and Food Chemistry, 61*, 3039–3046.
Müller, C. (2009). Interactions between glucosinolate- and myrosinase-containing plants and the sawfly *Athalia rosae*. *Phytochemistry Reviews, 8*, 121–134.

Mumm, R., Burow, M., Bukovinszkine'Kiss, G., Kazantzidou, E., Wittstock, U., Dicke, M., & Gershenzon, J. (2008). Formation of simple nitriles upon glucosinolate hydrolysis affects direct and indirect defense against the specialist herbivore, *Pieris rapae*. *Journal of Chemical Ecology, 34*, 1311–1321.

Nagano, A. J., Matsushima, R., & Hara-Nishimura, I. (2005). Activation of an ER-body-localized β-glucosidase via a cytosolic binding partner in damaged tissues of *Arabidopsis thaliana*. *Plant & Cell Physiology, 46*, 1140–1148.

Nikiforova, V., Freitag, J., Kempa, S., Adamik, M., Hesse, H., & Hoefgen, R. (2003). Transcriptome analysis of sulfur depletion in *Arabidopsis thaliana*: interlacing of biosynthetic pathways provides response specificity. *Plant Journal, 33*, 633–650.

Palmieri, S., Iori, R., & Leoni, O. (1986). Myrosinase from *Sinapis alba* L.: a new method of purification for glucosinolate analyses. *Journal of Agricultural and Food Chemistry, 34*, 138–140.

Pastorczyk, M., & Bednarek, P. (2016). The function of glucosinolates and related metabolites in plant innate immunity. In S. Kopriva (Ed.), *Glucosinolates: Vol. 80* (in this volume).

Pessina, A., Thomas, R. M., Palmieri, S., & Luisi, P. L. (1990). An improved method for the purification of myrosinase and its physicochemical characterization. *Archives of Biochemistry and Biophysics, 280*, 383–389.

Petersen, B. L., Chen, S. X., Hansen, C. H., Olsen, C. E., & Halkier, B. A. (2002). Composition and content of glucosinolates in developing *Arabidopsis thaliana*. *Planta, 214*, 562–571.

Piotrowski, M. (2008). Primary or secondary? Versatile nitrilases in plant metabolism. *Phytochemistry, 69*, 2655–2667.

Pontoppidan, B., Hopkins, R., Rask, L., & Meijer, J. (2003). Infestation by cabbage aphid (*Brevicoryne brassicae*) on oilseed rape (*Brassica napus*) causes a long lasting induction of the myrosinase system. *Entomologia Experimentalis et Applicata, 109*, 55–62.

Pontoppidan, P., Ekbom, B., Eriksson, S., & Meijer, J. (2001). Purification and characterization of myrosinase from the cabbage aphid (*Brevicoryne brassicae*), a *Brassica* herbivore. *European Journal of Biochemistry, 268*, 1041–1048.

Pope, T. W., Kissen, R., Grant, M., Pickett, J. A., Rossiter, J. T., & Powell, G. (2008). Comparative innate responses of the aphid parasitoid *Diaeretiella rapae* to alkenyl glucosinolate derived isothiocyanates, nitriles, and epithionitriles. *Journal of Chemical Ecology, 34*, 1302–1310.

Rask, L., Andréasson, E., Ekbom, B., Eriksson, S., Pontoppidan, B., & Meijer, J. (2000). Myrosinase: gene family evolution and herbivore defense in Brassicaceae. *Plant Molecular Biology, 42*, 93–113.

Ratzka, A., Vogel, H., Kliebenstein, D. J., Mitchell-Olds, T., & Kroymann, J. (2002). Disarming the mustard oil bomb. *Proceedings of the National Academy of Sciences of the United States of America, 99*, 11223–11228.

Reese, E. T., Clapp, R. C., & Mandels, M. (1958). A thioglucosidase in fungi. *Archives of Biochemistry and Biophysics, 75*, 228–242.

Rodman, J. E., Karol, K. G., Price, R. A., & Sytsma, K. J. (1996). Molecules, morphology, and Dahlgrens expanded order Capparales. *Systematic Botany, 21*, 289–307.

Rosa, E. A. S., & Rodrigues, P. M. F. (1998). The effect of light and temperature on glucosinolate concentration in the leaves and roots of cabbage seedlings. *Journal of the Science of Food and Agriculture, 78*, 208–212.

Rossiter, J. T., Pickett, J. A., Bennett, M. H., Bones, A. M., Powell, G., & Cobb, J. (2007). The synthesis and enzymic hydrolysis of (*E*)-2-[2,3-$^{2}h_{2}$]propenyl glucosinolate: confirmation of the rearrangement of the thiohydroximate moiety. *Phytochemistry, 68*, 1384–1390.

Sarsby, J., Towers, M. W., Stain, C., Cramer, R., & Koroleva, O. A. (2012). Mass spectrometry imaging of glucosinolates in *Arabidopsis* flowers and siliques. *Phytochemistry, 77*, 110–118.

Schnug, E., & Haneklaus, S. (2016). Glucosinolates – the agricultural story. In S. Kopriva (Ed.), *Glucosinolates: Vol. 80* (in this volume).

Schramm, K., Vassão, D. G., Reichelt, M., Gershenzon, J., & Wittstock, U. (2012). Metabolism of glucosinolate-derived isothiocyanates to glutathione conjugates in generalist lepidopteran herbivores. *Insect Biochemistry and Molecular Biology, 42*, 174–182.

Shirakawa, M., Ueda, H., Nagano, A. J., Shimada, T., Kohchi, T., & Hara-Nishimura, I. (2014). FAMA is an essential component for the differentiation of two distinct cell types, myrosin cells and guard cells, in *Arabidopsis*. *Plant Cell, 26*, 4039–4052.

Shirakawa, M., Ueda, H., Shimada, T., Kohchi, T., & Hara-Nishimura, I. (2014). Myrosin cell development is regulated by endocytosis machinery and *PIN1* polarity in leaf primordia of *Arabidopsis thaliana*. *Plant Cell, 26*, 4448–4461.

Shroff, R., Vergara, F., Muck, A., Svatoš, A., & Gershenzon, J. (2008). Nonuniform distribution of glucosinolates in *Arabidopsis thaliana* leaves has important consequences for plant defense. *Proceedings of the National Academy of Sciences of the United States of America, 105*, 6196–6201.

Stauber, E. J., Kuczka, P., van Ohlen, M., Vogt, B., Janowitz, T., Piotrowski, M., … Wittstock, U. (2012). Turning the 'mustard oil bomb' into a 'cyanide bomb': aromatic glucosinolate metabolism in a specialist insect herbivore. *PLoS One, 7*, e35545.

Stein, R. J., & Waters, B. M. (2012). Use of natural variation reveals core genes in the transcriptome of iron-deficient *Arabidopsis thaliana* roots. *Journal of Experimental Botany, 63*, 1039–1055.

Stotz, H. U., Sawada, Y., Shimada, Y., Hirai, M. Y., Sasaki, E., Krischke, M., … Kamiya, Y. (2011). Role of camalexin, indole glucosinolates, and side chain modification of glucosinolate-derived isothiocyanates in defense of *Arabidopsis* against *Sclerotinia sclerotiorum*. *Plant Journal, 67*, 81–93.

Sun, J. Y., Sønderby, I. E., Halkier, B. A., Jander, G., & de Vos, M. (2009). Non-volatile intact indole glucosinolates are host recognition cues for ovipositing *Plutella xylostella*. *Journal of Chemical Ecology, 35*, 1427–1436.

Taipalensuu, J., Andréasson, E., Eriksson, S., & Rask, L. (1997). Regulation of the wound-induced myrosinase-associated protein transcript in *Brassica napus* plants. *European Journal of Biochemistry, 247*, 963–971.

Taipalensuu, J., Eriksson, S., & Rask, L. (1997). The myrosinase-binding protein from *Brassica napus* seeds possesses lectin activity and has a highly similar vegetatively expressed wound-inducible counterpart. *European Journal of Biochemistry, 250*, 680–688.

Taipalensuu, J., Falk, A., Ek, B., & Rask, L. (1997). Myrosinase-binding proteins are derived from a large wound-inducible and repetitive transcript. *European Journal of Biochemistry, 243*, 605–611.

Taipalensuu, J., Falk, A., & Rask, L. (1996). A wound- and methyl jasmonate-inducible transcript coding for a myrosinase-associated protein with similarities to an early nodulin. *Plant Physiology, 110*, 483–491.

Takeda, M., Sugimori, N., Torizawa, T., Terauchi, T., Ono, A. M., Yagi, H., … Kainosho, M. (2008). Structure of the putative 32 kDa myrosinase-binding protein from *Arabidopsis* (At3g16450.1) determined by SAIL-NMR. *FEBS Journal, 275*, 5873–5884.

Tani, N., Ohtsuru, M., & Hata, T. (1974). Purification and general characteristics of bacterial myrosinase produced by *Enterobacter cloacae*. *Agricultural and Biological Chemistry, 38*, 1623–1630.

Thangstad, O. P., Evjen, K., & Bones, A. (1991). Immunogold-EM localization of myrosinase in Brassicaceae. *Protoplasma, 161*, 85–93.

Thangstad, O. P., Gilde, B., Chadchawan, S., Seem, M., Husebye, H., Bradley, D., & Bones, A. M. (2004). Cell specific, cross-species expression of myrosinases in *Brassica napus, Arabidopsis thaliana* and *Nicotiana tabacum*. *Plant Molecular Biology, 54*, 597–611.

Thangstad, O. P., Iversen, T. H., Slupphaug, G., & Bones, A. (1990). Immunocytochemical localization of myrosinase in *Brassica napus* L. *Planta, 180*, 245–248.

Thangstad, O. P., Winge, P., Husebye, H., & Bones, A. (1993). The myrosinase (thioglucoside glucohydrolase) gene family in Brassicaceae. *Plant Molecular Biology, 23*, 511–524.
Tookey, H. L. (1973). Crambe thioglucoside glucohydrolase (EC 3.2.3.1): separation of a protein required for epithiobutane formation. *Canadian Journal of Biochemistry, 51*, 1654–1660.
Traka, M. H. (2016). Health benefits of glucosinolates. In S. Kopriva (Ed.), *Glucosinolates: Vol. 80* (in this volume).
Travers-Martin, N., Kuhlmann, F., & Müller, C. (2008). Revised determination of free and complexed myrosinase activities in plant extracts. *Plant Physiology and Biochemistry, 46*, 506–516.
Tsuruo, I., Yoshida, M., & Hata, T. (1967). Studies on the myrosinase in mustard seed. *Agricultural and Biological Chemistry, 31*, 18–32.
Ueda, H., Nishiyama, C., Shimada, T., Koumoto, Y., Hayashi, Y., Kondo, M., ... Hara-Nishimura, I. (2006). *AtVAM3* is required for normal specification of idioblasts, myrosin cells. *Plant & Cell Physiology, 47*, 164–175.
Urzica, E. I., Casero, D., Yamasaki, H., Hsieh, S. I., Adler, L. N., Karpowicz, S. J., ... Merchant, S. S. (2012). Systems and trans-system level analysis identifies conserved iron deficiency responses in the plant lineage. *Plant Cell, 24*, 3921–3948.
Vaughn, S. F., & Berhow, M. A. (2005). Glucosinolate hydrolysis products from various plant sources: pH effects, isolation, and purification. *Industrial Crops & Products, 21*, 193–202.
Vergara, F., Svatoš, A., Schneider, B., Reichelt, M., Gershenzon, J., & Wittstock, U. (2006). Glycine conjugates in a lepidopteran insect herbivore – the metabolism of benzylglucosinolate in the cabbage white butterfly, *Pieris rapae. ChemBioChem: A European Journal of Chemical Biology, 7*, 1982–1989.
Virtanen, A. I. (1965). Studies on organic sulphur compounds and other labile substances in plants. *Phytochemistry, 4*, 207–228.
Wang, M., Li, D., Sun, X., Zhu, Y. J., Nong, H., & Zhang, J. (2009). Characterization of a root-specific β-thioglucoside glucohydrolase gene in *Carica papaya* and its recombinant protein expressed in *Pichia pastoris. Plant Science, 177*, 716–723.
Waser, J., & Watson, W. H. (1963). Crystal structure of sinigrin. *Nature, 198*, 1297–1298.
Wentzell, A. M., & Kliebenstein, D. J. (2008). Genotype, age, tissue, and environment regulate the structural outcome of glucosinolate activation. *Plant Physiology, 147*, 415–428.
Wheat, C. W., Vogel, H., Wittstock, U., Braby, M. F., Underwood, D., & Mitchell-Olds, T. (2007). The genetic basis of a plant-insect coevolutionary key innovation. *Proceedings of the National Academy of Sciences of the United States of America, 104*, 20427–20431.
Winde, I., & Wittstock, U. (2011). Insect herbivore counteradaptations to the plant glucosinolate-myrosinase system. *Phytochemistry, 72*, 1566–1575.
Wittstock, U., Agerbirk, N., Stauber, E. J., Olsen, C. E., Hippler, M., Mitchell-Olds, T., ... Vogel, H. (2004). Successful herbivore attack due to metabolic diversion of a plant chemical defense. *Proceedings of the National Academy of Sciences of the United States of America, 101*, 4859–4864.
Wittstock, U., & Burow, M. (2007). Tipping the scales - specifier proteins in glucosinolate hydrolysis. *IUBMB Life, 59*, 744–751.
Wittstock, U., & Burow, M. (2010). Glucosinolate breakdown in *Arabidopsis* – mechanism, regulation and biological significance. In R. Last, C. Chang, G. Jander, D. Kliebenstein, R. McClung, H. Millar, et al. (Eds.), *The* Arabidopsis *book* (p. e0134). The American Society of Plant Biologists.
Wittstock, U., & Gershenzon, J. (2002). Constitutive plant toxins and their role in plant defense. *Current Opinion in Plant Biology, 5*, 300–307.
Wittstock, U., Kliebenstein, D. J., Lambrix, V., Reichelt, M., & Gershenzon, J. (2003). Glucosinolate hydrolysis and its impact on generalist and specialist insect herbivores.

In J. T. Romeo (Ed.), *Integrative phytochemistry: From ethnobotany to molecular ecology: Vol. 37. Recent advances in phytochemistry* (pp. 101–125). Amsterdam: Elsevier.

Witzel, K., Hanschen, F. S., Klopsch, R., Ruppel, S., Schreiner, M., & Grosch, R. (2015). *Verticillium longisporum* infection induces organ-specific glucosinolate degradation in *Arabidopsis thaliana*. *Frontiers in Plant Science, 6.*

Xu, Z., Escamilla-Trevino, L. L., Zeng, L., Lalgondar, M., Bevan, D. R., Winkel, B. S. J., … Esen, A. (2004). Functional genomic analysis of *Arabidopsis thaliana* glycoside hydrolase family 1. *Plant Molecular Biology, 55*, 343–367.

Xue, J., Pihlgren, U., & Rask, L. (1993). Temporal, cell-specific, and tissue-preferential expression of myrosinase genes during embryo and seedling development in *Sinapis alba*. *Planta, 191*, 95–101.

Zhang, H., Schonhof, I., Krumbein, A., Gutezeit, B., Li, L., Stützel, H., & Schreiner, M. (2008). Water supply and growing season influence glucosinolate concentration and composition in turnip root (*Brassica rapa* ssp. *rapifera* L.). *Journal of Plant Nutrition and Soil Science, 171*, 255–265.

Zhang, Z., Ober, J. A., & Kliebenstein, D. J. (2006). The gene controlling the quantitative trait locus *EPITHIOSPECIFIER MODIFIER1* alters glucosinolate hydrolysis and insect resistance in *Arabidopsis*. *Plant Cell, 18*, 1524–1536.

Zhao, Y., Wang, J., Liu, Y., Miao, H., Cai, C., Shao, Z., … Wang, Q. (2015). Classic myrosinase-dependent degradation of indole glucosinolate attenuates fumonisin B1-induced programmed cell death in *Arabidopsis*. *Plant Journal, 81*, 920–933.

Zhao, Z., Zhang, W., Stanley, B. A., & Assmann, S. M. (2008). Functional proteomics of *Arabidopsis thaliana* guard cells uncovers new stomatal signaling pathways. *Plant Cell, 20*, 3210–3226.

Zhou, C., Tokuhisa, J. G., Bevan, D. R., & Esen, A. (2012). Properties of β-thioglucoside hydrolases (TGG1 and TGG2) from leaves of *Arabidopsis thaliana*. *Plant Science, 191–192*, 82–92.

CHAPTER SEVEN

The Function of Glucosinolates and Related Metabolites in Plant Innate Immunity

M. Pastorczyk, P. Bednarek[1]
Institute of Bioorganic Chemistry, Polish Academy of Sciences, Poznań, Poland
[1]Corresponding author: E-mail: bednarek@ibch.poznan.pl

Contents

Advances in Botanical Research, Volume 80
ISSN 0065-2296
http://dx.doi.org/10.1016/bs.abr.2016.06.007

Abstract

Numerous studies showed that glucosinolate metabolism products support broad spectrum immunity in the model plant *Arabidopsis thaliana* and possibly other Brassicaceae species. According to the experimental evidence, in response to pathogenic infection, methionine-derived aliphatic glucosinolates are metabolized similarly as during interactions with insects, and resulting products may restrict the in planta growth of at least same nonadapted microbial pathogens. Opposite to aliphatic glucosinolates, a specialized metabolic pathway evolved in the Brassicaceae species to metabolize the tryptophan-derived indole glucosinolates (IGs) in response to pathogen recognition. In addition to the activation of glucosinolate metabolism, microbial infection of Brassicaceae species triggers the biosynthesis of indole-type phytoalexins that are interconnected with IGs at the biosynthetic level. Unfortunately, despite years of studies the mode of action of glucosinolate-related phytochemicals in plant immunity remains unknown. Originally, due to their antimicrobial activity in vitro, these compounds were classified as in planta antibiotic. However, recent experimental evidence indicate that some molecules released during pathogen-triggered glucosinolate metabolism may activate evolutionarily conserved immune responses, possibly through affecting glutathione redox state in the infected cell.

1. INTRODUCTION

The plant immune system is developed as an effect of long-term coevolutionary processes between host plants and their pathogens and is constituted by two independent branches (Chisholm, Coaker, Day, & Staskawicz, 2006; Jones & Dangl, 2006). The more evolutionary ancient primary immune response, referred to as pathogen-associated molecular pattern (PAMP)-triggered immunity (PTI), relies on transmembrane-resident pattern recognition receptors that detect highly conserved and common microbial elicitors, such as flagellin or chitin, known as PAMPs or MAMPs (microbe-associated molecular patterns). During the coevolution process, pathogens acquired the ability to suppress PTI by delivering effector proteins into the host cell. However, the presence of effectors can be recognized and monitored by polymorphic resistance (R) proteins, which are the executors of effector-triggered immunity (ETI) (Dodds & Rathjen, 2010; Jones & Dangl, 2006).

Pathogenic microorganisms that are capable of invading plant cells exhibit a huge variability in their lifestyle and infection strategies. Germinating spores of plant pathogenic fungi differentiate into functional structures that attach to the cuticle of the epidermis. After successful penetration, nutritional structures allow the fungus to directly benefit from the active metabolism of the host, and the pathogen completes its life cycle either by keeping the host cells alive during colonization (biotrophic strategy) or inducing their disintegration after infection (necrotrophic strategy). Alternatively, some pathogenic fungi and oomycetes pass through a biotrophic phase and later switch to a necrotrophic phase (hemibiotrophic strategy) (Horbach, Navarro-Quesada, Knogge, & Deising, 2011; Huckelhoven & Panstruga, 2011). Pathogenic bacteria usually employ other infection strategies than those used by filamentous pathogens. For instance, the hemibiotrophic *Pseudomonas syringae* takes advantage of wounds or natural openings to colonize host tissue and lives both on the surface and in the apoplast, transferring its effector proteins directly inside host cells by a type III secretion system (Cunnac, Lindeberg, & Collmer, 2009). Importantly, pathogenic microorganisms belonging to distant phylogenetic groups and representing different infection strategies activate similar defence mechanisms in host plants across the plant kingdom. These highly conserved mechanisms include cell wall reinforcement, which is linked with the biosynthesis and deposition of β-1,3-glucan polymer callose (Ellinger & Voigt, 2014; Nishimura et al., 2003); a hypersensitive response (HR) that is usually accompanied with rapid and localized cell death at the infection sites (Coll, Epple, & Dangl, 2011; Heath, 2000); stomatal closure (Melotto, Underwood, Koczan, Nomura, & He, 2006; Zeng, Melotto, & He, 2010) and the biosynthesis of low-molecular weight secondary metabolites (Ahuja, Kissen, & Bones, 2012; Piasecka, Jedrzejczak-Rey, & Bednarek, 2015). Collectively, higher plants have the capacity to synthesize thousands of structurally diverse bioactive natural products constitutively or in response to pathogen detection. According to the frequently used classification system based on biosynthesis and mode of action, these defensive metabolites are classified as phytoanticipins, which are constitutively stored in plant tissues or released from constitutive precursors (VanEtten, Mansfield, Bailey, & Farmer, 1994), or phytoalexins, which are synthesized de novo upon pathogen recognition (Ahuja et al., 2012; Paxton, 1981). In plants belonging to the Brassicaceae family, two types of sulphur-containing secondary metabolites are implicated in its innate immunity. These include sulphur-containing indole-type phytoalexins and glucosinolates, which are constitutively present in plant tissue and function as phytoanticipins (Halkier & Gershenzon, 2006; Pedras, Yaya, & Glawischnig, 2011).

2. ALIPHATIC GLUCOSINOLATES IN PLANT IMMUNITY

2.1 Glucosinolate-Derived Isothiocyanates Can Limit Pathogen Growth in Planta

Myrosinase-mediated glucosinolate hydrolysis may lead to the generation of many types of products including nitriles, thiocyanates and isothiocyanates (ITCs). Among the glucosinolate breakdown products, ITCs possess the highest chemical reactivity, which is attributed to the central carbon of the isothiocyanate group (−N=C=S). This reactivity correlates with strong antimicrobial activity in vitro, even at low concentrations, and has negative effects on the growth of pathogenic fungal species (Calmes et al., 2015; Dufour, Stahl, & Baysse, 2015). While the generation of nitriles or epithionitriles requires the presence of specifier proteins, ITCs form rapidly without the involvement of additional enzymes. However, the generation of ITCs in planta is usually triggered by tissue damage, which is rather limited and often does not occur, during interactions with pathogenic microorganisms. Despite relatively low tissue damage, several reports indicate that activation of methionine-derived aliphatic glucosinolates (AGs) to antibiotic ITCs may be efficient in the interactions of the model plant *Arabidopsis thaliana* with necrotrophic pathogens. A study with the *myb28 myb29* line, in which AG biosynthesis is blocked at the transcriptional level and consequently these plants do not accumulate any compounds belonging to this class of glucosinolates (Sonderby et al., 2007), revealed that one out of two tested *Botrytis cinerea* strains took advantage of the AG-deficiency of the host and grew more than in the wild-type (WT) plants (Buxdorf, Yaffe, Barda, & Levy, 2013). Enhanced disease symptoms observed upon inoculation with the same *B. cinerea* strain (*B.05.10*) in the *tgg1 tgg2* double mutant and in a transgenic line overexpressing the epithiospecifier protein (*35S:ESP*), which redirects AG-degradation products from ITCs to nitriles, suggested that TGG-catalysed AG-hydrolysis, together with the spontaneous decomposition of the resulting aglycones to ITCs, is required in the immunity towards this particular pathogen (Buxdorf et al., 2013). Interestingly, infection of WT *A. thaliana* with *B. cinerea B05.10*, but not with the second tested isolate, strongly induced the accumulation of AGs and indole glucosinolates (IGs). This suggests that the AG contribution to plant immunity may be dependent on the pathogen-triggered induction of their biosynthesis and that successful pathogens may acquire ability to suppress the activation of the AG biosynthetic pathway. In contrast with this assumption, during the same study, an *Alternaria brassicicola* isolate was also tested, and it was found that it caused stronger disease symptoms in the *myb28 myb29* and *tgg1 tgg2* mutants as well as in the *35S:ESP* line, compared to the WT plants, but did not induce AG and IG biosynthesis. The *myb28 myb29* double mutant plants showed also enhanced diseased symptoms after inoculation of their leaves with *Sclerotinia sclerotiorum*, a necrotrophic pathogen closely related to *B. cinerea*, possibly due to a deficiency in the AG-derived ITCs (Stotz et al., 2011). Interestingly, the enhanced virulence of this aggressive ascomycete was also observed in the *myb28* but not in the *myb29* single knockout line. *myb29* is defective only in the accumulation of short chain AGs, while *myb28* accumulates reduced levels of both short and long chain AGs (Sonderby et al., 2007). This suggests that long-chain AGs may make a stronger contribution to the resistance towards *S. sclerotiorum*. In accordance with this assumption, in vitro assays revealed that the antimicrobial activity of ITCs towards this pathogen positively correlates with side chain length, and from the five ITCs tested during this study, 8-methylsulfinyloctyl ITC showed the highest inhibition of fungal growth (Stotz et al., 2011).

Apart from the *myb* and *tgg* knockout lines contribution of AGs to *A. thaliana* immunity was also addressed in the *gsm1* mutant (Tierens et al., 2001). This line carries a mutation in the *MAM1* gene, which encodes an enzyme in the methionine elongation pathway and controls the variation between C_3 and C_4 glucosinolates (Kroymann et al., 2001). Consequently, the *gsm1* mutant accumulates strongly reduced levels of 4-methyl-sulfinylbutyl, 4-methylthiobutyl, 5-methylsulfinylpentyl, and 6-methylsulfinyl glucosinolates and produces higher amounts of 3-methylsulfinylpropyl and 3-methylthiopropyl glucosinolates (Haughn, Davin, Giblin, & Underhill, 1991; Kroymann et al., 2001). The inoculation of this line with *A. brassicicola* and *B. cinerea* revealed infection symptoms similar to those observed in the WT plants (Tierens et al., 2001). However, as the fungal isolates tested in this study differed from those used by Buxdorf et al. (2013), it is difficult to conclude if the discrepant impact of *myb28 myb29* and *gsm1* mutations on the immunity towards *A. brassicicola* and *B. cinerea* results from the contribution of the particular AGs that are still present in the *gsm1* line or from the different adaptations of tested fungal strains. The *gsm1* line also revealed WT-like susceptibility to two other filamentous pathogens, *Plectosphaerella cucumerina* (necrotroph) and *Hyaloperonospora parasitica* (biotroph), and to the necrotrophic bacterium *Erwinia carotovora*. However, the hemibiotrophic fungus *Fusarium oxysporum* was found to be considerably more aggressive in *gsm1* plants, growing to a higher biomass and causing excessive chlorosis upon infection (Tierens et al., 2001).

2.2 Detoxification of Isothiocyanates Supports Pathogenesis in *Arabidopsis thaliana*

Collectively, the results obtained with destructive fungal hemibiotrophic and necrotrophic pathogens suggested that in these interactions, AGs can be degraded to bioactive ITCs, which in turn contribute to pathogen restriction. Opposite to filamentous pathogens, the hemibiotrophic bacterium *P. syringae* uses wounds or natural openings as gateways for infection and colonizes the apoplastic spaces between cells. Assays with the *gsm1* mutant line and the adapted *P. syringae* pv. *tomato* DC3000 strain did not reveal any defects in the immunity of this AG-deficient plant line (Tierens et al., 2001). Interestingly, results obtained during a study on the interactions between different *P. syringae* strains and *myb28 myb29* plants indicated that AGs may restrict the in planta growth of the nonadapted, but not the virulent, isolates of this pathogen (Fan et al., 2011). Strains that are pathogenic to *A. thaliana*, for instance, *P. syringae* pv. *tomato* DC3000, developed resistance mechanism to these antibiotics, which is based on the *sax* (*survival in Arabidopsis extracts*) operon. The contribution of this operon to ITC tolerance is highlighted by the elevated in vitro resistance of *sax*-containing *P. syringae* strains for these phytochemicals (Fan et al., 2011). However, how the *sax* genes mediate ITC resistance in virulent *P. syringae* is still elusive. The products of the *saxA* and *saxB* genes possess putative enzymatic activity related to the β-lactamase and isochorismatase enzyme families, respectively. The *saxC* gene encodes a transcriptional regulator from the highly conserved AraC/XylS family found in diverse prokaryotes that is involved in carbon metabolism, stress response and pathogenesis. Finally, the *saxF/D/G* products represent a family of putative multidrug resistance proteins and seem to have distinct but complementary roles to *saxA/B/C* in resistance for ITCs. Overall, it is assumed that the *sax* operon encodes proteins that enable the detoxification and extrusion of ITC, but the function of each component is not necessarily indicated by sequence homology. For instance, although the product of *saxA*, which is found only in the genome of pathogenic *P. syringae* strains, is related to class B of β-lactamases, it does not confer resistance to any conventional β-lactam antibiotics (Fan et al., 2011).

2.3 Mechanisms of Aliphatic Glucosinolate Activation in Plant Immunity Remain Obscure

Overall, the results obtained from the *myb28 myb29*, *tgg1 tgg2* and *gsm1* mutants indicated that the contribution of AGs and ITCs to the immunity of *A. thaliana* may vary among particular pathogen species or even isolates. This is explained by the ability of the adapted strains to detoxify ITCs and/or to suppress the induction of AG biosynthesis (Buxdorf et al., 2013; Fan et al., 2011). Unfortunately, as the tested mutant sets differ between reported studies (Table 1), it is difficult to obtain comprehensive information on the mechanisms of AG function. For instance, only one study included the *tgg1 tgg2* mutant and the *35S:ESP* line (Buxdorf et al., 2013). Thus, the idea that the contribution of AGs to *A. thaliana* immunity against *S. sclerotiorum, F. oxysporum* or *P. syringae* is based on mechanisms of activation different from the direct TGG1/2 mediated hydrolysis to ITCs cannot be excluded (Fan et al., 2011; Stotz et al., 2011; Tierens et al., 2001). Similarly, as *myb28* and *myb29* single mutants were only tested with *S. sclerotiorum*, it is unclear if the significant contribution of the AGs with long side chain is unique for the *A. thaliana* response against this particular fungi or if this is a common phenomenon (Stotz et al., 2011).

3. PATHOGEN-TRIGGERED INDOLE GLUCOSINOLATE METABOLISM

In contrast to necrotrophs and hemibiotrophs, biotrophic plant pathogens rely exclusively on nutrient uptake from living host cells. These microbes are capable of reorganizing the host cell architecture for their own benefit; however, they do not disturb the plant cell and tissue integrity (Huckelhoven & Panstruga, 2011). Interestingly, it was shown that *A. thaliana* in response to the attempted penetration of biotrophic powdery mildews, activates IG metabolism autonomously from cell damage. This process involves the atypical myrosinase PENETRATION2 (PEN2), anchored in the peroxisomal and mitochondrial membranes, which is actively delivered to the cell periphery at fungal entry sites after pathogen detection (Fuchs et al., 2016; Lipka et al., 2005). The enhanced penetration of the nonadapted powdery mildews *Blumeria graminis f.* sp. *hordei* and *Erysiphe pisi* into epidermal cells of *pen2* mutant leaves indicated that the PEN2

Table 1 Infection Phenotypes Observed in *Arabidopsis thaliana* Mutant Lines Impaired in Aliphatic and Indole Glucosinolate Biosynthesis and Metabolism Observed During Colonization With Various Microorganisms

	Aliphatic Glucosinolate Mutants			Indole Glucosinolate Mutants					
	myb28/29	***tgg1/2***	***gsm1***	***pen2***	***cyp81F2***	***pen2 cyp81F2***	***cyp79B2/B3***	***myb51***	***myb34/51/121***
Alternaria alternata				B[1]C[1]					
Alternaria brassicicola	D[2]	D[2]	E[3]	D[2]	D[2]	D[2]	D[2]		
Blumeria gramminis f. sp *hordei*				A[4;5;6;7]	A[6]	A[6]	A[6]		
Botritis cinerea	D[2;8]	E[2]	E[3]	D[2]	D[2]	D[2]	D[2;9]		
Colletotrichum gloeosporioides				A[10;11]	A[10]		B[10;11]		
Colletotrichum tofieldiae★				E[12]			C[12]		
Erwinia carotovora			E[3]						
Erysiphe pisi				A[4;6]	A[6]	A[6]	B[6]		
Fusarium oxysporum			C[3] D[3]						
Hyaloperonospora parasitica			E[3]	E[7]					
Leptosphaeria maculans				A[13]					
Magnaporthe oryzae				A[14;15]					
Phakospora pachyrhizi				A[16;17]					
Phytophthora brassicae				A[18]	A[18]		B[18]	A[18]	
Phytophthora capsici	D[19]			D[19]	D[19]		D[19]	D[19]	
Phytophthora infestans				B[4]					
Piriformospora indica★				C[20]	E[21]		C[22]		C[21]
Plectosphaerella cucumerina			E[3]	C[6;23;24] D[23;24]	C[23] D[23]	C[6]	C[23;24] D[23;24]	C[24] D[24]	C[24] D[24]
Pseudomonas syringae	C[25]			E[7]					
Sclerotonia sclerotiorum	D[8]						D[8]		
Sebacina vermifera★					E[21]		C[21]		C[21]

A, defect in preinvasive immunity (elevated entry rates); *B*, defect in pre- and postinvasive immunity; *C*, elevated pathogen biomass; *D*, increased disease symptoms; *E*, wild type-like susceptibility. ★nonpathogenic growth-promoting fungi.

[1] Egusa et al. (2013), [2] Buxdorf et al. (2013), [3] Tierens et al. (2001), [4] Lipka et al. (2005), [5] Stein et al. (2006), [6] Bednarek et al. (2009), [7] Johansson et al. (2014), [8] Stotz et al. (2011), [9] Kliebenstein, Rowe, and Denby (2005), [10] Hiruma et al. (2010), [11] Hiruma et al. (2013), [12] Hiruma et al. (2016), [13] Elliott et al. (2008), [14] Nakao et al. (2011), [15] Maeda et al. (2009), [16] Langenbach et al. (2013), [17] Loehrer et al. (2008), [18] Schlaeppi et al. (2010), [19] Wang et al. (2013), [20] Jacobs et al. (2011), [21] Lahrmann et al. (2015), [22] Nongbri et al. (2012), [23] Sanchez-Vallet et al. (2010), [24] Frerigmann et al. (2016), [25] Fan et al. (2011).

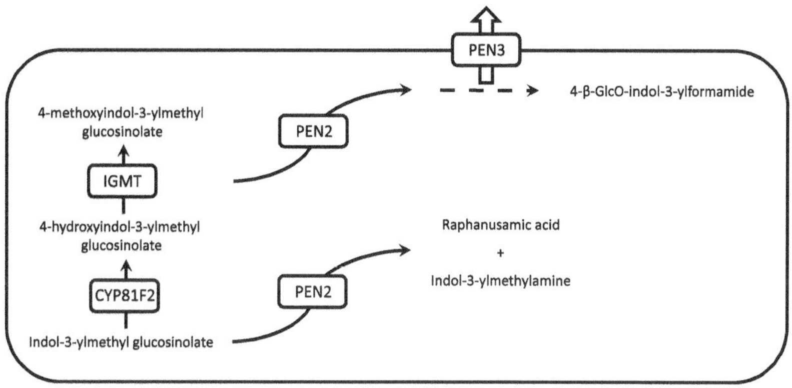

Figure 1 ***Contribution of CYP81F2 monooxygenase, PEN2 myrosinase and PEN3 ABC transporter to the pathogen-triggered indole glucosinolate metabolism in* Arabidopsis thaliana.** CYP81F2 converts indol-3-ylmethyl glucosinolate to 4-hydroxy-indol-3-ylmethyl glucosinolate, which is further metabolised by *O*-methyltransferase (IGMT) to 4-methoxyindol-3-ylmethyl glucosinolate; however, contribution of IGMT to the immune function of indole glucosinolates has not been so far investigated. PEN2 can hydrolyse different types of IGs; however, only hydrolysis of IGs substituted at the position C-4 leads to the formation of the critical for plant immunity end product that is transferred by the plasma membrane localized PEN3 transporter to the apoplast for preinvasive defence. In the case of PEN3-deficiency, this bioactive compound is converted to the 4-OGlc-indol-3-ylformamide. Unknown biosynthetic steps linking PEN2 product, PEN3 substrate and 4-OGlc-indol-3-ylformamide are indicated with *dashed lines*.

pathway is crucial for preinvasive resistance in *A. thaliana* (Lipka et al., 2005). Apart from IG hydrolysis, pathogen recognition activates CYP81F2 monooxygenase, which substitutes indol-3-ylmethyl glucosinolate (I3G) at the C-4 position in the indole ring to yield 4-hydroxy-I3G, and this compound is subsequently converted by respective *O*-methyltransferases to 4-methoxy-I3G (4MI3G; Fig. 1) (Bednarek et al., 2009; Pfalz et al., 2011; Pfalz, Vogel, & Kroymann, 2009). Together with in vitro enzymatic tests, the metabolic profiles of pathogen-inoculated *A. thaliana* WT and *cyp81F2* leaves indicated that, apart from I3G, PEN2 myrosinase mediates also hydrolyses IGs with modifications on the indole ring (Bednarek et al., 2009). As both *pen2* and *cyp81F2* single mutant plants revealed similar defects in their preinvasive resistance and the infection phenotype of the *pen2 cyp81F2* double knockout resembled that of the single mutants, it was assumed that the bioactive end products crucial for penetration restriction are produced in *A. thaliana* exclusively through the hydrolysis of IGs substituted at the C-4 position (Bednarek et al., 2009). In addition to these two enzymatic components, the immune function of the IG metabolism pathway is dependent on the pleiotropic drug resistance-type ATP-binding cassette transporter PEN3/PDR8/ABCG36 (Stein et al., 2006). Under attempted penetration, this plasma membrane protein is relocated to the exact penetration site, where it was proposed to translocate the bioactive end product(s) of the CYP81F2/PEN2 pathway to the extracellular space to terminate invasive pathogen growth (Fig. 1) (Bednarek et al., 2009; Stein et al., 2006; Underwood & Somerville, 2013).

3.1 Biological Significance

3.1.1 Immunity Against Filamentous Pathogens

The defence pathway involving PEN2 myrosinase was first identified as an element of the preinvasive resistance of *A. thaliana* for the nonadapted powdery mildews *B. graminis f.* sp. *hordei* and *E. pisi* (Bednarek et al., 2009; Lipka et al., 2005). Further studies demonstrated that the same enzymatic components, together with the respective end products, are essential to the effective defence response to a wide range of fungal and oomycete pathogens (Table 1). For instance, the enhanced colonization of *pen2* mesophyll by the Asian soybean rust agent *Phakopsora pachyrhizi* indicated that IG metabolism can control the invasive growth of other biotrophic fungal pathogens in *A. thaliana* (Langenbach, Campe, Schaffrath, Goellner, & Conrath, 2013; Loehrer, Langenbach, Goellner, Conrath, & Schaffrath, 2008). The nonadapted hemibiotrophic oomycete *Phytophthora infestans* also revealed frequent initiation of invasive growth on the *pen2* single knockout line, which resulted in a localized plant cell death response (Lipka et al., 2005). Similarly, the enhanced penetration of *pen2* and *cyp81F2* epidermal cells was observed for an isolate of a closely related species, *Phytophthora brassicae*, which potently colonizes *A. thaliana* (Schlaeppi, Abou-Mansour, Buchala, & Mauch, 2010). Both mutant lines revealed also stronger disease symptoms in response to inoculation with *Phytophthora capsici* compared to WT plants; however, in this case, the entry rates were not evaluated (Wang, Bouwmeester, van de Mortel, Shan, & Govers, 2013).

In accordance with the results obtained for *Phytophthora* spp., the rates of entry into epidermal cells of *pen2* and *cyp81F2* mutant plants were enhanced for two *Colletotrichum* spp., *Colletotrichum gloeosporioides* and *C. orbiculare*, which are fungal hemibiotrophic pathogens nonadapted to *A. thaliana* (Hiruma et al., 2010, 2013). Similarly, rice blast fungus, *Magnaporthe oryzae*, is capable of penetrating the epidermal cells of *pen2* mutants but not WT plants (Maeda et al., 2009; Nakao, Nakamura, Kita, Inukai,

& Ishikawa, 2011). In addition, experiments with the hemibiotrophic cabbage blackleg disease agent *Leptosphaeria maculans* indicated that PEN2 myrosinase can also control penetration through the stomata aperture (Elliott, Harjono, & Howlett, 2008). Importantly, the function of PEN2-mediated IG metabolism in controlling plant pathogens is not limited to biotrophic and hemibiotrophic microorganisms. Assays with the necrotrophic ascomycete *P. cucumerina* revealed the enhanced susceptibility of the *pen2* and *cyp81F2* mutant lines towards both adapted and nonadapted strains of this pathogenic fungus (Bednarek et al., 2009; Lipka et al., 2005; Sanchez-Vallet et al., 2010). Compared to WT plants, both mutants showed increased disease symptoms and supported enhanced hyphal growth, which lead to increased pathogen biomass. The nonadapted strain of the necrotrophic fungus *Alternaria alternata* efficiently penetrated and triggered cell death in *pen2* epidermal cells, which was not the case for the WT plants (Egusa, Miwa, Kaminaka, Takano, & Kodama, 2013). Likewise, the selected isolates of *B. cinerea*, upon inoculation, formed necrotic lesions on the leaves of *pen2* and *cyp81F2* knockout plants that were relatively larger than those observed on WT plants (Buxdorf et al., 2013).

3.1.2 Control of Colonization With Endophytic and Symbiotic Fungi

Interestingly, apart from the interactions with filamentous pathogens, PEN2 myrosinase-mediated IG metabolism was also shown to contribute to the growth control of plant-colonizing microorganisms that are not necessarily pathogenic. The sebacinoid endophytic fungus *Piriformospora indica* and the orchid mycorrhizal fungus *Sebacina vermifera* are considered nonobligate root symbionts that establish beneficial interactions with different hosts, including the model plant *A. thaliana*. Both fungi induce root modification and enhance shoot growth, which may result in increased biomass production. The colonization of *A. thaliana* roots by *P. indica* was shown to be controlled by PEN2 activity, as indicated by elevated fungal biomass in the roots of *pen2* knockout plants (Jacobs et al., 2011). According to these results, later studies including an IG-deficient *myb34 myb51 myb122* line revealed the enhanced colonization of this triple mutant by *P. indica* and *S. vermifera* (Lahrmann et al., 2015). However, reported in the same study, the WT-like phenotype of the *cyp81F2* mutant suggested that opposite to the PEN2-mediated immunity towards leaf-infecting pathogens, the function of the PEN2 pathway in the restriction of root colonization of sebacinoid fungi does not require CYP81F2 activity.

A recent study on microorganisms that colonize *A. thaliana* roots in natural soils identified *Colletotrichum tofieldiae* as an endophytic fungus that promotes plant growth under limited phosphate supply (Hiruma et al., 2016). Interestingly, the growth promotion mediated by this endophyte was abolished in *pen2* and *myb34 myb51 myb122* mutant lines. However, opposite to *P. indica* and *S. vermifera*, the biomass of *C. tofieldiae* in roots of WT and *pen2* plants did not differ significantly. This suggested that the fungal biomass assay was not sensitive enough to detect slightly elevated fungal growth or that PEN2 and IGs may have a positive role in the promotion of plant growth (Hiruma et al., 2016). In any case, the results obtained with *P. indica*, *S. vermifera* and *C. tofieldiae* clearly indicated that the functions of PEN2 myrosinase and IGs go beyond interactions with plant pathogens.

3.2 Myrosinases Responsible for Indole Glucosinolate Metabolism During Immune Responses

Despite the proven role of IG metabolism in *A. thaliana* defence responses, it remains unclear if myrosinases other than PEN2 can support the immune function of IGs. For instance, published results indicate that the TGG1 and TGG2 enzymes are capable of converting IGs (Barth & Jander, 2006). However, despite this substrate adaptation, *tgg1 tgg2* double knockout plants have only slightly elevated susceptibility for the *B. cinerea B05.10* strain, which causes clearly enhanced disease symptoms in *pen2* and *cyp81F2* mutants (Buxdorf et al., 2013). This suggests that TGG1 and TGG2 rather do not act in parallel with PEN2 in innate immunity. In accordance with this finding, recently, IG-deficient *myb34 myb51* and *myb34 myb51 myb122* mutants have been found to be similarly susceptible towards adapted and nonadapted strains of *P. cucumerina* as *pen2* plants (Frerigmann et al., 2016).

These results indicated that PEN2 is the major, or even the only, leaf myrosinase that can activate IGs for immune responses. It seems likely that this exclusive function of PEN2 in immunity results not from its unique substrate specificity but rather from the subcellular localization of this enzyme, which, upon pathogen recognition, is recruited to the attempted entry sites (Fuchs et al., 2016; Lipka et al., 2005). However, it should be noted that most of the experimental data on the immune function of IGs were obtained in leaves. Therefore, it cannot be excluded that additional enzymes may metabolize IGs during defence responses in other plant organs. In accordance with this suspicion, the colonization of *A. thaliana* roots by the soil-born fungal pathogen *Verticillium longisporum* triggered a high expression of *PEN2-LIKE1*, the closest *PEN2* homologue (Iven et al., 2012).

In addition, the roots of the *pyk10* mutants, which lack another root-specific homologue of PEN2, were colonized with higher efficiency than WT roots by the endophytic fungus *P. indica* (Sherameti et al., 2008). These results suggest that in opposite to leaves, in the roots other myrosinases may work together with PEN2 in the immune function of IGs.

3.3 Biochemical Pathway for Pathogen-Triggered Indole Glucosinolate Metabolism

As indicated by the metabolic profiles of pathogen-inoculated WT and *pen2* leaves, PEN2-mediated hydrolysis of I3G leads to the formation of at least two end products, raphanusamic acid (RA) and indol-3-ylmethyl amine (I3A; Fig. 1) (Bednarek et al., 2009; Hiruma et al., 2013; Sanchez-Vallet et al., 2010). Notably, these two metabolites do not fit into the compound classes reported as the products of glucosinolate metabolism formed upon cell disruption (Bones & Rossiter, 2006). This suggests that additional enzymatic or metabolic components contribute to glucosinolate metabolism in intact cells during preinvasive immune responses. In accordance with this assumption, two independent mutant alleles, *pad2* and *cad2*, partially impaired in glutathione biosynthesis (Parisy et al., 2007) accumulated reduced, as compared to WT plants, amounts of RA and I3A after pathogen inoculation, suggesting that the formation of both products is glutathione dependent (Bednarek et al., 2009). This observation fits with the RA chemical structure and suggests that glutathione acts as a cysteine donor during pathogen-triggered IG metabolism. Based on these findings, it was hypothesized that indol-3-ylmethyl ITC generated after pathogen-triggered I3G hydrolysis is immediately conjugated to glutathione, leading to the formation of an intermediate dithiocarbamate adduct that is metabolized to an S-cysteinyl isothiocyanate conjugate, which finally releases RA and I3A (Fig. 2). However, as none of the respective intermediates and enzymatic components have been identified, this pathway is hypothetical.

Despite the production of I3A, the increased accumulation of 4-OH-I3A or 4-methoxy-I3A upon pathogen challenge was not reported for *A. thaliana*, suggesting that the PEN2-mediated metabolism of IGs with substitutions at the position C-4 of the indole ring leads to other types of end products. According to this assumption, 4-O-β-D-glucosyl-indol-3-ylformamid (4OGlcI3F) was identified as a metabolite whose biosynthesis is induced with infection and dependent on the activity of CYP81F2 and PEN2 (Fig. 2). In addition, upon inoculation with *B. graminis* spores, 4OGlcI3F hyperaccumulated, as compared to WT plants, in several *pen3* alleles,

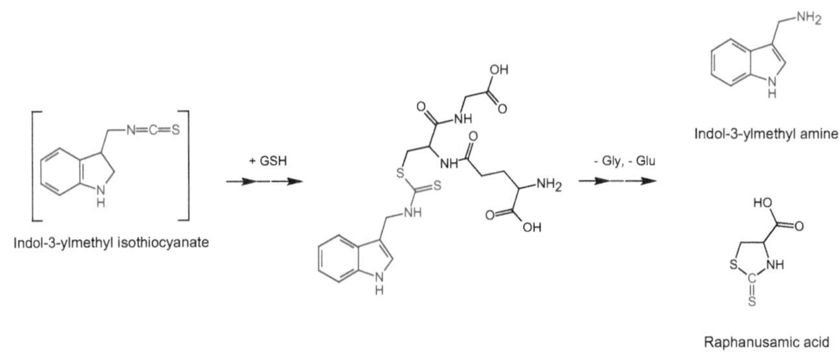

Figure 2 ***Putative contribution of glutathione (GSH) to the pathogen triggered IG metabolism in*** **Arabidopsis thaliana.** Indol-3-ylmethyl isothiocyanate generated after PEN2-dependent indol-3-ylmethyl glucosinolate hydrolysis is conjugated to GSH and forms dithiocarbamate adduct. This product is further processed to releases indol-3-ylmethyl amine and raphanusamic acid.

including the *pen3-5* mutant that uncouples the contribution of PEN3 to IG metabolism from other functions of this ABC transporter (Lu et al., 2015). This observed hyperaccumulation suggests that in WT plants, 4OGlcI3F is translocated to the apoplast by the PEN3 transporter upon pathogen challenge. Alternatively, as glycosylation is usually considered as a means to deactivate phytochemicals (Morant et al., 2008), 4OGlcI3F could be a disarmed derivative of the biologically active PEN3 substrate. Interestingly, a recent study showed that PEN2-mediated IG hydrolysis may lead to the generation of the thiocyanate ion that is released out of the cell in a PEN3-dependent manner (Xu et al., 2016). This ion could be generated as a consequence of the breakdown of the unstable indolic isothiocyanates. However, the precise mechanism underlying generation of the thiocyanate ion during interactions of *A. thaliana* with fungal pathogens is not known.

3.4 Microbe-Triggered Indole Glucosinolate Metabolism in Other Brassicaceae Species

The function of IG metabolism in plant immunity has been proven in *A. thaliana,* and it remains an open question to what extent the PEN2 pathway is conserved in other Brassicaceae species. Notably, a study with several species representing this plant family indicated that the PEN2 protein seems to be conserved in the Brassicaceae family (Bednarek et al., 2011). Moreover, most of the tested species enhanced the production of 4MI3G in response to pathogen inoculation, suggesting that CYP81F2 monooxygenase constitutes an important immunity component in species representing this family. The only exemption from this rule was *Capsella rubella*, which is not only IG-deficient but also lacks orthologues of *PEN2* and *CYP81F2* in its genome (Bednarek et al., 2011). Interestingly, this particular species was found to be colonized to a higher extent and ultimately killed by the same *C. tofieldiae* strain that promotes growth in *A. thaliana* and *Cardamine hirsuta*, another closely related Brassicaceae species (Hiruma et al., 2016). These findings suggest that in Brassicaceae, the PEN2 pathway for IG metabolism plays a key role in controlling plant colonization with different types of filamentous microorganisms.

4. FUNCTIONAL AND BIOSYNTHETIC RELATIONSHIPS BETWEEN INDOLE GLUCOSINOLATES AND BRASSICACEAE PHYTOALEXINS

In addition to the activation of the IG metabolic pathway, in the Brassicaceae species, the recognition of potential pathogens triggers the biosynthesis of sulphur-containing indole-type phytoalexins, whose biosynthesis, at least in some cases, is linked with the IG pathway (Bednarek, 2012b; Pedras et al., 2011). Despite the identification of many phytoalexin structures, their precursors, along with the intermediates and enzymatic components of the respective biosynthetic pathways, in most cases, remain elusive.

4.1 Camalexin

Among the exceptions of this rule is camalexin (Fig. 3), which is synthesized in *A. thaliana* and species belonging to the Camelinae tribe (Bednarek et al., 2011). Although camalexin biosynthesis has not yet been fully deciphered, the main components of the respective metabolic pathway in *A. thaliana* have been characterized using genetic tools and mutant lines available for this model plant (Böttcher et al., 2009; Glawischnig, Hansen, Olsen, & Halkier, 2004; Nafisi et al., 2007; Zhou, Tootle, & Glazebrook, 1999). These studies indicated that similar to IGs, camalexin is synthesized from tryptophan using indole-3-acetaldoxime (IAOx), which is the key branching point between these two metabolic pathways (Glawischnig et al., 2004). Isolated in a genetic screen, a *pad3* (*phytoalexin deficient 3*) mutant, which is defective in camalexin production, displayed increased susceptibility to a wide range of microbial pathogens, providing evidence for camalexin function in *A. thaliana* immunity (Thomma, Nelissen, Eggermont, &

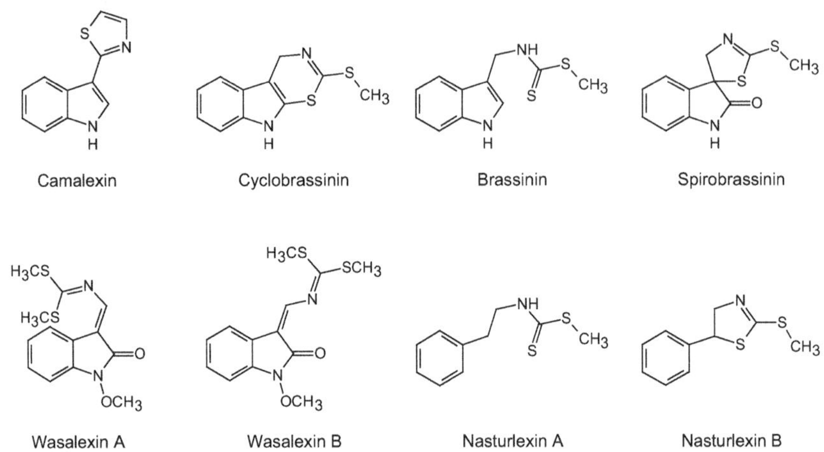

Figure 3 Chemical structures of selected Brassicaceae phytoalexins.

Broekaert, 1999). Interestingly, the infection phenotypes of *pen2* and *pad3* single and double mutant lines indicated that despite a common amino acid precursor and a shared initial step of biosynthesis, IGs and camalexin are crucial for plant immunity at different stages of pathogen development. According to the obtained results, the products of IG metabolism play a role in controlling the entry of microbial pathogens into epidermal cells, whereas camalexin restricts subsequent pathogen development and hyphae spreading to the cells surrounding the infection site. For example, the nonadapted powdery mildew *E. pisi* penetrates the epidermal cells of *pen2* single and *pen2 pad3* double mutants with similar efficiency; however, the hyphae growth evidently increased in the *pen2 pad3* leaves compared to that of the *pen2* single knockout (Bednarek et al., 2009). Additionally, in the case of *P. brassicae*, the simultaneous deactivation of the PEN2 metabolic pathway and camalexin biosynthesis did not escalate entry rates, but it lead to more severe disease symptoms than those observed in both *pen2* and *pen3* single mutants (Schlaeppi et al., 2010). Thus, IGs and camalexin are thought to have complementary but autonomous functions during the infection time course, contributing to the pre- and postinvasive plant resistance, respectively (Bednarek et al., 2009; Lipka et al., 2005). In addition to camalexin, in response to infection, *A. thaliana* produces additional IAOx-derived compounds, including indole-3-carboxylic acids and indole-3-carbonyl nitriles (Bednarek, Schneider, Svatos, Oldham, & Hahlbrock, 2005; Böttcher et al., 2014; Hagemeier, Schneider, Oldham, & Hahlbrock, 2001; Rajniak, Barco, Clay, & Sattely, 2015). As indicated by the severe infection phenotypes of *cyp79B2 cyp79B3* mutants, which are depleted from all IAOx-derived metabolites, these compounds may function together with camalexin in the postinvasive immunity against necrotrophic and hemibiotrophic filamentous pathogens (Hiruma et al., 2013; Sanchez-Vallet et al., 2010; Schlaeppi et al., 2010). Similarly, the beneficial endophytic fungi *P. indica*, *S. vermifera* and *C. tofieldiae* colonized *cyp79B2 cyp79B3* plants more efficiently than *pen2* or *myb34 myb51 myb121* mutants, causing clear disease symptoms or even killing the plants (Hiruma et al., 2016; Lahrmann et al., 2015). This indicates that IGs, along with other IAOx-derived compounds, are critical for maintaining the proper interaction between *A. thaliana* and endophytic fungi.

4.2 Brassinin and Related Phytoalexins

In addition to camalexin, some insights have been obtained into the biosynthesis of brassinin and biosynthetically linked compounds (Fig. 3). For instance, a recent study identified two P450 monooxygenases that are involved in the *S*-heterocyclizations of brassinin into cyclobrassinin and spirobrassinin in *Brassica rapa*. Interestingly, feeding the plants with labelled precursors indicated that some of the brassinin-type phytoalexins may be more tightly linked to the biosynthetic pathway of IGs than camalexin, which branches at IAOx. This biosynthetic link was first proposed based on experiments, which indicated that the C-2 carbon of the labelled tryptophan was retained as the iminocarbon of spirobrassinin (Monde, Takasugi, & Ohnishi, 1994). As a similar rearrangement occurs during the spontaneous generation of ITCs released during myrosinase-mediated glucosinolate hydrolysis, it was assumed that brassinin and its derivatives can be synthesized from IGs via ITCs (Fig. 4A). Because of the high chemical instability of indolyl ITCs, this hypothetical biosynthetic step could not be directly confirmed by tracking labelled precursors. Nevertheless, further feeding experiments supported the existence of biosynthetic interconnections between IGs and indole-type phytoalexins. Exogenously applied labelled I3G was shown to be incorporated into wasalexins A and B (Fig. 3) in *Thellungiella salsuginea*, confirming the biosynthetic link between IGs and these two metabolites (Pedras, Yaya, & Hossain, 2010). As wasalexins belong to a class of compounds biosynthetically linked with brassinin, it appears possible that I3G may also be the precursor of many other sulphur-containing phytoalexins (Bednarek, 2012b). Recently, some Brassicaceae species, including *Nasturtium officinale*, *Barbarea vulgaris* and *Barbarea verna*, have been reported to produce a series of related phenyl-containing phytoalexins named

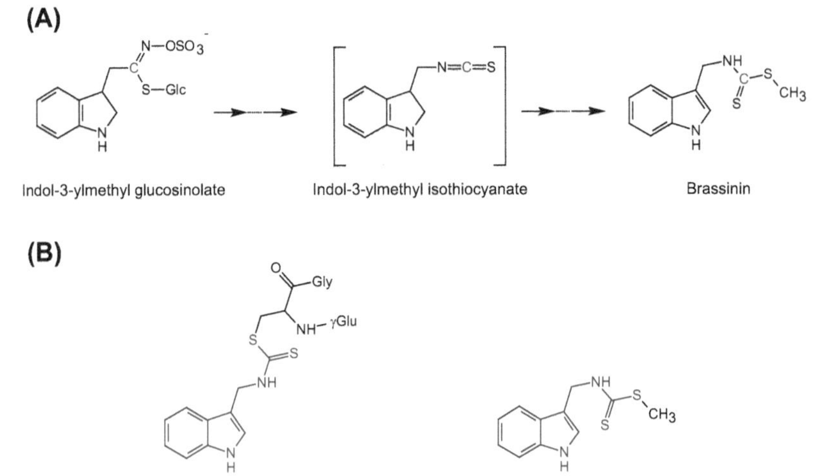

Figure 4 ***The biosynthetic link between IG metabolism and indole-type Brassicaceae phytoalexins.*** (A) Retention of the C-2 carbon (indicated in *red*) as the iminocarbon of spirobrassinin supports biosynthetic link between indol-3-ylmethyl glucosinolate, brassinin and other related Brassicaceae phytoalexins (Monde et al., 1994). (B) The structural similarity between the dithiocarbamate adduct (left) and brassinin (right).

nasturlexins (Fig. 3) (Pedras, Alavi, & To, 2015). Interestingly, despite differences in the core structure (phenyl vs. indole), the side chains of nasturlexins are analogous to those of the brassinin-type phytoalexins. Because of these structural features, it was proposed that nasturlexins, similarly to brassinin-type phytoalexins, are derived from benzyl glucosinolates via the respective ITCs (Pedras & To, 2015).

4.3 PEN2 Pathway and Brassicaceae Phytoalexins

Concerning the biosynthetic origin of Brassicaceae phytoalexins, it should be mentioned that a fragment of the chemical structure of the hypothetical dithiocarbamate adduct, which is thought to be generated as a consequence of the PEN2-mediated hydrolysis of IGs, is identical to the chemical structure of brassinin (Fig. 4B) (Bednarek et al., 2009). As the PEN2 protein is conserved across the Brassicaceae family, it is possible that brassinin and the related compounds originate from IGs via the PEN2 pathway (Bednarek et al., 2011). Similarly, the formation of nasturlexins could be linked to the activity of a PEN2-like myrosinase that converts benzyl glucosinolates to their respective ITCs. Evolutionarily, glucosinolates present in the order Brassicales seem to be more ancient than the indole-type or phenyl-containing phytoalexins that have only been identified in Brassicaceae species so far. It is likely that the PEN2 metabolic pathway, which probably evolved after the glucosinolate innovation, was later acquired for phytoalexin production, thus increasing the biochemical diversity of the secondary metabolites in this plant family (Bednarek, 2012b).

5. MODE OF ACTION OF GLUCOSINOLATE-DERIVED PRODUCTS

The significant contribution of glucosinolates to plant immunity is strongly supported by experimental evidence. Recent years have brought significant progress in the characterization of their respective metabolic pathways, including the identification of their enzymatic components and their intermediate or end products. However, little is known regarding the principles underlying the mechanisms of their action that ultimately result in the restriction of pathogen development.

5.1 Direct Antimicrobial Activity

Among the compound classes that could be formed during myrosinase-catalysed glucosinolate degradation, ITCs possess the highest chemical reactivity and may affect the biological function of various proteins, including enzymes, by modifying cysteine, selenocysteine or amine residues. *In vitro* and in planta bioassays addressing the effect of several glucosinolate-derived ITCs, including allyl-, benzyl- and 4-methylsulfinylbutyl ITC, confirmed the antimicrobial activity of these compounds for a wide range of bacterial and fungal plant pathogens (Sellam, Iacomi-Vasilescu, Hudhomme, & Simoneau, 2007; Tierens et al., 2001). In addition to aliphatic ITCs, the brassinin-type IG-derived phytoalexins possess clear antimicrobial activity (Pedras et al., 2011). Collectively, these observations suggest that at least some products of glucosinolate metabolism may contribute to plant immunity as in planta antibiotics that directly restrict pathogen growth (Fig. 5).

5.2 Impact on Conserved Plant Immune Responses

5.2.1 Stomatal Closure

Interestingly, recent studies have indicated that glucosinolate hydrolytic products may affect pathogenic microbes through their chemical properties; however, alternatively, they may be also involved in controlling immune responses that are highly evolutionarily conserved in the plant kingdom (Bednarek, 2012a; Maag, Erb, Köllner, & Gershenzon, 2015). For instance,

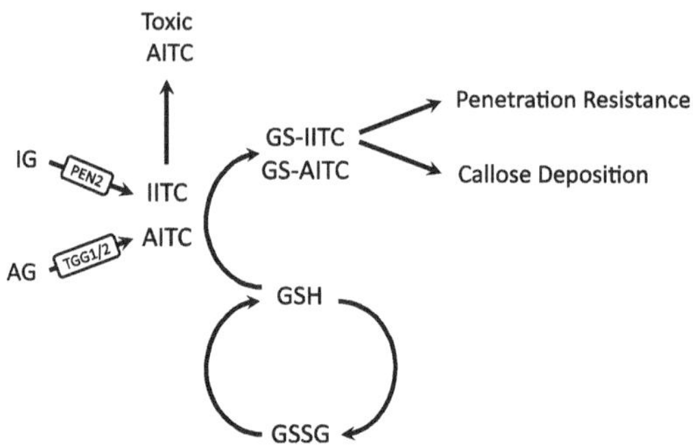

Figure 5 ***Joint contribution of isothiocyanates (ITCs) and glutathione (GSH) to different immune responses in*** **Arabidopsis thaliana.** Aliphatic glucosinolates (AGs) are hydrolysed by TGG1 and TGG2 myrosinases isoforms to toxic aliphatic ITCs (AITCs), which can directly restrict pathogen development. Indole glucosinolates (IGs) are hydrolysed by PEN2 myrosinase to highly unstable indolic ITCs (IITCs). Conjugation of IITCs and AITCs with GSH lead to the formation of GS-IITCs and GS-AITCs, respectively. Downstream product(s) of GS-IITC metabolism control(s) penetration resistance and act(s) as a signal in callose deposition. Generation of ITC-GS alters the ratio between reduced (GSH) and oxidized glutathione (GSSG) in favour of GSSG. This in turn increases levels of reactive oxygen species that can stimulate stomatal closure and initiate programmed cell death.

the TGG1 myrosinase was identified as one of the most abundant proteins in *A. thaliana* guard cells (Zhao, Zhang, Stanley, & Assmann, 2008). In accordance with this finding, the FAMA transcription factor has been shown recently to control differentiation of guard cells and myrosin cells in *A. thaliana* (Li & Sack, 2014; Shirakawa et al., 2014). These findings pointed to a link between glucosinolate metabolism and stomatal closure in response to attempted invasion. As many pathogenic microorganisms base their infection strategies on these natural openings that are present on the leaf surface, stomatal closure in response to microbial colonization is considered as an immune mechanism conserved in the plant kingdom (Zeng et al., 2010). The detailed analysis of stomatal closure in *A. thaliana* single *tgg1* and double *tgg1 tgg2* knockouts indicated that both enzymes may contribute to guard cell function. The single *tgg1* mutant is defective in wound-induced stomatal closure and has reduced sensitivity to ABA inhibition of stomatal opening (Zhao et al., 2008), while the *tgg1 tgg2* double mutant displayed impaired ABA-induced stomatal closure (Islam et al., 2009). In accordance with the phenotypes of the *tgg* mutants, several glucosinolate-derived ITCs have been shown to induce stomatal closure in *A. thaliana* in a dose-dependent manner (Hossain et al., 2013; Khokon et al., 2011). Taken together, these results indicate interconnections among myrosinase activity, ITCs and the regulation of stomatal closure (Fig. 5). This function of ITCs may be linked to their ability to deplete reduced glutathione, which was shown to regulate reactive oxygen species–mediated ABA signalling in guard cells (Fig. 5) (Munemasa et al., 2013).

5.2.2 Programmed Cell Death

Apart from the defects in guard cell function, the *tgg1 tgg2* mutants and AG-deficient *myb28 myb29* plants had reduced or delayed programmed cell death (PCD) upon inoculation with incompatible strains of the bacterial pathogen *P. syringae* and the downy mildew *Hyaloperonospora arabidopsidis* (Andersson et al., 2015). As PCD is a component of the effector-triggered HR, these results implicated the involvement of AG-derived ITCs in the effector-triggered HR, which is a conserved plant immune mechanism (Andersson et al., 2015). Similar to stomatal closure, the results of the conducted experiments suggested that the function of ITCs in PCD likely results from their ability to decrease the pool of reduced glutathione (Fig. 5), which has been reported to affect many cellular processes, including cell death (Andersson et al., 2015; Noctor, Queval, Mhamdi, Chaouch, & Foyer, 2011). Interestingly, *pen2* and *pen3* mutants also had HR-associated cell death in response to *P. syringae* and *H. arabidopsidis* infection (Johansson et al., 2014). As glutathione is one of the substrates in the PEN2-metabolic pathway, it seems likely that the decreased PCD in *pen2* plants results from elevated levels of reduced glutathione at the interaction sites (Fig. 5).

5.2.3 Callose Deposition

Another conserved defence response in the plant kingdom that can be controlled by glucosinolate metabolism is callose deposition, which was proposed to function as a physical barrier against pathogens in the cell wall (Aist & Bushnell, 1991). The microscopic observations of callose deposits in *A. thaliana* mutant seedlings impaired in IG biosynthesis and metabolism revealed defects in the deposition of this β-1,3-glucan polymer in response to flg22, a synthetic 22-amino acid polypeptide that corresponds to a highly conserved region of eubacterial flagellin (Clay, Adio, Denoux, Jander, & Ausubel, 2009). The exogenous application of 4MI3G restored flg22-triggered callose deposition in the *cyp81F2* mutant but not in the *pen2* loss-of-function mutant seedlings defective in the pathogen-triggered IG hydrolysis. These results indicate that the compounds generated by the CYP81F2/PEN2 pathway for IG metabolism may act as signalling molecules that trigger callose deposition (Fig. 5) (Clay et al., 2009). Similarly, CYP81F2 and PEN2 have been found to be required for callose deposition in *A. thaliana* roots (Millet et al., 2010). The proposed contribution of IG-derived products in signalling processes was further addressed by experiments performed with other MAMPs, such as chitin, a major component of the fungal cell wall (Millet et al., 2010), and its derivative chitosan (Luna et al., 2011). The callose depositions in the roots of *pen2* and *cyp81F2* mutants are impaired after chitin elicitation (Millet et al., 2010). However, after chitosan treatment, the cotyledons of *pen2* seedlings showed callose levels similar to those observed in WT plants (Luna et al., 2011).

5.2.4 Biosynthesis of Tryptophan-Derived Metabolites

Interestingly, the putative signalling function of IG-metabolism products is likely not only restricted to callose deposition. Recently, *myb34 myb51 myb122* and *pen2* mutants have been reported to be defective in the flg22-and *P. syringae*-triggered activation of indole-3-carboxylic acid and camalexin biosynthesis (Frerigmann et al., 2016; Stahl et al., 2016). As the biosynthetic pathways leading to these tryptophan derivatives neither use IGs as precursors nor involve PEN2 activity, these results suggest that IG metabolic products may contribute to the MAMP-mediated induction of biosynthetic pathways leading to indole-3-carboxylic acids and camalexin. However, in this case, the observed phenotype was not dependent on CYP81F2 monooxygenase, suggesting that, in contrast to callose deposition, a substitution at the C-4 position of the indole ring is not required for IGs to control the expression of genes linked to tryptophan metabolism (Stahl et al., 2016). In addition, *myb34 myb51 myb122* and *pen2* plants produced WT-like amounts of indole-3-carboxylic acids and camalexin upon inoculation with *P. cucumerina* (Frerigmann et al., 2016). This indicated that the requirement of IG-derived products for the activation of tryptophan metabolism can be overridden by signalling pathways triggered by fungal pathogens.

6. CONCLUDING REMARKS

Experimental results published during the last decade vastly improved our knowledge of the immune function of glucosinolates. Currently, it seems that out of the two classes of glucosinolates present in *A. thaliana*, IGs play a more important role than AGs in plant immunity. However, it should be noted that the function of IGs has been investigated in a greater detail than the contribution of AGs (Table 1). Therefore, it seems possible that the immune role of AGs has been underestimated because of insufficient experimental evidence. A more detailed analysis of the respective mutant lines is still necessary to address this point. It should also be noted that the immune function of glucosinolates has been mainly addressed in leaves and much less is known regarding their function in roots. In this context, of particular interest are the recent results obtained with endophytic fungi indicating that the immune role of IGs goes beyond interactions with pathogenic microorganisms. Finally, the metabolic interconnections of glucosinolates and other defensive metabolites, together with the precise molecular mechanisms of action of glucosinolate-derived products in immunity, remain elusive and should be examined in greater detail.

ACKNOWLEDGEMENTS

P.B. acknowledges the National Science Centre SONATA BIS grant (2012/07/E/NZ2/04098) and Polish Ministry of Science and Higher Education (KNOW program).

REFERENCES

Ahuja, I., Kissen, R., & Bones, A. M. (2012). Phytoalexins in defense against pathogens. *Trends in Plant Science, 17*(2), 73–90.

Aist, J. R., & Bushnell, W. R. (1991). Invasion of plants by powdery mildew fungi, and cellular mechanisms of resistance. In G. T. Cole, & H. C. Hoch (Eds.), *The fungal spore and disease initiation in plants and animals* (pp. 321–345). Boston, MA: Springer US.

Andersson, M. X., Nilsson, A. K., Johansson, O. N., Boztaş, G., Adolfsson, L. E., Pinosa, F., … Ellerström, M. (2015). Involvement of the electrophilic isothiocyanate sulforaphane in Arabidopsis local defense responses. *Plant Physiology, 167*(1), 251–261.

Barth, C., & Jander, G. (2006). Arabidopsis myrosinases TGG1 and TGG2 have redundant function in glucosinolate breakdown and insect defense. *Plant Journal, 46*(4), 549–562.

Bednarek, P. (2012a). Chemical warfare or modulators of defence responses - the function of secondary metabolites in plant immunity. *Current Opinion in Plant Biology, 15*(4), 407–414.

Bednarek, P. (2012b). Sulfur-containing secondary metabolites from *Arabidopsis thaliana* and other Brassicaceae with function in plant immunity. *ChemBioChem: a European Journal of Chemical Biology, 13*(13), 1846–1859.

Bednarek, P., Pislewska-Bednarek, M., Svatos, A., Schneider, B., Doubsky, J., Mansurova, M., … Schulze-Lefert, P. (2009). A glucosinolate metabolism pathway in living plant cells mediates broad-spectrum antifungal defense. *Science, 323*(5910), 101–106.

Bednarek, P., Pislewska-Bednarek, M., Ver Loren van Themaat, E., Maddula, R. K., Svatos, A., & Schulze-Lefert, P. (2011). Conservation and clade-specific diversification of pathogen-inducible tryptophan and indole glucosinolate metabolism in *Arabidopsis thaliana* relatives. *New Phytologist, 192*(3), 713–726.

Bednarek, P., Schneider, B., Svatos, A., Oldham, N. J., & Hahlbrock, K. (2005). Structural complexity, differential response to infection, and tissue specificity of indolic and phenylpropanoid secondary metabolism in Arabidopsis roots. *Plant Physiology, 138*(2), 1058–1070.

Bones, A. M., & Rossiter, J. T. (2006). The enzymic and chemically induced decomposition of glucosinolates. *Phytochemistry, 67*(11), 1053–1067.

Böttcher, C., Chapman, A., Fellermeier, F., Choudhary, M., Scheel, D., & Glawischnig, E. (2014). The biosynthetic pathway of indole-3-carbaldehyde and indole-3-carboxylic acid derivatives in Arabidopsis. *Plant Physiology, 165*(2), 841–853.

Böttcher, C., Westphal, L., Schmotz, C., Prade, E., Scheel, D., & Glawischnig, E. (2009). The multifunctional enzyme CYP71B15 (PHYTOALEXIN DEFICIENT3) converts cysteine-indole-3-acetonitrile to camalexin in the indole-3-acetonitrile metabolic network of *Arabidopsis thaliana*. *The Plant Cell, 21*(6), 1830–1845.

Buxdorf, K., Yaffe, H., Barda, O., & Levy, M. (2013). The effects of glucosinolates and their breakdown products on necrotrophic fungi. *PLoS One, 8*(8), e70771.

Calmes, B., N'Guyen, G., Dumur, J., Brisach, C. A., Campion, C., Iacomi, B., … Simoneau, P. (2015). Glucosinolate-derived isothiocyanates impact mitochondrial function in fungal cells and elicit an oxidative stress response necessary for growth recovery. *Frontiers in Plant Science, 6*, 14.

Chisholm, S. T., Coaker, G., Day, B., & Staskawicz, B. J. (2006). Host-microbe interactions: shaping the evolution of the plant immune response. *Cell, 124*(4), 803–814.

Clay, N. K., Adio, A. M., Denoux, C., Jander, G., & Ausubel, F. M. (2009). Glucosinolate metabolites required for an *Arabidopsis* innate immune response. *Science, 323*(5910), 95–101.

Coll, N. S., Epple, P., & Dangl, J. L. (2011). Programmed cell death in the plant immune system. *Cell Death and Differentiation, 18*(8), 1247–1256.

Cunnac, S., Lindeberg, M., & Collmer, A. (2009). *Pseudomonas syringae* type III secretion system effectors: repertoires in search of functions. *Current Opinion in Microbiology, 12*(1), 53–60.

Dodds, P. N., & Rathjen, J. P. (2010). Plant immunity: towards an integrated view of plant–pathogen interactions. *Nature Reviews Genetics, 11*(8), 539–548.

Dufour, V., Stahl, M., & Baysse, C. (2015). The antibacterial properties of isothiocyanates. *Microbiology, 161*, 229–243.

Egusa, M., Miwa, T., Kaminaka, H., Takano, Y., & Kodama, M. (2013). Nonhost resistance of *Arabidopsis thaliana* against *Alternaria alternata* involves both pre- and postinvasive defenses but is collapsed by AAL-toxin in the absence of LOH2. *Phytopathology, 103*(7), 733–740.

Ellinger, D., & Voigt, C. A. (2014). Callose biosynthesis in arabidopsis with a focus on pathogen response: what we have learned within the last decade. *Annals of Botany, 114*(6), 1349–1358.

Elliott, C. E., Harjono, & Howlett, B. J. (2008). Mutation of a gene in the fungus *Leptosphaeria maculans* allows increased frequency of penetration of stomatal apertures of *Arabidopsis thaliana*. *Molecular Plant, 1*(3), 471–481.

Fan, J., Crooks, C., Creissen, G., Hill, L., Fairhurst, S., Doerner, P., & Lamb, C. (2011). *Pseudomonas sax* genes overcome aliphatic isothiocyanate-mediated non-host resistance in *Arabidopsis*. *Science, 331*(6021), 1185–1188.

Frerigmann, H., Piślewska-Bednarek, M., Sánchez-Vallet, A., Molina, A., Glawischnig, E., Gigolashvili, T., & Bednarek, P. (2016). Regulation of pathogen triggered tryptophan metabolism in *Arabidopsis thaliana* by MYB transcription factors and indole glucosinolate conversion products. *Molecular Plant, 9*(5), 682–695.

Fuchs, R., Kopischke, M., Klapprodt, C., Hause, G., Meyer, A. J., Schwarzländer, M., … Lipca, V. (2016). Immobilized subpopulations of leaf epidermal mitochondria mediate PENETRATION2-Dependent Pathogen Entry Control in Arabidopsis. *The Plant Cell, 28*(1), 130–145.

Glawischnig, E., Hansen, B. G., Olsen, C. E., & Halkier, B. A. (2004). Camalexin is synthesized from indole-3-acetaldoxime, a key branching point between primary and secondary metabolism in *Arabidopsis*. *Proceedings of the National Academy of Sciences of the United States of America, 101*(21), 8245–8250.

Hagemeier, J., Schneider, B., Oldham, N. J., & Hahlbrock, K. (2001). Accumulation of soluble and wall-bound indolic metabolites in *Arabidopsis thaliana* leaves infected with virulent or avirulent *Pseudomonas syringae* pathovar tomato strains. *Proceedings of the National Academy of Sciences of the United States of America, 98*(2), 753–758.

Halkier, B. A., & Gershenzon, J. (2006). Biology and biochemistry of glucosinolates. *Annual Review of Plant Biology, 57*, 303–333.

Haughn, G. W., Davin, L., Giblin, M., & Underhill, E. W. (1991). Biochemical genetics of plant secondary metabolites in *Arabidopsis thaliana* : the Glucosinolates. *Plant Physiology, 97*(1), 217–226.

Heath, M. C. (2000). Hypersensitive response-related death. *Plant Molecular Biology, 44*(3), 321–334.

Hiruma, K., Fukunaga, S., Bednarek, P., Piślewska-Bednarek, M., Watanabe, S., Narusaka, Y., … Takano, Y. (2013). Glutathione and tryptophan metabolism are required for *Arabidopsis* immunity during the hypersensitive response to hemibiotrophs. *Proceedings of the National Academy of Sciences, 110*(23), 9589–9594.

Hiruma, K., Gerlach, N., Sacristán, S., Nakano, R. T., Hacquard, S., Kracher, B., … Schulze-Lefert, P. (2016). Root endophyte *Colletotrichum tofieldiae* confers plant fitness benefits that are phosphate status dependent. *Cell, 165*(2), 464–474.

Hiruma, K., Onozawa-Komori, M., Takahashi, F., Asakura, M., Bednarek, P., Okuno, T., … Takano, Y. (2010). Entry mode-dependent function of an indole glucosinolate pathway in Arabidopsis for nonhost resistance against anthracnose pathogens. *The Plant Cell, 22*(7), 2429–2443.

Horbach, R., Navarro-Quesada, A. R., Knogge, W., & Deising, H. B. (2011). When and how to kill a plant cell: infection strategies of plant pathogenic fungi. *Journal of Plant Physiology, 168*(1), 51–62.

Hossain, M. S., Ye, W., Hossain, M. A., Okuma, E., Uraji, M., Nakamura, Y., … Murata, Y. (2013). Glucosinolate degradation products, isothiocyanates, nitriles, and thiocyanates, induce stomatal closure accompanied by peroxidase-mediated reactive oxygen species production in *Arabidopsis thaliana. Bioscience, Biotechnology, and Biochemistry,* 77(5), 977–983.

Huckelhoven, R., & Panstruga, R. (2011). Cell biology of the plant-powdery mildew interaction. *Current Opinion in Plant Biology, 14*(6), 738–746.

Islam, M. M., Tani, C., Watanabe-Sugimoto, M., Uraji, M., Jahan, M. S., Masuda, C., … Murata, Y. (2009). Myrosinases, TGG1 and TGG2, redundantly function in ABA and MeJA signaling in Arabidopsis guard cells. *Plant & Cell Physiology, 50*(6), 1171–1175.

Iven, T., König, S., Singh, S., Braus-Stromeyer, S. A., Bischoff, M., Tietze, L. F., … Dröge-Laser, W. (2012). Transcriptional activation and production of tryptophan-derived secondary metabolites in *Arabidopsis* roots contributes to the defense against the fungal vascular pathogen *Verticillium longisporum. Molecular Plant, 5*(6), 1389–1402.

Jacobs, S., Zechmann, B., Molitor, A., Trujillo, M., Petutschnig, E., Lipka, V., … Schäfer, P. (2011). Broad-spectrum suppression of innate immunity is required for colonization of Arabidopsis roots by the fungus *Piriformospora indica. Plant Physiology, 156*(2), 726–740.

Johansson, O. N., Fantozzi, E., Fahlberg, P., Nilsson, A. K., Buhot, N., Tor, M., & Andersson, M. X. (2014). Role of the penetration-resistance genes PEN1, PEN2 and PEN3 in the hypersensitive response and race-specific resistance in *Arabidopsis thaliana. Plant Journal, 79*(3), 466–476.

Jones, J. D. G., & Dangl, J. L. (2006). The plant immune system. *Nature, 444*(7117), 323–329.

Khokon, M. A. R., Jahan, M. S., Rahman, T., Hossain, M. A., Muroyama, D., Minami, I., … Murata, Y. (2011). Allyl isothiocyanate (AITC) induces stomatal closure in *Arabidopsis. Plant, Cell & Environment, 34*(11), 1900–1906.

Kliebenstein, D. J., Rowe, H. C., & Denby, K. J. (2005). Secondary metabolites influence Arabidopsis/*Botrytis* interactions: variation in host production and pathogen sensitivity. *Plant Journal, 44*(1), 25–36.

Kroymann, J., Textor, S., Tokuhisa, J. G., Falk, K. L., Bartram, S., Gershenzon, J., & Mitchell-Olds, T. (2001). A gene controlling variation in Arabidopsis glucosinolate composition is part of the methionine chain elongation pathway. *Plant Physiology, 127*(3), 1077–1088.

Lahrmann, U., Strehmel, N., Langen, G., Frerigmann, H., Leson, L., Ding, Y., … Zuccaro, A. (2015). Mutualistic root endophytism is not associated with the reduction of saprotrophic traits and requires a noncompromised plant innate immunity. *New Phytologist, 207*(3), 841–857.

Langenbach, C., Campe, R., Schaffrath, U., Goellner, K., & Conrath, U. (2013). UDP-glucosyltransferase UGT84A2/BRT1 is required for Arabidopsis nonhost resistance to the Asian soybean rust pathogen *Phakopsora pachyrhizi*. *New Phytologist, 198*(2), 536–545.

Li, M., & Sack, F. D. (2014). Myrosin idioblast cell fate and development are regulated by the *Arabidopsis* transcription factor FAMA, the auxin pathway, and vesicular trafficking. *The Plant Cell, 26*(10), 4053–4066.

Lipka, V., Dittgen, J., Bednarek, P., Bhat, R., Wiermer, M., Stein, M., … Schulze-Lefert, P. (2005). Pre- and postinvasion defenses both contribute to nonhost resistance in *Arabidopsis*. *Science, 310*(5751), 1180–1183.

Loehrer, M., Langenbach, C., Goellner, K., Conrath, U., & Schaffrath, U. (2008). Characterization of nonhost resistance of *Arabidopsis* to the Asian soybean rust. *Molecular Plant–Microbe Interactions, 21*(11), 1421–1430.

Lu, X., Dittgen, J., Pislewska-Bednarek, M., Molina, A., Schneider, B., Svatos, A., … Schulze-Lefert, P. (2015). Mutant allele-specific uncoupling of PENETRATION3 functions reveals engagement of the ATP-binding cassette transporter in distinct tryptophan metabolic pathways. *Plant Physiology, 168*(3), 814–827.

Luna, E., Pastor, V., Robert, J., Flors, V., Mauch-Mani, B., & Ton, J. (2011). Callose deposition: a multifaceted plant defense response. *Molecular Plant–Microbe Interactions, 24*(2), 183–193.

Maag, D., Erb, M., Köllner, T. G., & Gershenzon, J. (2015). Defensive weapons and defense signals in plants: some metabolites serve both roles. *BioEssays: News and Reviews in Molecular, Cellular and Developmental Biology, 37*(2), 167–174.

Maeda, K., Houjyou, Y., Komatsu, T., Hori, H., Kodaira, T., & Ishikawa, A. (2009). AGB1 and PMR5 contribute to PEN2-mediated preinvasion resistance to *Magnaporthe oryzae* in *Arabidopsis thaliana*. *Molecular Plant–Microbe Interactions, 22*(11), 1331–1340.

Melotto, M., Underwood, W., Koczan, J., Nomura, K., & He, S. Y. (2006). Plant stomata function in innate immunity against bacterial invasion. *Cell, 126*(5), 969–980.

Millet, Y. A., Danna, C. H., Clay, N. K., Songnuan, W., Simon, M. D., Werck-Reichhart, D., & Ausubel, F. M. (2010). Innate immune responses activated in *Arabidopsis* roots by microbe-associated molecular patterns. *The Plant Cell, 22*(3), 973–990.

Monde, K., Takasugi, M., & Ohnishi, T. (1994). Biosynthesis of cruciferous phytoalexins. *Journal of the American Chemical Society, 116*(15), 6650–6657.

Morant, A. V., Jørgensen, K., Jørgensen, C., Paquette, S. M., Sánchez-Pérez, R., Møller, B. L., & Bak, S. (2008). β-Glucosidases as detonators of plant chemical defense. *Phytochemistry, 69*(9), 1795–1813.

Munemasa, S., Muroyama, D., Nagahashi, H., Nakamura, Y., Mori, I. C., & Murata, Y. (2013). Regulation of reactive oxygen species-mediated abscisic acid signaling in guard cells and drought tolerance by glutathione. *Frontiers in Plant Science, 4*.

Nafisi, M., Goregaoker, S., Botanga, C. J., Glawischnig, E., Olsen, C. E., Halkier, B. A., & Glazebrook, J. (2007). *Arabidopsis* cytochrome P450 monooxygenase 71A13 catalyzes the conversion of indole-3-acetaldoxime in camalexin synthesis. *The Plant Cell, 19*(6), 2039–2052.

Nakao, M., Nakamura, R., Kita, K., Inukai, R., & Ishikawa, A. (2011). Non-host resistance to penetration and hyphal growth of *Magnaporthe oryzae* in *Arabidopsis*. *Scientific Reports, 1*, 171.

Nishimura, M. T., Stein, M., Hou, B. H., Vogel, J. P., Edwards, H., & Somerville, S. C. (2003). Loss of a callose synthase results in salicylic acid-dependent disease resistance. *Science, 301*(5635), 969–972.

Noctor, G., Queval, G., Mhamdi, A., Chaouch, S., & Foyer, C. H. (2011). Glutathione. *The Arabidopsis Book*, 1–32.

Nongbri, P. L., Johnson, J. M., Sherameti, I., Glawischnig, E., Halkier, B. A., & Oelmuller, R. (2012). Indole-3-acetaldoxime-derived compounds restrict root colonization in the beneficial interaction between *Arabidopsis* roots and the endophyte *Piriformospora indica*. *Molecular Plant–Microbe Interactions, 25*(9), 1186–1197.

Parisy, V., Poinssot, B., Owsianowski, L., Buchala, A., Glazebrook, J., & Mauch, F. (2007). Identification of PAD2 as a gamma-glutamylcysteine synthetase highlights the importance of glutathione in disease resistance of Arabidopsis. *Plant Journal, 49*(1), 159–172.

Paxton, J. D. (1981). Phytoalexins — A working redefinition. *Journal of Phytopathology, 101*(2), 106–109.

Pedras, M. S. C., Alavi, M., & To, Q. H. (2015). Expanding the nasturlexin family: nasturlexins C and D and their sulfoxides are phytoalexins of the crucifers *Barbarea vulgaris* and *B. verna*. *Phytochemistry, 118*, 131–138.

Pedras, M. S. C., & To, Q. H. (2015). Non-indolyl cruciferous phytoalexins: nasturlexins and tridentatols, a striking convergent evolution of defenses in terrestrial plants and marine animals? *Phytochemistry, 113*, 57–63.

Pedras, M. S. C., Yaya, E. E., & Glawischnig, E. (2011). The phytoalexins from cultivated and wild crucifers: chemistry and biology. *Natural Product Reports, 28*(8), 1381–1405.

Pedras, M. S. C., Yaya, E. E., & Hossain, S. (2010). Unveiling the phytoalexin biosynthetic puzzle in salt cress: unprecedented incorporation of glucobrassicin into wasalexins A and B. *Organic & Biomolecular Chemistry, 8*(22), 5150–5158.

Pfalz, M., Mikkelsen, M. D., Bednarek, P., Olsen, C. E., Halkier, B. A., & Kroymann, J. (2011). Metabolic engineering in *Nicotiana benthamiana* reveals key enzyme functions in *Arabidopsis* indole glucosinolate modification. *The Plant Cell, 23*(2), 716–729.

Pfalz, M., Vogel, H., & Kroymann, J. (2009). The gene controlling the *indole glucosinolate modifier1* quantitative trait locus alters indole glucosinolate structures and aphid resistance in *Arabidopsis*. *The Plant Cell, 21*(3), 985–999.

Piasecka, A., Jedrzejczak-Rey, N., & Bednarek, P. (2015). Secondary metabolites in plant innate immunity: conserved function of divergent chemicals. *New Phytologist, 206*(3), 948–964.

Rajniak, J., Barco, B., Clay, N. K., & Sattely, E. S. (2015). A new cyanogenic metabolite in *Arabidopsis* required for inducible pathogen defence. *Nature, 525*(7569), 376–379.

Sanchez-Vallet, A., Ramos, B., Bednarek, P., Lopez, G., Pislewska-Bednarek, M., Schulze-Lefert, P., & Molina, A. (2010). Tryptophan-derived secondary metabolites in *Arabidopsis thaliana* confer non-host resistance to necrotrophic *Plectosphaerella cucumerina* fungi. *Plant Journal, 63*(1), 115–127.

Schlaeppi, K., Abou-Mansour, E., Buchala, A., & Mauch, F. (2010). Disease resistance of Arabidopsis to *Phytophthora brassicae* is established by the sequential action of indole glucosinolates and camalexin. *Plant Journal, 62*(5), 840–851.

Sellam, A., Iacomi-Vasilescu, B., Hudhomme, P., & Simoneau, P. (2007). In vitro antifungal activity of brassinin, camalexin and two isothiocyanates against the crucifer pathogens *Alternaria brassicicola* and *Alternaria brassicae*. *Plant Pathology, 56*(2), 296–301.

Sherameti, I., Venus, Y., Drzewiecki, C., Tripathi, S., Dan, V. M., Nitz, I., ... Oelmüller, R. (2008). PYK10, a β-glucosidase located in the endoplasmatic reticulum, is crucial for the

beneficial interaction between *Arabidopsis thaliana* and the endophytic fungus *Piriformospora indica*. *Plant Journal, 54*(3), 428–439.
Shirakawa, M., Ueda, H., Nagano, A. J., Shimada, T., Kohchi, T., & Hara-Nishimura, I. (2014). FAMA is an essential component for the differentiation of two distinct cell types, myrosin cells and guard cells, in *Arabidopsis*. *The Plant Cell, 26*(10), 4039–4052.
Sonderby, I. E., Hansen, B. G., Bjarnholt, N., Ticconi, C., Halkier, B. A., & Kliebenstein, D. J. (2007). A systems biology approach identifies a R2R3 MYB gene subfamily with distinct and overlapping functions in regulation of aliphatic glucosinolates. *PLoS One, 2*(12), 16.
Stahl, E., Bellwon, P., Huber, S., Schlaeppi, K., Bernsdorff, F., Vallat-Michel, A., … Zeier, J. (2016). Regulatory and functional aspects of indolic metabolism in plant systemic acquired resistance. *Molecular Plant, 9*(5), 662–681.
Stein, M., Dittgen, J., Sanchez-Rodriguez, C., Hou, B. H., Molina, A., Schulze-Lefert, P., … Somerville, S. (2006). *Arabidopsis* PEN3/PDR8, an ATP binding cassette transporter, contributes to nonhost resistance to inappropriate pathogens that enter by direct penetration. *The Plant Cell, 18*(3), 731–746.
Stotz, H. U., Sawada, Y., Shimada, Y., Hirai, M. Y., Sasaki, E., Krischke, M., … Kamiya, Y. (2011). Role of camalexin, indole glucosinolates, and side chain modification of glucosinolate-derived isothiocyanates in defense of Arabidopsis against *Sclerotinia sclerotiorum*. *Plant Journal, 67*(1), 81–93.
Thomma, B. P. H. J., Nelissen, I., Eggermont, K., & Broekaert, W. F. (1999). Deficiency in phytoalexin production causes enhanced susceptibility of *Arabidopsis thaliana* to the fungus *Alternaria brassicicola*. *Plant Journal, 19*(2), 163–171.
Tierens, K. F., Thomma, B. P., Brouwer, M., Schmidt, J., Kistner, K., Porzel, A., … Broekaert, W. F. (2001). Study of the role of antimicrobial glucosinolate-derived isothiocyanates in resistance of Arabidopsis to microbial pathogens. *Plant Physiology, 125*(4), 1688–1699.
Underwood, W., & Somerville, S. C. (2013). Perception of conserved pathogen elicitors at the plasma membrane leads to relocalization of the *Arabidopsis* PEN3 transporter. *Proceedings of the National Academy of Sciences of the United States of America, 110*(30), 12492–12497.
VanEtten, H. D., Mansfield, J. W., Bailey, J. A., & Farmer, E. E. (1994). Two classes of plant antibiotics: phytoalexins versus "Phytoanticipins". *The Plant Cell, 6*(9), 1191–1192.
Wang, Y., Bouwmeester, K., van de Mortel, J. E., Shan, W., & Govers, F. (2013). A novel Arabidopsis-oomycete pathosystem: differential interactions with *Phytophthora capsici* reveal a role for camalexin, indole glucosinolates and salicylic acid in defence. *Plant, Cell & Environment, 36*(6), 1192–1203.
Xu, J., Meng, J., Meng, X., Zhao, Y., Liu, J., Sun, T., … Zhang, S. (2016). Pathogen-responsive MPK3 and MPK6 reprogram the biosynthesis of indole glucosinolates and their derivatives in Arabidopsis immunity. *The Plant Cell, 28*(5), 1144–1162.
Zeng, W., Melotto, M., & He, S. Y. (2010). Plant stomata: a checkpoint of host immunity and pathogen virulence. *Current Opinion in Biotechnology, 21*(5), 599–603.
Zhao, Z., Zhang, W., Stanley, B. A., & Assmann, S. M. (2008). Functional proteomics of *Arabidopsis thaliana* guard cells uncovers new stomatal signaling pathways. *The Plant Cell, 20*(12), 3210–3226.
Zhou, N., Tootle, T. L., & Glazebrook, J. (1999). Arabidopsis PAD3, a gene required for camalexin biosynthesis, encodes a putative cytochrome P450 monooxygenase. *The Plant Cell, 11*(12), 2419–2428.

CHAPTER EIGHT

Insect Detoxification of Glucosinolates and Their Hydrolysis Products

V. Jeschke, J. Gershenzon[1], D.G. Vassão[1]
Max Planck Institute for Chemical Ecology, Jena, Germany
[1]Corresponding authors: E-mail: gershenzon@ice.mpg.de; vassao@ice.mpg.de

Contents

Abstract

The glucosinolate–myrosinase system is the main activated chemical defence present among plants of the order Brassicales, which includes agriculturally important crucifer crops such as cabbage, mustard and rapeseed. Upon plant damage, glucosinolates (GLS) are activated by myrosinase to form hydrolysis products, such as toxic isothiocyanates (ITCs). Yet many insects readily feed and perform well on Brassicaceae species. The avoidance of glucosinolate toxicity is due at least partly to detoxification processes. General detoxification strategies include the deactivation of toxic ITCs formed upon hydrolysis via conjugation to nucleophilic substrates, such as glutathione. On the other hand, specialized strategies avoid the formation of toxic glucosinolate breakdown products. Instead, GLS hydrolysis may be diverted to form less toxic metabolites, such as nitriles instead of ITCs, or GLS may be modified by desulphation to prevent activation altogether. Lastly, some insects sequester intact GLS and employ them in their own defence against predators. We further discuss how these adaptations help insects with different feeding strategies thrive on GLS-containing plants, and consider new directions for future research.

Advances in Botanical Research, Volume 80
ISSN 0065-2296
http://dx.doi.org/10.1016/bs.abr.2016.06.003

List of Abbreviations

ESP Epithiospecifier protein
GLS Glucosinolates
GSH Glutathione
GSS Glucosinolate sulphatase
GST Glutathione *S*-transferase
ITC Isothiocyanate
NSP Nitrile-specifier protein
TFP Thiocyanate-forming protein

1. INTRODUCTION: GLUCOSINOLATE HYDROLYSIS AND ITS EFFECTS ON INSECT HERBIVORES

Plants produce an exceptional diversity of specialized (secondary) metabolites that act in defence against insect herbivores, including phenolics, terpenes, alkaloids and other amino acid-derived compounds with toxic or deterrent properties (Fraenkel, 1959; Goyal, Lambert, Cluzet, Mérillon, & Ramawat, 2012; Hartmann, 2007; Kliebenstein, 2004; Kliebenstein, Kroymann, & Mitchell-Olds, 2005; Schoonhoven et al., 2005; Wink, 1988; Wittstock & Gershenzon, 2002). These chemical defences vary extensively among different plant lineages. The main defences of the order Brassicales are the glucosinolates (GLS), which are sulphur- and nitrogen-containing, amino acid-derived metabolites that have long been the subject of research (Ahuja, Rohloff, & Bones, 2010; Halkier & Gershenzon, 2006).

Belonging to the group of activated, two-component plant defence systems, GLS are stored constitutively in high quantities and can be quickly activated by the thioglucoside hydrolysing enzyme myrosinase upon destruction of tissue and cellular integrity, for example, when a herbivore attacks (Wittstock, Kliebenstein, Lambrix, Reichelt, & Gershenzon, 2003). Although intact GLS have some biological activity, for example, serving as oviposition cues for female moths (Shroff et al., 2015; Städler, Renwick, Radke, & Sachdev-Gupta, 1995; Sun, Sønderby, Halkier, Jander, & de Vos, 2009), the activities of GLS are conveyed mostly by their hydrolysis products. GLS hydrolysis products include isothiocyanates (ITCs), simple nitriles, thiocyanates, epithionitriles and others; the composition of the resulting mix depends on the side chain of the parent GLS, the presence of specifier proteins and other conditions such as pH and metal cation concentrations (Fig. 1) (Agerbirk, De Vos, Kim, & Jander, 2009; Kissen, Rossiter, & Bones, 2009; Wittstock & Burow, 2010).

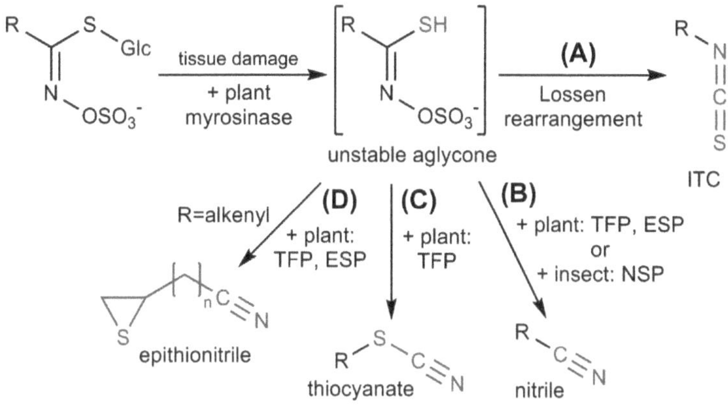

Figure 1 The breakdown of glucosinolates (GLS) upon tissue damage. Plant myrosinase cleaves the glucose moiety releasing an unstable aglycone which can be converted to a variety of metabolites: (A) isothiocyanates (ITCs) form by spontaneous Lossen rearrangement, (B) nitrile formation is mediated by the presence of specifier proteins, which differ between insects and plants, (C) thiocyanate formation requires TFPs, (D) epithionitriles are the breakdown products formed when the side chain of the parent glucosinolate has an alkenyl function and a plant specifier protein is present. *ESP*, epithiospecifier protein; *NSP*, nitrile-specifier protein, *TFP*, thiocyanate-forming protein.

ITCs are the most characteristic and reactive GLS-derived metabolites and have been intensely studied for their biological activities. While the reported anti-carcinogenic activity of ITCs in mammalian cells is of advantage for humans (Jeffery & Araya, 2009; Traka & Mithen, 2009), nonadapted insects suffer detrimental consequences from ITCs (Agrawal & Kurashige, 2003; Beekwilder et al., 2008; Hopkins, van Dam, & van Loon, 2009; Jeschke, Gershenzon, & Vassão, 2016; Li, Eigenbrode, Stringham, & Thiagarajah, 2000; Müller et al., 2010; Schramm, Vassão, Reichelt, Gershenzon, & Wittstock, 2012; Seo & Tang, 1982) and the toxicities of natural ITCs are comparable to those of synthetic insecticides (Lichtenstein, Strong, & Morgan, 1962). The lipophilic properties of ITCs facilitate passive diffusion through membranes into the cell, where their electrophilic functional core (−N=C=S) reacts with biological nucleophiles. Potential nucleophilic targets under physiological conditions include glutathione (GSH), some amino acid residues of proteins (mostly cysteines and lysines) (Hanschen et al., 2012; Jeschke et al., 2016; Kawakishi & Namiki, 1982; Nakamura, Kawai, Kitamoto, Osawa, & Kato, 2009; Schramm et al., 2012) and nucleic acids, although the extent of reaction with nucleic acids is not fully clear (Baasanjav-Gerber et al., 2011; Xiao, Mi, Chung, &

Veenstra, 2012). The conjugation of ITCs to nucleophilic residues can structurally and functionally disrupt proteins resulting in impaired catalytic activity, increases in oxidative stress and the initiation of stress-related signalling cascades (Brown & Hampton, 2011; Calmes et al., 2015; Cross et al., 2009; Hu, Eggler, Mesecar, & van Breemen, 2011; Kawakishi & Kaneko, 1987; Kawakishi & Namiki, 1982; Mi, Di Pasqua, & Chung, 2011). ITCs may also directly deplete GSH and disrupt redox balance within the organism (Jeschke et al., 2016).

The targets, modes of action and toxic activities of other hydrolysis products are less studied than those of ITCs. For example, simple nitriles have lower direct toxicity than ITCs with analogous side chains (Burow, Müller, Gershenzon, & Wittstock, 2006; Lambrix, Reichelt, Mitchell-Olds, Kliebenstein, & Gershenzon, 2001; Wittstock et al., 2003). These nitriles do not seem to play an important role in direct defence, but are more involved in indirect defence, e.g., attracting parasitoids of lepidopteran larvae (Burow, Müller, et al., 2006; Mumm et al., 2008; Wittstock & Burow, 2010; Wittstock et al., 2003). Other non-ITC hydrolysis products are derived from indolic GLS whose ITCs are less stable in aqueous solution than those derived from aliphatic or benzenic GLS, with the −N=C=S group being quickly displaced by water to form the corresponding indole-3-carbinols or by other nucleophiles, as discussed later (Agerbirk et al., 2009). The biological activities of GLS hydrolysis products are heavily dependent on the surrounding physiological conditions and the presence of other hydrolysis products or plant specialized metabolites (Bidart-Bouzat & Kliebenstein, 2008; Burow, Müller, et al., 2006; Wittstock & Burow, 2007, 2010; Wittstock et al., 2003).

The GLS–myrosinase system has been well studied in the current era of plant–insect research because of the agricultural importance of GLS-containing crucifers (Brassicaceae), as well as the long history of *Arabidopsis thaliana* (thale cress) as a model research organism (Ahuja et al., 2010; Bohinc, Ban, Ban, & Trdan, 2012; Hopkins et al., 2009; Kliebenstein et al., 2005; Textor & Gershenzon, 2009; Wittstock & Halkier, 2002). The enzymes and genes involved in the biosynthesis of GLS are mostly elucidated (Benderoth, Pfalz, & Kroymann, 2009; Geu-Flores, Olsen, & Halkier, 2009; Halkier & Du, 1997; Halkier & Gershenzon, 2006; Sønderby, Geu-Flores, & Halkier, 2010), allowing for the engineering of transgenic plants with modified GLS and myrosinase profiles to study the importance of side chain chemistry, the role of myrosinase activation and the effects of the breakdown products in insect defence (Barth & Jander, 2006;

Beekwilder et al., 2008; Müller et al., 2010; Schlaeppi, Bodenhausen, Buchala, Mauch, & Reymond, 2008; Schramm et al., 2012; Sønderby et al., 2007; Sun et al., 2009; Zhao et al., 2002).

The structural diversity of GLS and corresponding hydrolysis products affects a wide range of herbivores, including birds, rabbits, gastropod mollusks, insects and plant pathogens (Bednarek & Osbourn, 2009; Falk et al., 2014; Giamoustaris & Mithen, 1995; Hopkins et al., 2009; Lipka, Fuchs, & Lipka, 2008; Lipka et al., 2005; Pedras & Hossain, 2011; Tierens et al., 2001; Winde & Wittstock, 2011). However, the different GLS classes (aliphatic GLS derived from Met or other aliphatic amino acids; indolic GLS derived from Trp and benzenic GLS derived from Phe or Tyr (Fahey, Zalcmann, & Talalay, 2001)) exert different effects and toxicities depending on the herbivore studied. To compare the differential effects of GLS, transgenic plant lines devoid of one or more classes of GLS can be employed (Table 1). These experiments have shown, for example, that the generalist chewing larvae of the tobacco hornworm, *Manduca sexta* (Lepidoptera) and the cabbage looper, *Trichoplusia ni* (Lepidoptera), are negatively affected only by the presence of aliphatic GLS and not other types. Indolic GLS negatively influence the growth and development of the beet armyworm, *Spodoptera exigua* (Lepidoptera), albeit less strongly than aliphatic GLS (Müller et al., 2010). On the other hand, both aliphatic (Beekwilder et al., 2008) and indolic GLS (Schlaeppi et al., 2008) strongly affect some other generalist feeders, such as the African cotton leafworm, *Spodoptera littoralis* (Lepidoptera), and the cabbage moth, *Mamestra brassicae* (Lepidoptera). Benzenic GLS cause weight reduction and feeding deterrence for *S. littoralis* larvae reared on overexpressing lines (Bejai, Fridborg, & Ekbom, 2012). The detrimental effects of GLS experienced during larval development on a plant of the Brassicaceae family also affect choices in adult *S. littoralis* moths, which then avoid Brassicaceae for mating and oviposition (Proffit, Khallaf, Carrasco, Larsson, & Anderson, 2015; Thöming, Larsson, Hansson, & Anderson, 2013). Surprisingly, larvae of the cotton bollworm, *Helicoverpa armigera* (Lepidoptera), and of the crucifer-specialist diamondback moth, *Plutella xylostella* (Lepidoptera), were inhibited from feeding on an aliphatic GLS-deficient *A. thaliana* line (*apk1 apk2*) compared to wild-type plants (Badenes-Perez, Reichelt, Gershenzon, & Heckel, 2013). However, other changes in the biochemistry of this mutant plant, resulting, for example, in the accumulation of metabolites such as auxin and cysteine, were suggested to cause the deterrent effects. In summary, all classes of GLS can defend against generalist chewing herbivores, with aliphatic and

Table 1 Effects of Some *Arabidopsis thaliana* Lines With Altered GLS Profiles on Growth of Insect Herbivores

Insect Herbivore	*A. thaliana* Line Fed Upon	Metabolite Decreased in Line	Weight Gain in Comparison to Wild-type (Fold)	Age of Larvae, Duration of Experiment	References
Spodoptera littoralis	*pad2-1*	GSH	×2.4	Neonate 8 days	Schlaeppi et al. (2008)
	cyp79B2 cyp79B3	Indolic GLS	×1.5		
	gsm1-1	Aliphatic GLS	×1.8		
	coi1-1	JA perception	×5.2		
Spodoptera exigua	*myb28 myb29*	Aliphatic GLS	×2.5	Neonate 10 days	Müller et al. (2010)
	cyp79B2 cyp79B3	Indolic GLS	×2.2		
	myb28/29 cyp79B2/3	Aliphatic + indolic GLS	×9.1		
Trichoplusia ni	*myb28 myb29*	Aliphatic GLS	×1.7	9 days	
	cyp79B2 cyp79B3	Indolic GLS	×1		
	myb28/29 cyp79B2/3	Aliphatic + indolic GLS	×2.1		
Manduca sexta	*myb28 myb29*	Aliphatic GLS	×2.6	8 days	
	cyp79B2 cyp79B3	Indolic GLS	×1.2		
	myb28/29 cyp79B2/3	Aliphatic + indolic GLS	×3.1		
Mamestra brassicae	*myb28 myb29*	Aliphatic GLS	×2.6	Neonate 12 days	Beekwilder et al. (2008)

Insect Herbivore	*A. thaliana* Line Fed Upon	Metabolite Increased in Line	Weight Change in Comparison to Wild-type (Fold)	Age of Larvae, Duration of Experiment	References
Spodoptera littoralis	35S:*CYP79A1*	4-Hydroxybenzyl GLS	0.6–1.0	First instar 10 days	Bejai et al. (2012)
			0.8–0.9	Second instar	
	35S:*CYP79A2*	Benzyl GLS	0.5–0.7	First instar	
			0.7–0.8	Second instar	
	35S:*CYP79D2*	Isopropyl + 1-methylpropyl GLS	1.0–1.2	First instar	
			0.9–1.0	Second instar	

GLS, glucosinolates; *GSH*, glutathione.

benzenic GLS, which produce long-lasting ITCs upon hydrolysis, having a stronger effect. Accordingly, young generalist caterpillars adapt their feeding behaviour and minimize the negative effects of a Brassicaceae diet by selectively avoiding leaf areas with high constitutive or induced levels of GLS (Perkins et al., 2013; Shroff, Vergara, Muck, Svatoš, & Gershenzon, 2008).

While specialist herbivores are also negatively affected by consumption of GLS-derived ITCs or by certain classes of GLS (Agrawal & Kurashige, 2003; Åhman, 1986; Frisch et al., 2014; Kos, Houshyani, Wietsma, et al., 2012; Li et al., 2000; Müller et al., 2015; Santolamazza-Carbone, Sotelo, Velasco, & Cartea, 2015), they often successfully prevent the formation of toxic ITCs in nature (see Section 3), and their performance is generally not negatively affected by the GLS profile of the host plant (Chaplin-Kramer et al., 2011; Gols et al., 2008; Harvey, Gols, Wagenaar, & Bezemer, 2007; Hopkins et al., 2009; Kroymann, Donnerhacke, Schnabelrauch, & Mitchell-Olds, 2003; Nielsen, Hansen, Agerbirk, Petersen, & Halkier, 2001; Sarfraz, Dosdall, & Keddie, 2006; Sarosh, Wittstock, Halkier, & Ekbom, 2010; Schlaeppi et al., 2008; Schweizer et al., 2013; Van Geem, Harvey, Cortesero, Raaijmakers, & Gols, 2015). Some specialists even use intact GLS and their breakdown products as cues to locate suitable host plants for oviposition (Barker, Molotsane, Müller, Schaffner, & Städler, 2006; Hopkins et al., 2009; Marazzi, Patrian, & Städler, 2004; Marazzi & Städler, 2004; Mumm et al., 2008; Shroff et al., 2015; Städler et al., 1995; Sun et al., 2009) and perceive GLS as feeding stimulants (Badenes-Perez et al., 2013; David & Gardiner, 1966; Hicks, 1974; Huang & Renwick, 1993; Müller et al., 2015).

In contrast to chewing herbivores, piercing-sucking insects such as aphids do not severely damage plant tissue upon feeding, and are thus thought to largely avoid setting off the 'mustard oil bomb' (Kim & Jander, 2007). However, indolic GLS were shown to be unstable under the physiological conditions of the aphid gut and to form breakdown products independently of myrosinase activation, including nitriles, alcohols and unstable ITCs that are metabolized further (Agerbirk et al., 2009). Thus, indolic GLS appear to be more important for plant resistance against piercing-sucking insect species, although some indolic breakdown products also act as oviposition deterrents against GLS specialists (De Vos, Kriksunov, & Jander, 2008). The presence of both generalist and specialist representatives of either the chewing or piercing-sucking guilds in the insect herbivore community also selects for different GLS profiles (Giamoustaris & Mithen, 1995; Gols et al., 2008; Kliebenstein et al., 2001; Lankau, 2007; Mosleh Arany et al.,

2008; Newton, Bullock, & Hodgson, 2009). For example, higher allyl-GLS (sinigrin) concentrations in *Brassica nigra* are thought to have been selected for by generalist herbivores, which preferentially avoid this compound, but not by specialist herbivores, which feed more on high sinigrin foliage (Lankau, 2007). Allyl-GLS concentration was shown to be a heritable trait in natural *Brassica oleracea* populations, and the responses of different herbivore species to the presence of these GLS differ significantly (Newton et al., 2009). The induction of GLS upon herbivory by chewing herbivores seems to be higher than after infestation by piercing-sucking herbivore species (reviewed in Textor & Gershenzon, 2009), in agreement with the low amount of tissue damaged by piercing-sucking herbivores and the active manipulation of plant defences by these insects. However, indolic GLS are generally induced in higher amounts in comparison to aliphatic GLS regardless of feeding guild or degree of specialization.

The complexity of the GLS—myrosinase system, including its chemical diversity (Fahey et al., 2001) and tissue-specific expression (Brown, Tokuhisa, Reichelt, & Gershenzon, 2003), allows plants to fend off a large diversity of herbivores, and even affects predators and parasitoids of GLS-consuming insects (Chaplin-Kramer et al., 2011; Hopkins et al., 2009; Kos, Houshyani, Achhami, et al., 2012; Kos, Houshyani, Wietsma, et al., 2012; Kos et al., 2011; Pratt, Pope, Powell, & Rossiter, 2008). But despite the presence of GLS and other specialized metabolites, some herbivores can feed and survive on these toxic plants (Fig. 2). Several biochemical counter-adaptations by insect herbivores to plant specialized metabolites have been described (Dussourd & Eisner, 1987; Hartmann, 1999; Heckel, 2014; Holzinger, Frick, & Wink, 1992; Ivie, Bull, Beier, Pryor, & Oertli, 1983; Pentzold et al., 2015; Self, Guthrie, & Hodgson, 1964), and such information often facilitates the understanding of insect resistance to commercial insecticides (Després, David, & Gallet, 2007; Furlong & Wright, 1994; Iqbal & Wright, 1997; Yu, 1984). The toxicity of a two-component activated defence such as the GLS—myrosinase system could in theory be circumvented at different stages of the activation process. Herbivores could either avoid the activation of the glucosides or minimize the toxicity of the breakdown products (Pentzold, Zagrobelny, Rook, & Bak, 2014). If the hydrolysis of GLS occurs, toxic metabolites such as ITCs can be neutralized by spontaneous chemical and/or enzymatic reactions (Kim, Lee, Schroeder, & Jander, 2008; Schramm et al., 2012), or GLS hydrolysis may be controlled by the insect to form less toxic metabolites (Wittstock et al., 2004). In addition, the GLS core structure itself may be modified to prevent activation by myrosinase (Falk & Gershenzon, 2007; Malka et al., 2016; Ratzka, Vogel, Kliebenstein, Mitchell-Olds, & Kroymann, 2002). Finally, herbivores may adopt feeding strategies that circumvent GLS activation altogether, and the intact GLS may even be selectively taken up and utilized for the insect's own defence (Aliabadi, Renwick, & Whitman, 2002; Beran et al., 2014; Francis, Lognay, Wathelet, & Haubruge, 2001; Müller et al., 2001; Nishida, 2002). In the following sections, we will discuss these strategies, divided into general detoxification mechanisms (Section 2) and specialized adaptations (Section 3).

2. GENERAL DETOXIFICATION STRATEGIES

Generalist herbivores feed on a wide range of host plants often from different orders, encountering a diverse range of specialized defence metabolites. Thus they require efficient biochemical and behavioural adaptations towards many different xenobiotics. Many plant chemical defences are lipophilic compounds that can undergo passive noncontrolled diffusion through cell membranes before exerting toxicity intracellularly. Broad-range xenobiotic detoxification mechanisms target especially such nonpolar compounds, aiming to increase their polarity for easier and safer excretion. Such general detoxification typically has three phases. Phase I involves the functionalization of xenobiotics via chemical modification to insert or expose polar groups and is generally catalysed by cytochrome P450 enzymes, peroxidases, esterases or hydrolases. Phase II encompasses the transferase-catalysed conjugation of reactive groups, such as GSH, phosphate, sulphate, sugars and amino acids, to form polar or charged derivatives. In phase III, these polar derivatives are excreted from the cell and eventually from the organism, usually by means of transmembrane transporters.

When feeding on GLS-containing plants, the major lipophilic toxins encountered by generalist chewing herbivores, as we have seen in the first section, are ITCs formed from aliphatic and benzenic GLS. The ITC group is chemically very reactive, and the metabolic pathways described for their generalized detoxification consist mostly of phase II conjugation reactions. On the other hand, piercing-sucking insects, which avoid myrosinase-catalysed hydrolysis, encounter the breakdown products of indolic GLS, which include nitriles, ITCs and derived alcohols that can undergo both phase I and II reactions. In the following two sections, we will discuss the

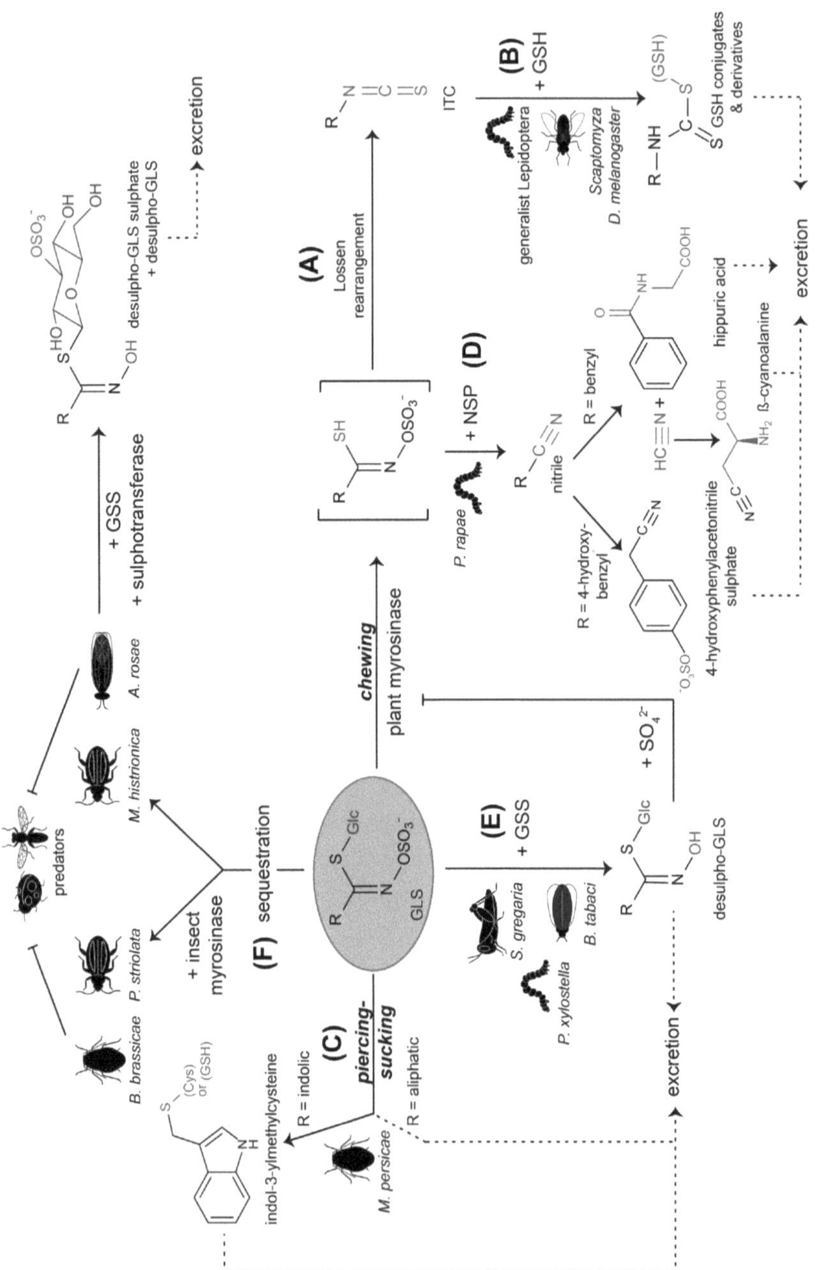
predators
B. brassicae
P. striolata
M. histrionica
A. rosae
+ GSS
+ sulphotransferase
desulpho-GLS sulphate
+ desulpho-GLS
excretion
+ insect
myrosinase
(F) sequestration
GLS
chewing
plant myrosinase
(A)
Lossen
rearrangement
ITC
(B)
+ GSH
generalist Lepidoptera
Scaptomyza
D. melanogaster
GSH conjugates
& derivatives
P. rapae
+ NSP (D)
nitrile
R = 4-hydroxy-
benzyl
R = benzyl
4-hydroxyphenylacetonitrile
sulphate
ß-cyanoalanine
hippuric acid
excretion
(C)
piercing-
sucking
R = indolic
R = aliphatic
M. persicae
indol-3-ylmethylcysteine
(E)
+ GSS
S. gregaria
P. xylostella
B. tabaci
desulpho-GLS
+ SO_4^{2-}
excretion

Figure 2 Insects have developed different means to circumvent the hydrolysis of glucosinolates (GLS) or detoxify the harmful products. (A) Chewing or piercing-sucking insects that have not developed specialized strategies must deal with toxic hydrolysis products (Section 2). (B) Chewing herbivores such as generalist Lepidoptera and larvae of *Scaptomyza* conjugate isothiocyanates (ITCs) with glutathione (GSH). (C) Piercing-sucking insects, such as generalist aphids, avoid contact with myrosinases, and thus aliphatic GLS are excreted intact while indolic GLS, which break down regardless of myrosinase, are detoxified via conjugation, e.g., to GSH or amino acids. Specialized strategies involve the utilization of proteins or enzymes designed to counteract the GLS—myrosinase system (Section 3). (D) Larvae of *Pieris rapae* express a nitrile-specifier protein (NSP) and redirect the aglycone to form less toxic nitriles. Depending on the side chain of the parent GLS, nitriles are further metabolized to sulphates, or hippuric acid and cyanide, which is further converted to β-cyanoalanine. (E) Some insects, such as *Plutella xylostella* larvae, *Schistocerca gregaria* and *Bemisia tabaci*, produce a GLS sulphatase (GSS) which yields desulpho-GLS. (F) A variety of specialist insects sequesters GLS and some even produce their own myrosinase, employing GLS and their breakdown products for their own defence against predators. Known sequestering insects include *Brevicoryne brassicae, Phyllotreta striolata, Murgantia histrionica* and *Athalia rosae*, which further metabolizes the sequestered GLS to desulpho-GLS sulphates.

mechanisms described for the generalized detoxification of GLS breakdown products in more detail.

2.1 Chewing Insects and the Conjugation of Isothiocyanate Hydrolysis Products

Chewing insects such as caterpillars cause extensive plant tissue damage while feeding, and the resulting cell disintegration leads to the efficient mixing of GLS with myrosinases, immediately detonating the 'mustard oil bomb' (Matile, 1980) (Fig. 1). GLS-derived ITCs are not only lipophilic but also very reactive electrophiles. ITCs likely enter cells via passive diffusion, and the central electrophilic ITC carbon (R—N═C═S) readily reacts with many biological nucleophiles, for example, the thiol groups of cysteine and GSH and the amine group of lysine residues of proteins (Brown & Hampton, 2011). The dithiocarbamates resulting from reaction with thiols are semi-stable under intracellular pH conditions (Fig. 3). A common detoxification of electrophilic metabolites is conjugation with GSH, which is an important biological nucleophile and reducing agent and is typically present in cells in mM concentrations. GSH is a Glu-Cys-Gly tripeptide with a γ-peptide bond between the carboxyl group of the Glu side chain and the amine of Cys. The addition of the thiol at the Cys residue to the electrophilic centres of ITCs can happen spontaneously or be catalysed by glutathione *S*-transferases (GSTs). The GSH conjugates formed are then sequentially hydrolysed via the mercapturic acid pathway to give the GlyCys- and Cys-conjugates, with the latter further *N*-acetylated (Boyland & Chasseaud, 1969; Habig, Pabst, & Jakoby, 1974) (Fig. 3). The conjugation of GSH to GLS-derived ITCs has been demonstrated or suggested to occur in several insect species, molluscs and mammals including humans (Al Janobi et al., 2006; Falk et al., 2014; Gloss et al., 2014; Kassahun, Davis, Hu, Martin, & Baillie, 1997; Schramm et al., 2012; Wadleigh & Yu, 1988). The resulting ITC-GSH conjugates and mercapturic acid derivatives have increased polarity relative to the parent ITC and can be more easily excreted from the cell and ultimately the organism; however, little is currently known about the transporters involved in this process. GLS-derived nitriles, on the other hand, lack the electrophilicity of ITCs and are not metabolized in this way.

Conjugation to GSH has been shown to be involved in ITC detoxification in several generalist lepidopteran herbivores with strongly varying GLS preferences (Schramm et al., 2012), but its overall importance is not always clear. In following the metabolic fate of isotopically labelled [^{14}C]-4-methylsulfinylbutyl (4msob) GLS in *S. littoralis*, only 11% of the ingested

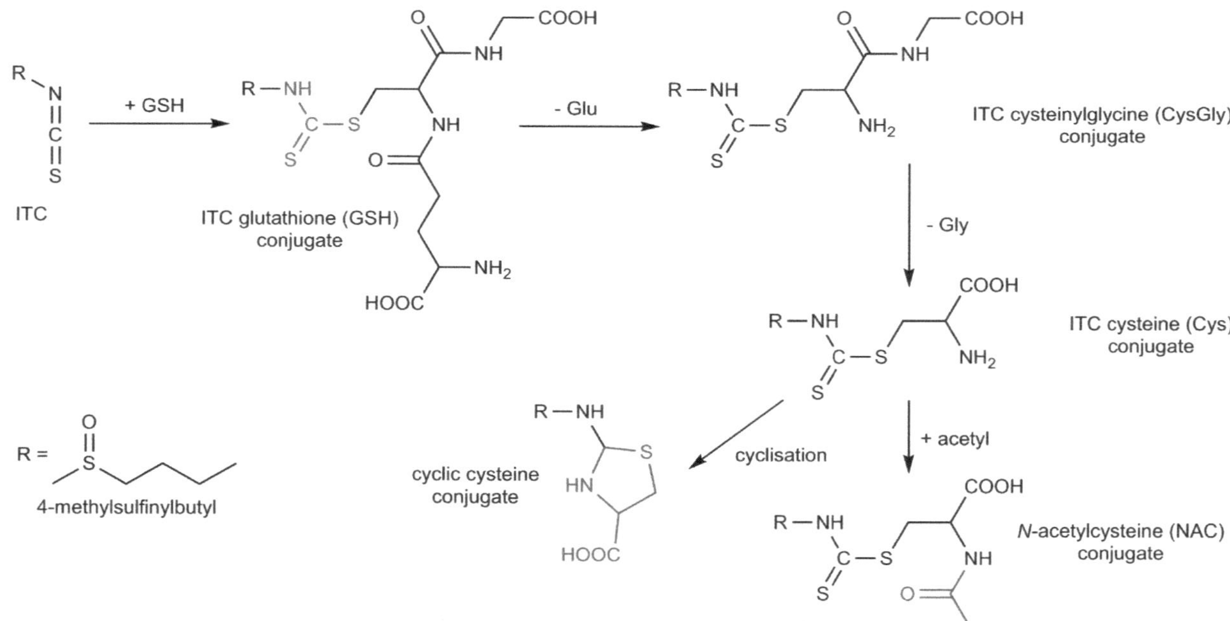

Figure 3 The conjugation of glutathione (GSH) to isothiocyanate (ITC). Generalist insects, including several lepidopteran species and *Scaptomyza*, detoxify ingested ITCs via conjugation to GSH. The resulting GSH conjugate undergoes further modification to its amino acid derivatives following the mercapturic acid pathway, which recovers some of the amino acids and produces additional derivatives.

GLS were found to be excreted as ITC-GSH conjugates (including its hydrolysis derivatives) and 10% as the nitrile, but more than 60% were excreted as the unmodified ITC (Schramm et al., 2012). A portion of this unmodified ITC pool could have been initially conjugated but then dissociated (deconjugated) to re-release the free ITC while passing through the gut, or may never have entered the midgut tissue cells where conjugation presumably takes place.

The conjugation of GSH with ITCs can progress spontaneously and nonenzymatically under the conditions present in insect midgut cells (~pH 7, high GSH concentrations) and gut lumen (~pH 7–10), but GSTs have long been suspected to be involved in the conjugation in vivo. In an analysis of the transcriptome of midgut cells in *T. ni*, a total of 28 GST-encoding genes were identified, two of which were upregulated in response to GLS consumption (Herde & Howe, 2014). GLS feeding also resulted in a strong general induction of genes encoding phase I and II detoxification enzymes, including UGTs and cytochrome P450s. Analysis of cDNA sequences of the midgut tissue of the common cutworm, *Spodoptera litura* (Lepidoptera) led to the identification of eight GSTs (Deng, Huang, Feng, & Zheng, 2009; Huang, Xu, Lin, Feng, & Zheng, 2011). One GST gene, encoding cytosolic GST epsilon 1 (SlGSTE1), was upregulated when larvae fed on leaves of mustard greens, *Brassica juncea*, or artificial diets containing indole-3-carbinol (I3C) and allyl-ITC (Zou et al., 2016). Larvae in which *SlGSTE1* was silenced by RNAi consumed less *B. juncea* leaves resulting in decreased body weight compared to a control group, demonstrating a role for the encoded enzyme in ITC detoxification. Interestingly, SlGSTE1 also showed higher activity towards allyl-ITC than to I3C in vitro (Zou et al., 2016). Thus, although insect GSTs were first linked to ITC detoxification almost 30 years ago (Yu, 1989, 1992), definitive proof of product formation (Schramm et al., 2012) and function in vivo (Zou et al., 2016) have come only recently.

While GST-dependent detoxification is critical for the ability of generalist lepidopteran caterpillars to feed on Brassicaceae, it also imposes a high metabolic cost. GSH levels in *S. littoralis* midgut tissues and haemolymph were observed to drop significantly upon ITC ingestion in a dose-dependent manner, suggesting that the available pool of GST is limited (Jeschke et al., 2016). Since caterpillars feeding on 4msob-ITC containing diet also had reduced levels of free cysteine, the biosynthesis of GSH seems likely to be limited by the availability of this amino acid precursor. These metabolic changes led to other metabolic consequences such as decreased weight and body protein levels, with some of these effects ameliorated by supplementation of cysteine (as its oxidized dimer cystine) (Jeschke et al., 2016). However, generalist herbivores can adjust their feeding behaviour upon encountering ITCs to avoid leaf areas of high constitutive (Shroff et al., 2008) or induced GLS (Perkins et al., 2013).

Although conjugation with GSH is an ITC detoxification strategy generally employed by nonadapted insects, some *Scaptomyza* species (Diptera) were also found to excrete the GSH conjugation products of ITCs (Gloss et al., 2014). These larvae are leaf-mining specialists on GLS-containing plants, but unlike other known GLS specialist herbivores (Section 3), *Scaptomyza* larvae do not prevent ITC formation but metabolize them intracellularly via GSH conjugation (Gloss et al., 2014). The enzyme GSTD1, which catalyses conjugation of several ITCs to GSH, seems to have acquired increased activity against ITCs in Brassicaceae-feeding *Scaptomyza* due to gene duplication and positive selection (Gloss et al., 2014). The products identified for *Scaptomyza flava* include the GSH conjugate and its amino acid derivatives (ITC-CysGly, ITC-Cys), as well as the *N*-acetylcysteine derivatives of a number of ITCs. The ITC *N*-acetylcysteine derivatives, the final products of the mercapturic acid pathway, are also found in humans (Al Janobi et al., 2006) but are not excreted in high amounts by generalist Lepidopterans (Schramm et al., 2012) and molluscs (Falk et al., 2014). In the faeces of molluscs, an additional downstream metabolite of 4msob-ITC-Cys was characterized that is formed by intramolecular cyclization of the cysteine moiety (Falk et al., 2014) (Fig. 3). This cyclic cysteine product is also formed from conjugates of phenylethyl- and allyl-ITC, and can be generated spontaneously in solution (Eklind, Morse, & Chung, 1990; Kawakishi & Namiki, 1982).

2.2 Piercing-Sucking Insects and Indolic Glucosinolates

In contrast to chewing herbivores, piercing-sucking insects such as aphids only cause minimal tissue damage during feeding and are thought to largely avoid the detonation of the mustard oil bomb. These herbivores guide their stylets between individual plant cells to the phloem sieve elements (Tjallingii & Hogen Esch, 1993), and therefore the GLS present in the phloem and apoplast are not brought into contact with myrosinases that are localized in separate myrosin cells (Andréasson & Jørgensen, 2003; Barth & Jander, 2006; Husebye, Chadchawan, Winge, Thangstad, & Bones, 2002; Koroleva et al., 2000; Thangstad et al., 2004). Accordingly, aliphatic GLS are excreted intact in the honeydew of the green peach aphid, *Myzus persicae* (Hemiptera)

(Kim & Jander, 2007), which is an extreme generalist infesting close to 400 host plant species. However, indolic GLS break down under the conditions present in the gut independently of myrosinases and so have a strong anti-feedant effect on *M. persicae* (Fig. 4A) (Kim & Jander, 2007; Kim et al., 2008). Thus, indolic GLS are suggested to play a more important role in plant–aphid interactions than aliphatic GLS. A defensive role of indolic GLS is further supported by their significant induction upon herbivory by *M. persicae* and by plant hormone treatment (Agerbirk et al., 2009; Bodnaryk, 1994; Kim & Jander, 2007).

Although indolic GLS are widespread in Brassicaceae plants, they do not typically dominate the constitutive GLS profile (Agerbirk et al., 2009; Fahey et al., 2001). Still, their abundance in *A. thaliana* has greatly facilitated research on their biology and function in plant defence (Agerbirk et al., 2009; Brown et al., 2003). Indol-3-ylmethyl (I3M) GLS is the most abundant indolic GLS in *A. thaliana* and other plants in the Brassicaceae family, while less abundant structural derivatives include 1-methoxy-, 4-methoxy- and 4-hydroxy-I3M (Fig. 4B) (Agerbirk et al., 2009). Hydrolysis of the thioglucose results in indoleacetonitrile (IAN) or the corresponding ITC, which is unstable and reacts with nucleophiles to form conjugates or indole-3-carbinol (I3C) in aqueous conditions (Fig. 4A). I3C itself undergoes a plethora of further modifications, being oxidized to the corresponding aldehyde and carboxylic acid, dimerized, or also forming conjugates with amino acids, GSH or ascorbate (reviewed in Agerbirk et al., 2009) (Fig. 4A).

Indolic GLS are known to be preferentially induced upon herbivore attack (Textor & Gershenzon, 2009), and a peptide (between 25 and 100 amino acids) has been identified in *M. persicae* saliva that locally induces plant genes encoding *O*-methyltransferases that could be involved in indolic GLS biosynthesis (De Vos & Jander, 2009). However, this *O*-methyltransferase gene induction did not correlate with an increase in 4-methoxy-I3M, which was previously shown to be elevated upon *M. persicae* infestation of *A. thaliana* (Kim & Jander, 2007). 4-methoxy-I3M and its isomer 1-methoxy-I3M exert stronger negative effects on feeding and reproduction than their biosynthetic precursor I3M when myrosinases are absent, while I3M was more toxic when combined with myrosinase (Barth & Jander, 2006; Kim & Jander, 2007). The upregulation of 4-methoxy-I3M biosynthesis thus demonstrates an effective plant response to attack by piercing-sucking insects.

It is not fully understood which conditions determine the breakdown of indolic GLS, but the rates of breakdown seem to depend upon the hydroxy

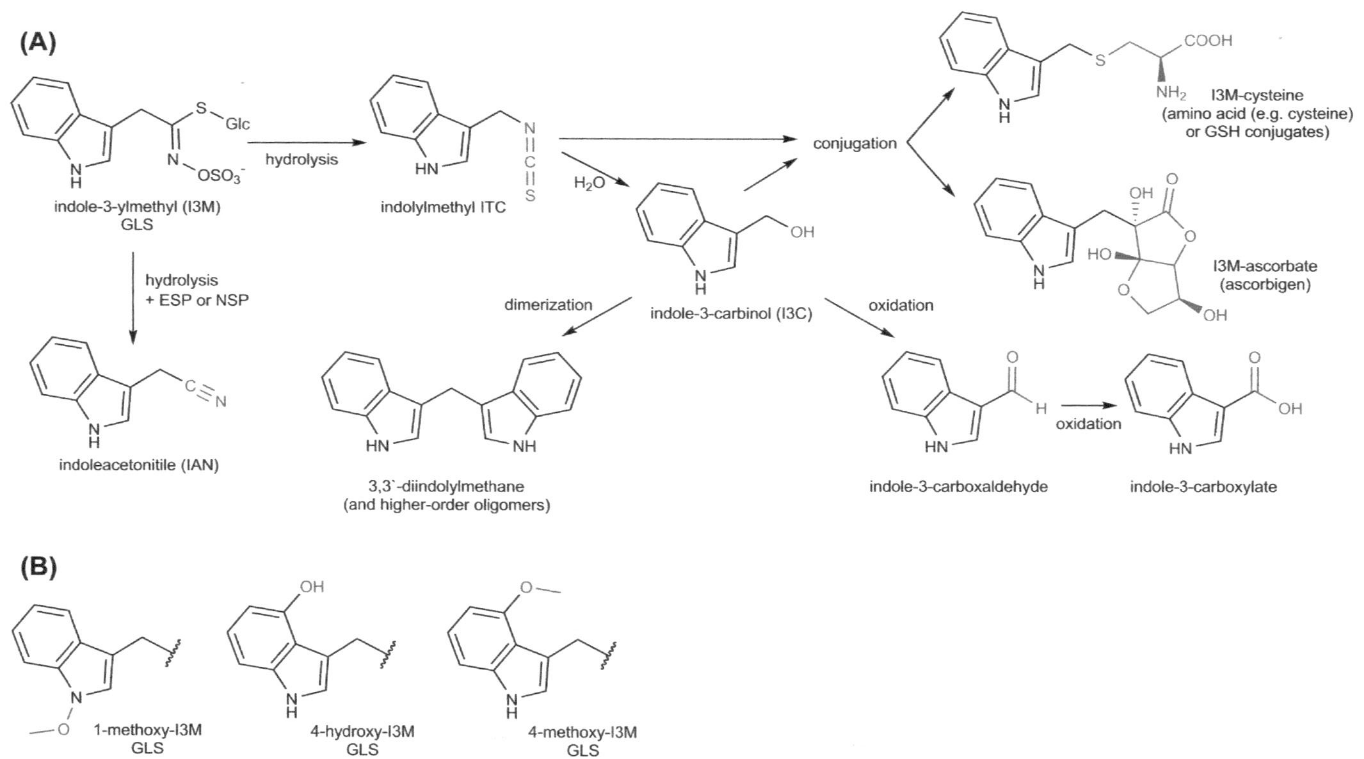

Figure 4 (A) Metabolic conversions of indolic glucosinolates (GLS). Upon myrosinase-dependent or myrosinase-independent hydrolysis of the glucose moiety, the corresponding nitrile or isothiocyanate (ITC) is produced. Indolylmethyl ITC reacts with water to form indole-3-carbinol (I3C) that has several possible fates: I3C can be conjugated with amino acids such as cysteine, and glutathione (GSH) or ascorbate. Indolylmethyl ITC can also react directly with these other nucleophiles, resulting in the corresponding conjugates. Oxidation of I3C yields the corresponding aldehyde and further the carboxylic acid. I3C may also dimerize. (B) Naturally occurring indolic GLS, derivatives of the I3M precursor.

and methoxy functional groups present (Barth & Jander, 2006). Interestingly, *M. persicae* aphids harbour a gut-expressed gene with similarity to plant myrosinases that could encode a protein that participates in indolic GLS breakdown (Ramsey et al., 2007). The breakdown products detected in the honeydew include amino acid and GSH conjugates, which could represent active detoxification products. An analysis of the cDNA of *M. persicae* identified between 14 and 21 GST-like genes, of which eight were grouped into the delta class but none in the epsilon class (Ramsey et al., 2010). Similarly, the pea aphid, *Acyrthosiphon pisum*, a Fabaceae specialist, harbours 20 putative GST genes, of which 10 belong to the delta class but none to the epsilon class (Ramsey et al., 2010). The delta and epsilon classes of GSTs appear to be insect-specific GSTs (Ranson et al., 2001), but the epsilon class seems to be absent in aphids. *M. persicae* was shown to increase its GST activity in response to feeding on GLS- or ITC-containing diets (Francis et al., 2001; Francis, Vanhaelen, & Haubruge, 2005), but these activities were assayed in vitro using artificial substrates, and not ITCs, and so could be a result of an increased antioxidant response of the insect rather than a specific detoxification of GLS-derived metabolites via GSH conjugation. Interestingly, some indolic-derived detoxification products of I3C, specifically its conjugates with GSH and cysteine, caused a significant reduction in aphid reproduction as well as feeding deterrence (Kim et al., 2008). However, other amino acid conjugates such as I3M-Pro, I3M-Ile and I3M-Leu did not have a negative effect on reproduction and were detected in more than 10-fold higher abundance in the honeydew than the inhibitory conjugates (Kim et al., 2008). Besides these influences on feeding behaviour and reproduction, a GLS-containing diet may also affect the primary metabolism of aphids. A proteomic study revealed that proteins involved in mitochondrial function, protein and lipid synthesis were downregulated in *M. persicae* upon feeding on rapeseed, *Brassica napus*, compared to a control group reared on *Vicia faba* (broad bean, Fabaceae) (Francis et al., 2006).

Recently, the influence of GLS transport and distribution on aphid infestation was investigated using an *A. thaliana* GLS transporter mutant (*gtr1 gtr2*) with reduced GLS content in the phloem sap (Madsen, Kunert, Reichelt, Gershenzon, & Halkier, 2015). Unexpectedly, *M. persicae* aphids performed worse on the *gtr1 gtr2* plants, producing less offspring and having higher adult mortality. Mutant *gtr1 gtr2* plants have a higher level of total GLS in leaf tissue, which may be encountered by aphids during the probing phase before the stylets reach the phloem, and so cause the observed detrimental effects (Madsen et al., 2015).

Although our understanding of the biosynthesis and biological activities of indolic GLS has improved in recent years, the modes of action and biological targets of the toxic indolic breakdown products in herbivores remain to be identified, as do any metabolic counter-adaptations on the part of herbivores. The chemical lability of indolic GLS, particularly of the methoxylated derivatives, also raises the question about how plants can avoid autotoxicity problems and manage to stabilize the compounds *in planta*.

3. SPECIALIZED DETOXIFICATION STRATEGIES

Specialized detoxification strategies of plant defensive metabolites are often thought to be characteristic of monophagous feeders that are adapted to one plant family or species, and so cope with a limited range of plant defences. Some such 'specialist' biochemical adaptations against the GLS–myrosinase system have been identified that avoid exposure to toxic ITCs (Müller et al., 2001; Ratzka et al., 2002; Wittstock et al., 2004). However, in recent years several polyphagous insects have been shown to exploit some of the same strategies previously found in specialist feeders (Falk & Gershenzon, 2007; Malka et al., 2016). Specialized insects generally do not suffer severe detrimental effects after consuming GLS-containing plants (such as the strong growth retardation seen in generalist caterpillars, for example) (Chaplin-Kramer et al., 2011; Gols et al., 2008; Harvey et al., 2007; Kroymann et al., 2003; Li et al., 2000; Müller et al., 2010; Sarosh et al., 2010; Schlaeppi et al., 2008; Schweizer et al., 2013; Van Geem et al., 2015). Some even require GLS for feeding or choosing oviposition sites (Badenes-Perez et al., 2013; Barker et al., 2006; David & Gardiner, 1966; Giamoustaris & Mithen, 1995; Hicks, 1974; Hilker & Meiners, 2002; Huang & Renwick, 1993; Isidoro, Bartlet, Ziesmann, & Williams, 1998; Marazzi et al., 2004; Marazzi & Städler, 2004; Müller et al., 2015; Mumm et al., 2008; Nielsen et al., 2001; Renwick, Radke, Sachdev-Gupta, & Städler, 1992; Roessingh, Städler, Baur, Hurter, & Ramp, 1997; Sarfraz et al., 2006; Shroff et al., 2015; Siemens & Mitchell-Olds, 1996; Städler et al., 1995; Sun et al., 2009). However, some GLS metabolites can act as deterrents to specialists, for example, the indolic IAN to *Pieris rapae* (small white, Lepidoptera) oviposition (De Vos et al., 2008) or the unusual ITC moringin to the feeding of *Pieris napi* (green-veined white, Lepidoptera), *Pieris brassicae* (large white, Lepidoptera) and *Athalia rosae* (Hymenoptera) (Müller et al., 2015). GLS specialists are still sensitive to the more toxic

GLS breakdown products (Agrawal & Kurashige, 2003; Åhman, 1986; Li et al., 2000), even though their biochemical adaptations to the GLS–myrosinase system allow them to feed successfully on many Brassicaceae plants, causing major agricultural damage on cabbage, broccoli and other important food crops (Ahuja et al., 2010). Interestingly, while most identified specialist adaptations to GLS arise directly from the insect, the specialist larvae of the cabbage fly *Delia radicum* (Diptera) were recently shown to recruit gut microbes capable of detoxifying GLS-derived ITCs (Welte et al., 2015). The microbial enzyme SaxA metabolizes the *Brassica* root volatile 2-phenylethyl ITC in vitro and in vivo (Welte et al., 2015); however, the end products of this transformation remain to be identified.

Two specialized GLS metabolic processes have been particularly well-characterized. The cabbage white butterfly, *P. rapae*, prevents the release of toxic ITCs by redirecting the spontaneous GLS hydrolysis towards less toxic nitriles (Wittstock et al., 2004). On the other hand, larvae of the diamondback moth, *P. xylostella*, employ a fundamentally different strategy, modifying the core GLS chemical structure to make them unrecognizable to plant myrosinases (Ratzka et al., 2002). While the former strategy is a key innovation so far only found in *Pieris* species (Wheat et al., 2007), the latter has also been identified in some generalist insect species (Falk & Gershenzon, 2007; Malka et al., 2016). Lastly, some insects sequester intact GLS and exploit them for their own defence, even harbouring their own endogenous myrosinases (Aliabadi et al., 2002; Beran et al., 2014; Francis, Lognay, Wathelet, & Haubruge, 2002; Müller et al., 2001). These processes are explored in more detail in the three following sections.

3.1 Specifier Proteins: Diverting Hydrolysis to Less Toxic Products

A classic example of the 'coevolutionary arms race' (Ehrlich & Raven, 1964) between herbivores and their host plants is the well-studied relationship between Pieridae butterflies (Lepidoptera) and Brassicales plants, which led to both increased speciation and GLS chemical diversity (Edger et al., 2015). Approximately 10 million years after the evolution of the GLS–myrosinase system, ancestral Pieridae insects developed a key biochemical adaptation that allowed them to utilize Brassicales plants as their food source (Beilstein, Nagalingum, Clements, Manchester, & Mathews, 2010; Wheat et al., 2007). The host shift to Brassicales plants was facilitated by the evolution of a nitrile-specifier protein (NSP), which directs the myrosinase-catalysed breakdown of GLS in the larval gut to form nitriles, which are less toxic and reactive, instead of ITCs (Wittstock et al., 2004). These nitriles may be further metabolized prior to excretion.

While aliphatic GLS-derived nitriles are excreted unmodified (Fig. 5A), benzenic GLS-derived nitriles undergo metabolism of the nitrile moiety and/or the side chain prior to excretion (Fig. 5B and C) (Agerbirk, Olsen, Poulsen, Jacobsen, & Hansen, 2010; Agerbirk, Olsen, Topbjerg, & Sørensen, 2007; Vergara et al., 2006; Wittstock et al., 2004). The 4-hydroxybenzyl GLS-derived nitrile is transformed into the phenolic sulphate ester, yielding 4-phenylacetonitrile sulphate (Fig. 5B) (Agerbirk, Müller, Olsen, & Chew, 2006; Agerbirk et al., 2010; Wittstock et al., 2004). Phenylacetonitrile derived from benzyl GLS is metabolized to hippuric acid and in smaller quantity to *N*-phenylacetylglycine and *N*-benzoylisoserine (Fig. 5C) (Vergara et al., 2006). Surprisingly, hippuric acid and the malonyl conjugate of benzylamine were also detected in the faeces of some nonadapted generalist caterpillars (*S. exigua*, *S. littoralis* and *T. ni*) that ingested benzyl GLS-containing plant material. Generalist insects do not harbour an NSP, so these metabolites were concluded to derive from nitriles generated from natural GLS hydrolysis in *A. thaliana* leaves (Winde, 2011).

N-phenylacetylglycine can be formed by conjugation of glycine to phenylacetic acid, which could arise by nitrilase action on phenylacetonitrile (Vergara et al., 2006). However, the formation of hippuric acid, the main metabolite and its minor derivative *N*-benzoylisoserine require the loss of a C1 unit, and therefore cannot proceed directly via the action of a nitrilase on phenylacetonitrile. Instead, phenylacetonitrile undergoes an enzymatic α-hydroxylation resulting in an unstable α-hydroxynitrile, which further decomposes to the respective aldehyde losing the C1 unit as cyanide (Stauber et al., 2012). Oxidation of the resulting aldehyde then yields benzoic acid (Fig. 5C). In analogy to the benzyl nitrile-derived amino acid conjugates in *P. rapae*, glycine and isoserine conjugates of indole-3-carboxylic acid, which presumably derives from I3M GLS (Fig. 4A), were detected in the faeces of *P. rapae* (Vergara et al., 2006). Additionally, some benzenic GLS-derived products are subjected to other complex metabolism in *P. rapae* including *O*-demethylation followed by sulphation of the phenol group thus exposed (Agerbirk et al., 2010, 2007). Plant-derived enzymes can also metabolize the benzenic GLS metabolites in herbivore guts, further contributing to the diversity of metabolites; this includes the formation of carboxylic acids, which are considered to be more toxic than the respective nitriles (Agerbirk et al., 2010).

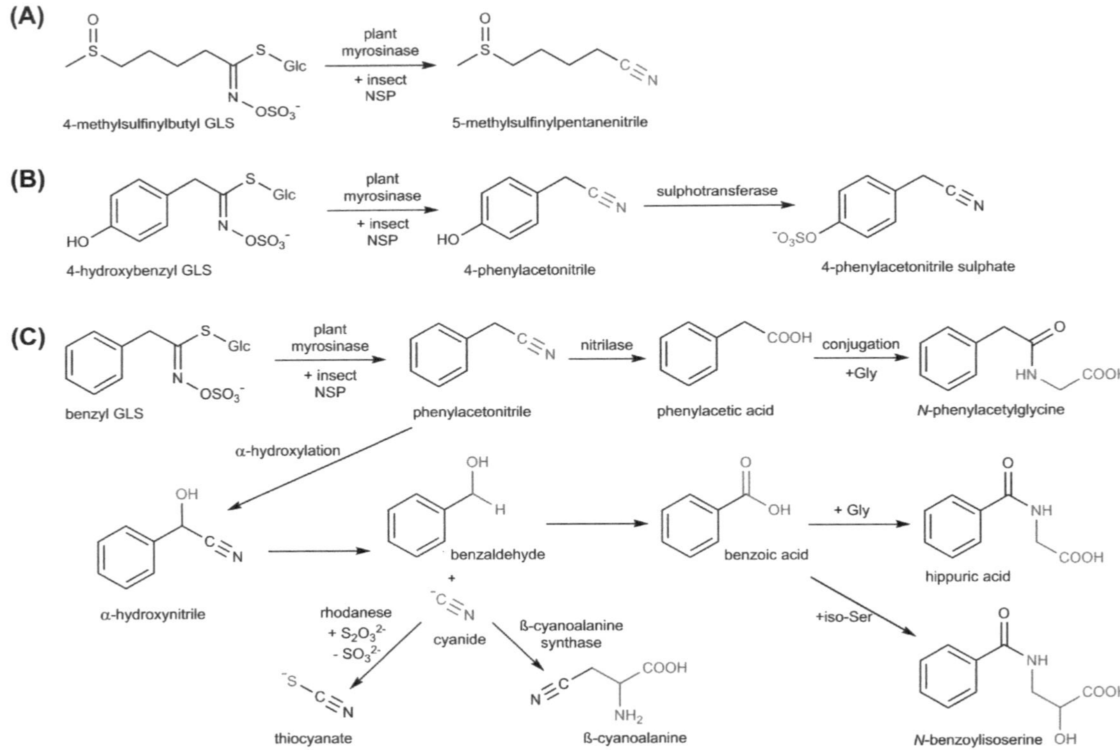

Figure 5 Metabolism of various glucosinolates (GLS) in *Pieris rapae*. (A) The hydrolysis of 4-methylsulfinylbutyl GLS (glucoraphanin) is redirected to the corresponding nitrile in the presence of the insect nitrile-specifier protein (NSP). (B) 4-Hydroxybenzyl GLS hydrolysis is redirected to the corresponding nitrile that is further converted to the sulphate ester. (C) The nitrile formed from benzyl GLS can have different fates. Following the action of a nitrilase, the phenylacetic acid produced may be further conjugated to glycine. Alternatively, phenylacetonitrile may undergo α-hydroxylation and the α-hydroxynitrile decomposes to benzaldehyde and cyanide. Cyanide is toxic and needs to be detoxified either by β-cyanoalanine synthase or rhodanese. Benzaldehyde can be further oxidized and conjugated to amino acids.

Although *Pieris* larvae evidently have an efficient strategy to avoid toxic ITCs, the formation of cyanide ('cyanide bomb') during metabolism of benzenic GLS could also result in toxicity (Stauber et al., 2012). Cyanide is universally toxic to aerobic cells, inhibiting the cytochrome c oxidase in cellular respiration (Ballhorn, Kautz, Heil, & Hegeman, 2009; Gleadow & Woodrow, 2002). However, *P. rapae* larvae that either ingested the cyanogenic glucoside dhurrin or were fumigated with cyanide did not show any ill-effects, as they very effectively detoxified cyanide enzymatically using β-cyanoalanine synthase and rhodanese to form the nontoxic β-cyanoalanine and thiocyanate (formerly rhodanide), respectively (Stauber et al., 2012) (Fig. 5C). Three constitutively active, gut-expressed β-cyanoalanine synthases were detected which are functionally different in their affinities for cysteine (Ohlen, Herfurth, Kerbstadt, & Wittstock, 2016). Additionally, *P. rapae* compensates for the costs of hippuric acid formation (loss of glycine) by recycling some of the cyanide formed via amino acid metabolism (Stauber et al., 2012).

These examples demonstrate that the metabolism of benzenic GLS in *Pieris* species is considerably more complex than that of aliphatic GLS, whose only described end products are the corresponding nitriles (Wittstock et al., 2004). In fact, the biochemical adaptations towards the benzenic GLS metabolite cyanide may have facilitated the initial colonization of cyanogenic glycoside-producing Fabales by the Pieridae (Conn, 1980; Stauber et al., 2012) long before they were a beneficial factor for the host plant shift to Brassicales that contained benzenic GLS (Edger et al., 2015; Fahey et al., 2001; Mithen, Bennett, & Marquez, 2010). Both the expansion of GLS structures in the Brassicales and counter-adaptations in the Pieridae were associated with increases in species diversification. The evolution of novel GLS biosynthetic routes in the Brassicales was facilitated by gene and whole genome duplications (Edger et al., 2015). To date, the only insect species identified to harbour NSP activity are members of the crucifer-feeding Pieridae, and NSP genes were lost in lineages that shifted to non-GLS plants (Edger et al., 2015; Wheat et al., 2007). Thus, successive cycles of coevolution between Brassicales plants and Pierid herbivores have resulted in the complexity of GLS metabolism in *Pieris* species.

Brassicales plants themselves have evolved specifier proteins that direct GLS hydrolysis away from ITCs and towards nitriles: an epithiospecifier protein (ESP) (Abdalsamee, Giampà, Niehaus, & Müller, 2014; Lambrix et al., 2001; Petroski & Kwolek, 1985) or towards thiocyanates: a thiocyanate-forming protein (TFP) (Burow, Bergner, Gershenzon, &

Wittstock, 2007). These plant and insect specifier proteins — NSP, ESP and TFP — are not capable of hydrolysing GLS by themselves but work together with myrosinases as enzymes rather than allosteric cofactors of myrosinases (Burow, Markert, Gershenzon, & Wittstock, 2006). Despite the biochemical similarity of the *A. thaliana* ESP and the *P. rapae* NSP in terms of mechanisms and products, the two proteins differ in their primary sequence, dependence on metal cofactors and substrate specificity (Burow, Markert, et al., 2006). Also, they have distinctive biological roles in plant–insect interactions. For example, the formation of nitriles in *A. thaliana* rosette leaves catalysed by the plant AtNSP1 and AtNSP2 is a part of constitutive and herbivore-induced defence (Burow et al., 2009). In response to herbivory, formation of simple nitriles in *A. thaliana* rosette leaves increases at the expense of ITC formation (Burow et al., 2009; Lambrix et al., 2001). Several observations help explain why these plants actively modify metabolism to produce defences that are less toxic. First, increased nitrile concentrations emitted by the plant can lower its attractiveness to ovipositing females (Mumm et al., 2008; Rothschild & Schoonhoven, 1977) by acting as a signal of larvae feeding on the plant, which (1) implies the presence of feeding competitors and (2) warns of induced defensive secondary metabolites (other than GLS) that would negatively affect larval fitness (Agrawal & Kurashige, 2003; Mewis, Appel, Hom, Raina, & Schultz, 2005; Mewis et al., 2006; Mumm et al., 2008). Second, aliphatic-derived nitriles attract natural enemies of the larvae such as the parasitic wasp *Cotesia rubecula* (Hymenoptera) (Mumm et al., 2008). Phenylacetonitrile from benzyl GLS breakdown acts as an indirect plant defence in two ways: It represses the mating of female Pierid moths (antiaphrodisiac effect) and attracts enemies such as the generalist egg parasitoid *Trichogramma brassicae* (Hymenoptera) (Andersson, Borg-Karlson, & Wiklund, 2003; Fatouros et al., 2008; Fatouros, Huigens, van Loon, Dicke, & Hilker, 2005). However, successful parasitism may depend on other factors, such as the qualitative GLS profile of the host plant, which influences the performance of *P. rapae* larvae (Kos, Houshyani, Wietsma, et al., 2012). Thus Brassicaceae plants appear to synthesize specifier proteins to switch between direct (ITC) and indirect (nitrile) defences, which can be effective against different types of herbivores. As a result, the same GLS breakdown metabolites — nitriles — formed by an insect for detoxification purposes are used by the plant for signalling indirect defences. In contrast to the simple nitriles, the biological functions of epithionitriles formed by the specifier protein ESP are still unknown.

3.2 Glucosinolate Sulphatases: Preventing Glucosinolate Activation

Another strategy in disarming the mustard oil bomb is to metabolize GLS to make them unrecognizable as substrates for the activating myrosinases. The enzyme GLS sulphatase (GSS) converts intact GLS into desulpho GLS (Fig. 2), a process that reverses the last step of core GLS biosynthesis. Unlike NSP, whose activity is only found in species of the Pieridae, GSS activity in vivo or in vitro has been identified in four different orders of Insecta and even one Gastropoda: in the diamondback moth, *P. xylostella* (Lepidoptera) (Ratzka et al., 2002); in the desert locust, *Schistocerca gregaria* (Orthoptera) (Falk & Gershenzon, 2007); in the silverleaf whitefly, *Bemisia tabaci* (Hemiptera) (Malka et al., 2016); in the turnip sawfly, *A. rosae* (Opitz, Mix, Winde, & Müller, 2011; discussed in Section 3.3) and in the Burgundy snail, *Helix pomatia* (Gastropoda: Stylommatophora) (Thies, 1979).

The GSS of *H. pomatia* was the first to be recognized (Thies, 1979), and is now routinely used in the purification of GLS from plant tissues for quantitative chemical analysis. After the extracted GLS are bound to an anion exchange column, they are treated with sulphatase to be released as desulpho GLS for further HPLC analysis. Although *H. pomatia* is a major herbivore of *A. thaliana*, it does not seem to use its GSS activity for the detoxification of GLS in vivo (Falk et al., 2014). Desulpho GLS were reported to be present in *H. pomatia* faeces after ingestion of filter paper soaked with allyl-GLS (sinigrin) (Ratzka et al., 2002), but they could not be detected when *H. pomatia* fed on cruciferous plants, suggesting that *H. pomatia* GSS cannot outcompete the plant myrosinase. Instead, *H. pomatia* and some other molluscan herbivores (Portuguese slug, *Arion lusitanicus*, and marsh slug, *Deroceras laeve*) detoxify GLS by conjugating ITCs resulting from hydrolysis to GSH as described earlier (Falk et al., 2014).

The first insect GSS was discovered in *P. xylostella* (Ratzka et al., 2002), which is to date also the only specialist feeder and lepidopteran insect known to use GSS to disarm the GLS—myrosinase system (Wheat et al., 2007). Conversion of the parent GLS to its desulphated derivative is effective in blocking hydrolysis since plant myrosinases specifically bind the sulphate moiety of GLS in their active site, but do not recognize desulpho GLS as substrates (Matile, 1980). Additionally, myrosinase activity is inhibited by the free sulphate released by the GSS-catalysed reaction (Shikita, Fahey, Golden, Holtzclaw, & Talalay, 1999). *P. xylostella* constitutively secretes the GSS protein into its gut lumen during all larval stages as long as it actively feeds on a diet of Brassicaceae species (Ratzka et al., 2002; You et al., 2013). The amount of protein secreted by a fourth instar larva has an activity of approximately 20–28 nmol/min, which is sufficient to degrade the GLS present in a typically ingested amount of *A. thaliana* Col-0 to desulpho GLS within minutes. Furthermore, *P. xylostella* GSS can efficiently desulphate GLS with varied side chains (Ratzka et al., 2002), thus minimizing the effects of host plant GLS variation (Kliebenstein, Pedersen, Barker, & Mitchell-Olds, 2002; Li et al., 2000; Sarosh et al., 2010). However, elevated myrosinase activity, which may outcompete GSS, can negatively affect *P. xylostella* larvae (Li et al., 2000).

Surprisingly, the polyphagous locust *S. gregaria* also harbours a GSS (Falk & Gershenzon, 2007). This insect was found to feed on the West African plant *Schouwia purpurea*, which contains extraordinarily high amounts of GLS, without suffering adverse effects (Ghaout, Louveaux, Mainguet, Deschamps, & Rahal, 1991; Mainguet, Louveaux, El Sayed, & Rollin, 2000). The GLS composition of *S. purpurea* is dominated by the short-chain aliphatic 3-butenyl GLS (64–92% of total GLS) followed by (*S*)-2-hydroxy-3-butenyl and indol-3-ylmethyl GLS (Falk & Gershenzon, 2007). The ingested GLS are 99% excreted as desulpho GLS in the faeces by *S. gregaria* (Falk & Gershenzon, 2007). The rest undergoes breakdown mainly to epithionitriles, e.g., 3-epithiobutyl cyanide (= 3-epithiopentanenitrile, >56% of total breakdown products) with minor amounts of 3-butenyl ITC (7%) (Falk & Gershenzon, 2007), suggesting *S. purpurea* also possesses an ESP (Burow, Müller, et al., 2006; Lambrix et al., 2001). *S. gregaria* does not constitutively exhibit high levels of GSS catalysis like *P. xylostella*, but induces this activity up to 10-fold upon feeding on GLS-containing plant material (Falk & Gershenzon, 2007). Induction of detoxification enzymes is a strategy found in other generalist herbivores that deal with a diversity of plant metabolites (Hoy, Head, & Hall, 1998; Terriere, 1984; Yu & Hsu, 1993), and avoids a waste of resources on detoxification capacity when a particular defence compound is not present in the diet.

Recently, a piercing-sucking insect, the silverleaf whitefly *B. tabaci*, was shown to utilize GSS activity for GLS detoxification (Malka et al., 2016). *B. tabaci* is a species complex with a very broad feeding range, infesting more than 600 host plants including some Brassicaceae. Since the feeding mode of piercing-sucking insects mostly avoids the detonation of the mustard oil bomb, these insects are generally considered to be less susceptible to the more abundant aliphatic GLS, while still being negatively affected by indolic GLS (as discussed in Section 2.2). Nonetheless, both types of GLS have adverse effects on the reproduction of *B. tabaci*, suggesting that this insect encounters significant amounts of GLS during host probing (Elbaz et al., 2012; Markovich et al., 2013). Notably, different species of the *B. tabaci* complex are differentially affected by aliphatic and indolic GLS: while indolic GLS seem to be more detrimental for the development of the MEAM1 species (formerly known as B), the MED Q2 species (formerly known as Q2) is more negatively affected by aliphatic GLS, likely due to differential expression of GST genes (Elbaz et al., 2012). Given the negative effects of aliphatic GLS on *B. tabaci*, it seems likely that it would possess adaptations to feed on plants of the Brassicaceae. Indeed, large amounts of desulphated as well as intact aliphatic GLS were excreted by *B. tabaci* (MED, Q2 subspecies) (Malka et al., 2016), but only traces of intact indolic GLS were recovered suggesting that these succumb to myrosinase-independent degradation. Future research should focus on the absolute quantification of these processes and the identification and biochemical characterization of the *B. tabaci* GSS to clarify substrate specificities, kinetic parameters and possible inducibility.

3.3 Sequestration: Herbivores Make Their Own Glucosinolate–Myrosinase Bomb

Sequestration of plant defence metabolites by insect herbivores after feeding is a widespread adaptation to exploit these compounds for their own defence against predators (Heckel, 2014). The sequestration of GLS has been described in insect species from different orders including the chewing specialist feeders *A. rosae* (turnip sawfly, Hymenoptera: Tenthredinidae) (Müller et al., 2001), *Murgantia histrionica* (harlequin cabbage bug, Hemiptera: Pentatomidae) (Aliabadi et al., 2002) and *Phyllotreta striolata* (striped flea beetle, Coleoptera: Chrysomelidae) (Beran et al., 2014), and the piercing-sucking specialist *Brevicoryne brassicae* (cabbage aphid, Hemiptera: Aphididae) (Francis et al., 2001). Successful sequestration requires efficient uptake of GLS before myrosinases break them down, which may involve the inhibition of myrosinase activity inside the insect during feeding. To avoid autotoxicity, a storage site is needed in which GLS breakdown is prevented until this is needed for defence (Müller, 2009). Some insect species have even evolved their own myrosinases, stored in a separate compartment from the sequestered GLS, which are utilized for activating defence against predators (Beran et al., 2014; Bridges et al., 2002; Francis et al., 2002). The mechanisms of uptake, storage, and metabolism of GLS and the evolution of insect myrosinases will be explored further in this section.

The sawfly *A. rosae* is a well-studied example of a GLS-sequestering insect whose larvae deploy this defence via the characteristic 'easy bleeding' phenomenon (Boevé & Schaffner, 2003): GLS are enriched in the haemolymph and readily released as droplets on the integument when the larva is injured or touched deterring predators (Müller et al., 2001; Müller & Brakefield, 2003; Ohara, Nagasaka, & Ohsaki, 1993; Opitz, Jensen, & Müller, 2010). Experiments employing [^{14}C]-labelled compounds showed that sequestered GLS are stored exclusively in the haemolymph in all larval stages and the adult (Müller & Wittstock, 2005) with the highest GLS concentrations present in the larval haemolymph (Opitz et al., 2010). Larvae can accumulate up to 5 μmol GLS/g fresh weight but the final concentration fluctuates depending on the side chain-specificity of sequestration and the composition of the host plant (Müller & Wittstock, 2005; Müller, Zwaan, de Vos, & Brakefield, 2003).

The GLS found in *A. rosae* haemolymph are the same as those in the host plant, suggesting that larvae do not possess the machinery for de novo GLS biosynthesis (Müller et al., 2001). However, the GLS classes are sequestered in different proportions with aliphatic and benzenic GLS detected in significantly higher amounts compared to their concentrations in the host plant (Abdalsamee & Müller, 2012; Müller et al., 2001; Opitz et al., 2010). Indolic GLS can reach haemolymph levels comparable with aliphatic and benzenic GLS if the host plant has a high indolic GLS content, but are usually present in minute amounts (Abdalsamee & Müller, 2012; Müller et al., 2001; Müller & Wittstock, 2005; Opitz et al., 2010), which suggests that they might undergo spontaneous degradation at least partially before uptake as discussed previously (Section 2.2) (Kim et al., 2008). Surprisingly, *A. rosae* larvae are apparently not sensitive to the less stable indolic GLS. Larvae reared on a wild *B. oleracea* population, which has indolic GLS in very high concentrations and a significantly higher total GLS content compared to a control population, showed improved adult physiological traits such as weight, fecundity and longevity (Abdalsamee & Müller, 2012). *A. rosae* larvae tolerate a high range of both GLS and myrosinase concentrations in their host plants (Müller & Sieling, 2006) suggesting a fast uptake mechanism and possibly a strategy to inhibit myrosinases. Analyses of GLS concentrations and myrosinase activities along the gut suggest that GLS uptake happens in the front part of the gut where myrosinase activity is also the lowest (Abdalsamee et al., 2014). 4-Hydroxybenzyl GLS (sinalbin) can be detected in haemolymph as soon as 5 min after feeding on *Sinapis alba* begins (Abdalsamee et al., 2014), and GLS concentrations in haemolymph increase steadily until reaching a maximum (Müller & Wittstock, 2005). However, sequestered GLS are rapidly turned over and leave the body within a day upon diet change (Müller & Wittstock, 2005). Thus, *A. rosae* larvae are required to continuously feed on GLS-containing plants to maintain their haemolymph GLS levels.

Besides the uptake and storage of dietary GLS, *A. rosae* larvae must also have a strategy for excretion that avoids autotoxicity. Intact GLS were only found in traces in the midgut and faeces, and minute concentrations of desulpho GLS could be found in haemolymph and faeces (Abdalsamee et al., 2014; Müller & Wittstock, 2005). Experiments with [^{14}C]-labelled benzyl GLS helped to clarify the metabolic fates of the GLS in the larval haemolymph (Opitz et al., 2011). Benzyl GLS was transformed mainly into desulphobenzyl-GLS-3″-sulphate, and minor amounts of its presumed precursor desulphobenzyl GLS were also detected (Fig. 6) (Opitz et al., 2011). GSS-catalysed GLS desulphation is a well-studied strategy in other insects of the orders Lepidoptera, Orthoptera and Hemiptera that feed on GLS-containing plants (Section 3.2) (Falk & Gershenzon, 2007; Malka et al., 2016; Ratzka et al., 2002; Thies, 1979). *A. rosae* likely also harbours one or more GSS, but these remain to be identified. Following desulphation, a sulphotransferase presumably catalyses the formation of desulphobenzyl GLS-3″-sulphate (Opitz et al., 2011). Sulphotransferases are commonly involved in xenobiotic detoxification (Smith, 1955; Yang & Wilkinson, 1973), and also participate in other GLS detoxification processes (Agerbirk et al., 2010; Vergara et al., 2006). The *O*-sulphation at C-3″ in the glucose moiety is advantageous for two reasons: (1) it makes the molecule more hydrophilic, thus aiding in excretion and (2) desulphobenzyl-GLS-3″-sulphate is also not a suitable substrate for plant myrosinases, avoiding further breakdown within the larva. The putative transporters that

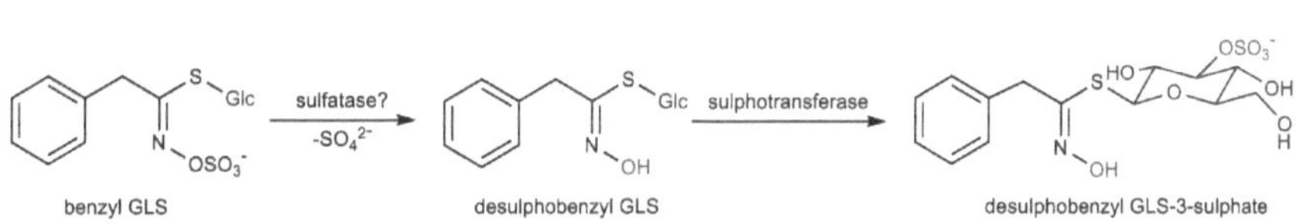

Figure 6 *Athalia rosae* metabolizes sequestered benzyl glucosinolate (GLS) to desulphobenzyl GLS, presumably by the action of a glucosinolate sulphatase (GSS). Further, a sulphotransferase adds a sulphate moiety to the C3 position of the glucose, and the resulting desulphobenzyl GLS-3″-sulphate is excreted.

facilitate the excretion of the sulphated GLS metabolites from the haemolymph into the gut lumen in *A. rosae* larvae are still unidentified. Future efforts in this area may be facilitated by the identification of transporters of conjugated plant metabolites in other insects (Discher et al., 2009; Homolya, Váradi, & Sarkadi, 2003; Kuhn et al., 2004; Liu et al., 2011; Strauss, Peters, Boland, & Burse, 2013).

Although the 'easy bleeding' phenomenon points to the use of GLS as a defence by *A. rosae*, an insect myrosinase activity was not found in larval tissue of this insect, suggesting the sequestered GLS remain inactivated (Müller & Wittstock, 2005). Other known GLS-sequestering herbivores, on the other hand, have been shown to harbour their own myrosinases, including some specialist aphids and the striped flea beetle *P. striolata*, which can be designated 'walking mustard oil bombs' (Kazana et al., 2007). The *P. striolata* and *B. brassicae* myrosinases are different from the Brassicaceae myrosinases, and phylogenetic analysis suggests that both insect myrosinases also evolved independently (Beran et al., 2014; Husebye et al., 2005; Pontoppidan, Ekbom, Eriksson, & Meijer, 2001). Both insect myrosinases displayed large differences in activity and kinetic parameters towards GLS with different side chains: Aliphatic GLS were hydrolysed more efficiently than benzenic GLS, followed by indolic GLS (Beran et al., 2014; Francis et al., 2002). The main hydrolysis products of the aphid myrosinase were the corresponding ITCs for all tested GLS, except for (*S*)-2-hydroxy-3-butenyl GLS (epiprogoitrin) whose main product is the cyclized oxazolidinethione (Francis et al., 2002). In aphids, the myrosinase is spatially separated in crystalline microbodies in the non-flight muscle, while the sequestered GLS are in the haemolymph (Bridges et al., 2002; Francis et al., 2002; Husebye et al., 2005; Jones, Bridges, Bones, Cole, & Rossiter, 2001; Pontoppidan et al., 2001). Furthermore, the abundance of the myrosinase protein increases steadily throughout the development (approximately twofold in total) in both wingless and winged aphids (Kazana et al., 2007).

The GLS are selectively sequestered in the flea beetle and cabbage aphid with the composition matching the hydrolysis preference of the corresponding myrosinases: aliphatic GLS are preferentially sequestered, while indolic GLS are only found in smaller amounts (Beran et al., 2014; Kos, Houshyani, Achhami, et al., 2012). Flea beetles are even stimulated to feed by aliphatic GLS like allyl-GLS (sinigrin) (Nielsen et al., 2001). In addition to the class-specific sequestration, the individual structures of GLS within classes influence sequestration preference. For example, when *B. brassicae* were reared on *B. oleracea* they selectively take up allyl-GLS (three carbon side chain), while (*R*)-2-hydroxy-3-butenyl GLS (progoitrin, four carbon side chain) is preferentially excreted (Goodey, Florance, Smirnoff, & Hodgson, 2015). In *P. striolata*, 4-methylthiobutyl GLS is sequestered in 100-fold higher amounts in the haemolymph in comparison to the host plant, while (*R*)-2-hydroxy-3-butenyl GLS is present in only 40-fold higher amounts (Beran et al., 2014).

P. striolata beetles sequester GLS up to 1.75% of their body weight and retain these for several weeks independently of the GLS present in the current host plant (Beran et al., 2014). This retention, as well as controlled utilization of its endogenous myrosinase, may enable *P. striolata* adults to release GLS hydrolysis products on demand. ITCs were proposed to have pheromone-like activities and induce aggregation behaviour in adult beetles at high concentrations (Beran et al., 2011). The GLS-sequestering *Lipaphis erysimi* (mustard aphid, Aphididae) releases ITCs as pheromones to synergize the action of the aphid alarm pheromone *E*-β-farnesene, which promotes the dispersion of the aphid colony after enemy attack (Dawson, Griffiths, Pickett, Wadhams, & Woodcock, 1987).

B. brassicae accumulate between 102 and 148 ng allyl-GLS in their haemolymph after feeding on *B. nigra* for one week (Bridges et al., 2002; Kazana et al., 2007). Similarly to the sawfly *A. rosae*, studies confirm that the aphids do not possess the machinery to biosynthesize GLS themselves (Francis, Haubruge, & Gaspar, 2000; Kazana et al., 2007; Pratt et al., 2008; Vanhaelen, Haubruge, Lognay, & Francis, 2001). Interestingly, GLS accumulation is three to four times higher in wingless aphids than in winged individuals, suggesting that GLS could be especially beneficial to wingless aphids which are less able to flee an approaching predator (Kazana et al., 2007). Different lines of evidence support the beneficial effects of GLS sequestration on aphids. Firstly, the performance, generation times and fecundity of *B. brassicae* aphids correlate positively with total host plant GLS concentrations (Chaplin-Kramer et al., 2011; Kos, Houshyani, Achhami, et al., 2012). Secondly, the accumulation of GLS in aphids has significant defensive value against a variety of predators. The survival of both hoverflies and ladybugs can be negatively affected by the GLS composition of the aphid host plants and the GLS amounts sequestered (Chaplin-Kramer et al., 2011; Francis et al., 2000, 2001; Kazana et al., 2007; Kos, Houshyani, Achhami, et al., 2012; Pratt et al., 2008). For example, survival of the ladybug *Adalia bipunctata* (Coleoptera) as well as the growth and development time of *Coccinella septempunctuata* (Coleoptera) are all negatively affected by increased allyl-GLS accumulation in the aphid (Pratt et al., 2008). However, *A. bipunctata* survival rates are high when feeding on the generalist aphid *M. persicae* reared on *B. nigra*, confirming that *M. persicae* does not sequester GLS (Pratt et al., 2008) and only carries small amounts of GLS in their bodies (Bridges et al., 2002). Interestingly, the hoverfly *Episyrphus balteatus* (Diptera) induces GSTs in reaction to the toxic GLS breakdown products ingested from its aphid prey (Vanhaelen et al., 2001) suggesting a GSH-dependent detoxification of ITCs in the third trophic level.

Although aphid predators are negatively influenced by GLS sequestration in their prey, the performance of the parasitic wasp *Diaeretiella rapae* (Hymenoptera) was positively influenced by aphids feeding on high GLS plants (Kos, Houshyani, Achhami, et al., 2012), with larval development time, adult longevity and parasitism rates additionally affected by changes in the GLS profiles of the aphid host plant (Bayhan, Ulusoy, & Bayhan, 2007). *D. rapae* even utilizes GLS volatile breakdown products as attractive cues in locating aphids (Najar-Rodriguez, Friedli, Klaiber, & Dorn, 2015). Interestingly, *B. brassicae* aphids are apparently able to suppress the emission of allyl-ITC by *B. oleracea* var. *gemmifera*, thus indirectly reducing parasitism by *D. rapae* (Najar-Rodriguez et al., 2015).

The harlequin bug *M. histrionica* is another example of a GLS-sequestering insect whose GLS composition is comparable to that of the host plant, and it retains intact GLS for several weeks mainly in the haemolymph (Aliabadi et al., 2002). GLS breakdown products (ITCs and nitriles) were detected in the prothoracic fluid, but it is not known whether this insect harbours its own myrosinase or how GLS hydrolysis is accomplished. The GLS breakdown products are emitted to the environment through the prothoracic fluid when *M. histrionica* is squeezed (Aldrich et al., 1996). These insects are unpalatable to several species of birds (Aliabadi et al., 2002), although direct proof for a defensive function of the sequestered GLS and corresponding breakdown products is still lacking. In this and all insect species discussed above, the transport processes that are involved in GLS uptake, accumulation and excretion are still poorly described and certainly merit further study taking advantage of recent advances in insect molecular biology.

4. CONCLUSION

The GLS–myrosinase system protects plants against many herbivore species, but both generalist and specialist feeders possess mechanisms to circumvent this defence system and feed on Brassicales plants to varying degrees. Some of these mechanisms involve specialized adaptations to the GLS—myrosinase system, while others utilize generalized detoxification processes for xenobiotic metabolism and transport. However, the examples highlighted in this chapter clearly show that a segregation between 'specialist' and 'generalist' GLS detoxification strategies does not necessarily correlate with corresponding insect feeding habits.

Herbivore mechanisms to cope with activated, two-component plant defences such as GLS could theoretically be directed at one or both components. We have seen that herbivores can metabolize GLS to nonactivate-able derivatives (desulpho-GLS) or divert hydrolysis to less toxic products (nitriles). Or, herbivores may allow GLS activation to occur and then detoxify the ITC hydrolysis products (via conjugation to GSH). However, these known examples concern mainly chewing rather than piercing-sucking herbivores and involve the detoxification of aliphatic and benzenic GLS rather than indolic GLS. There are still many gaps in our understanding.

GLS activation could also be prevented in other ways, for example, by inhibiting myrosinase activity, but there is as yet no evidence of this mechanism. Nevertheless, herbivores that sequester GLS from their diet must be able to block hydrolysis at least partially, and could employ this strategy. Even for the known mechanisms of GLS detoxification, we are missing many important details, such as how GLS and their metabolites are transported within the insect. And, our understanding of how GLS detoxification processes evolved is still very rudimentary. In addition to metabolic adaptations, we know little about whether herbivores avoid the toxicity of GLS by altering the target sites of ITCs or by behavioural adaptations, such as snipping leaves into large pieces, rather than chewing them, to minimize plant tissue damage and reduce GLS activation (Pentzold et al., 2014).

GLS possess a wide variety of side chains with different functional groups. In addition, there is diversity generated during hydrolysis that depends on the conditions present at the site of GLS breakdown. The effect of the natural diversity of GLS on herbivore detoxification is also unexplored. Understanding how insects of different orders or even different species are differentially affected by GLS and their hydrolysis products with varying chemistry might reveal why plants maintain such a large variety of GLS structures.

While there are still many open questions about herbivores and the GLS—myrosinase system, recent advances in analytical, chemical and molecular methods have improved our ability to answer them. For example, the increased sensitivity of MS and NMR methods allows identification of minute quantities of GLS metabolites, and MS imaging could determine their location in the insect. The availability of transgenic plant lines with modified GLS profiles permits researchers to control the GLS content of herbivore diets in a whole plant context. The increased availability of insect genomes and transcriptomes facilitates the isolation of genes involved in GLS detoxification and studies on their evolution and function. In particular, the ability to silence insect genes involved in detoxification processes will help to determine the effectiveness of these processes and what the costs and benefits are for herbivores. New developments that improve our basic knowledge of GLS detoxification by herbivores could also impact plant protection efforts. Information about how crop pests metabolize GLS or their hydrolysis products might be useful in engineering plants that are more resistant to insect damage.

REFERENCES

Abdalsamee, M. K., Giampà, M., Niehaus, K., & Müller, C. (2014). Rapid incorporation of glucosinolates as a strategy used by a herbivore to prevent activation by myrosinases. *Insect Biochemistry and Molecular Biology, 52*, 115–123.

Abdalsamee, M. K., & Müller, C. (2012). Effects of indole glucosinolates on performance and sequestration by the sawfly *Athalia rosae* and consequences of feeding on the plant defense system. *Journal of Chemical Ecology, 38*(11), 1366–1375.

Agerbirk, N., De Vos, M., Kim, J. H., & Jander, G. (2009). Indole glucosinolate breakdown and its biological effects. *Phytochemistry Reviews, 8*(1), 101–120.

Agerbirk, N., Müller, C., Olsen, C. E., & Chew, F. S. (2006). A common pathway for metabolism of 4-hydroxybenzylglucosinolate in *Pieris* and *Anthocaris* (Lepidoptera: Pieridae). *Biochemical Systematics and Ecology, 34*(3), 189–198.

Agerbirk, N., Olsen, C. E., Poulsen, E., Jacobsen, N., & Hansen, P. R. (2010). Complex metabolism of aromatic glucosinolates in *Pieris rapae* caterpillars involving nitrile formation, hydroxylation, demethylation, sulfation, and host plant dependent carboxylic acid formation. *Insect Biochemistry and Molecular Biology, 40*(2), 126–137.

Agerbirk, N., Olsen, C. E., Topbjerg, H. B., & Sørensen, J. C. (2007). Host plant-dependent metabolism of 4-hydroxybenzylglucosinolate in *Pieris rapae*: substrate specificity and effects of genetic modification and plant nitrile hydratase. *Insect Biochemistry and Molecular Biology, 37*(11), 1119–1130.

Agrawal, A. A., & Kurashige, N. S. (2003). A role for isothiocyanates in plant resistance against the specialist herbivore *Pieris rapae*. *Journal of Chemical Ecology, 29*(6), 1403–1415.

Åhman, I. (1986). Toxicities of host secondary compounds to eggs of the *Brassica* specialist *Dasineura brassicae*. *Journal of Chemical Ecology, 12*(6), 1481–1488.

Ahuja, I., Rohloff, J., & Bones, A. M. (2010). Defence mechanisms of Brassicaceae: implications for plant-insect interactions and potential for integrated pest management. A review. *Agronomy for Sustainable Development, 30*(2), 311–348.

Al Janobi, A. A., Mithen, R. F., Gasper, A. V., Shaw, P. N., Middleton, R. J., Ortori, C. A., & Barrett, D. A. (2006). Quantitative measurement of sulforaphane, iberin and their mercapturic acid pathway metabolites in human plasma and urine using liquid chromatography–tandem electrospray ionisation mass spectrometry. *Journal of Chromatography B, 844*(2), 223–234.

Aldrich, J. R., Avery, J. W., Lee, C. J., Graf, J. C., Harrisons, D. J., & Bin, F. (1996). Semiochemistry of cabbage bugs (Heteroptera: Pentatomidae: Eurydema and Murgantia). *Journal of Entomological Science, 31*(2), 172–182.

Aliabadi, A., Renwick, J. A. A., & Whitman, D. W. (2002). Sequestration of glucosinolates by harlequin bug *Murgantia histrionica. Journal of Chemical Ecology, 28*(9), 1749–1762.

Andersson, J., Borg-Karlson, A. K., & Wiklund, C. (2003). Antiaphrodisiacs in pierid butterflies: a theme with variation! *Journal of Chemical Ecology, 29*(6), 1489–1499.

Andréasson, E., & Jørgensen, L. B. (2003). Localization of plant myrosinases and glucosinolates. In J. T. Romeo (Ed.), *Integrative phytochemistry: From ethnobotany to molecular ecology: Vol. 37. Recent advances in phytochemistry* (pp. 79–99). Amsterdam: Elsevier.

Baasanjav-Gerber, C., Monien, B. H., Mewis, I., Schreiner, M., Barillari, J., Iori, R., & Glatt, H. (2011). Identification of glucosinolate congeners able to form DNA adducts and to induce mutations upon activation by myrosinase. *Molecular Nutrition & Food Research, 55*(5), 783–792.

Badenes-Perez, F. R., Reichelt, M., Gershenzon, J., & Heckel, D. G. (2013). Interaction of glucosinolate content of *Arabidopsis thaliana* mutant lines and feeding and oviposition by generalist and specialist lepidopterans. *Phytochemistry, 86*, 36–43.

Ballhorn, D. J., Kautz, S., Heil, M., & Hegeman, A. D. (2009). Cyanogenesis of wild lima bean (*Phaseolus lunatus* L.) is an efficient direct defence in nature. *PLoS One, 4*(5), e5450.

Barker, A. M., Molotsane, R., Müller, C., Schaffner, U., & Städler, E. (2006). Chemosensory and behavioural responses of the turnip sawfly, *Athalia rosae*, to glucosinolates and isothiocyanates. *Chemoecology, 16*(4), 209–218.

Barth, C., & Jander, G. (2006). Arabidopsis myrosinases TGG1 and TGG2 have redundant function in glucosinolate breakdown and insect defense. *Plant Journal, 46*(4), 549–562.

Bayhan, S.Ö., Ulusoy, M. R., & Bayhan, E. (2007). Is the parasitization rate of *Diaeretiella rapae* influenced when *Brevicoryne brassicae* feeds on *Brassica* plants? *Phytoparasitica, 35*(2), 146–149.

Bednarek, P., & Osbourn, A. (2009). Plant-microbe interactions: chemical diversity in plant defense. *Science, 324*(5928), 746–748.

Beekwilder, J., van Leeuwen, W., van Dam, N. M., Bertossi, M., Grandi, V., Mizzi, L., … Bovy, A. (2008). The impact of the absence of aliphatic glucosinolates on insect herbivory in Arabidopsis. *PLoS One, 3*(4), e2068.

Beilstein, M. A., Nagalingum, N. S., Clements, M. D., Manchester, S. R., & Mathews, S. (2010). Dated molecular phylogenies indicate a Miocene origin for *Arabidopsis thaliana. Proceedings of the National Academy of Sciences of the United States of America, 107*(43), 18724–18728.

Bejai, S., Fridborg, I., & Ekbom, B. (2012). Varied response of *Spodoptera littoralis* against *Arabidopsis thaliana* with metabolically engineered glucosinolate profiles. *Plant Physiology and Biochemistry, 50*, 72–78.

Benderoth, M., Pfalz, M., & Kroymann, J. (2009). Methylthioalkylmalate synthases: genetics, ecology and evolution. *Phytochemistry Reviews, 8*(1), 255–268.

Beran, F., Mewis, I., Srinivasan, R., Svoboda, J., Vial, C., Mosimann, H., … Reinecke, A. (2011). Male *Phyllotreta striolata* (F.) produce an aggregation pheromone: identification of male-specific compounds and interaction with host plant volatiles. *Journal of Chemical Ecology, 37*(1), 85–97.

Beran, F., Pauchet, Y., Kunert, G., Reichelt, M., Wielsch, N., Vogel, H., … Heckel, D. G. (2014). *Phyllotreta striolata* flea beetles use host plant defense compounds to create their own glucosinolate-myrosinase system. *Proceedings of the National Academy of Sciences of the United States of America, 111*(20), 7349–7354.

Bidart-Bouzat, M. G., & Kliebenstein, D. J. (2008). Differential levels of insect herbivory in the field associated with genotypic variation in glucosinolates in *Arabidopsis thaliana. Journal of Chemical Ecology, 34*(8), 1026–1037.

Bodnaryk, R. P. (1994). Potent effect of jasmonates on indole glucosinolates in oilseed rape and mustard. *Phytochemistry, 35*(2), 301–305.

Boevé, J. L., & Schaffner, U. (2003). Why does the larval integument of some sawfly species disrupt so easily? The harmful hemolymph hypothesis. *Oecologia, 134*(1), 104–111.

Bohinc, T., Ban, S. G., Ban, D., & Trdan, S. (2012). Glucosinolates in plant protection strategies: a review. *Archives of Biological Sciences, 64*(3), 821–828.

Boyland, E., & Chasseaud, L. F. (1969). The role of glutathione and glutathione *S*-transferases in mercapturic acid biosynthesis. *Advances in Enzymology and Related Areas of Molecular Biology, 32*, 173–219.

Bridges, M., Jones, A. M. E., Bones, A. M., Hodgson, C., Cole, R., Bartlet, E., … Rossiter, J. T. (2002). Spatial organization of the glucosinolate-myrosinase system in brassica specialist aphids is similar to that of the host plant. *Proceedings of the Royal Society B-Biological Sciences, 269*(1487), 187–191.

Brown, K. K., & Hampton, M. B. (2011). Biological targets of isothiocyanates. *Biochimica et Biophysica Acta (BBA) – General Subjects, 1810*(9), 888–894.

Brown, P. D., Tokuhisa, J. G., Reichelt, M., & Gershenzon, J. (2003). Variation of glucosinolate accumulation among different organs and developmental stages of *Arabidopsis thaliana*. *Phytochemistry, 62*, 471–481.

Burow, M., Bergner, A., Gershenzon, J., & Wittstock, U. (2007). Glucosinolate hydrolysis in *Lepidium sativum* - identification of the thiocyanate-forming protein. *Plant Molecular Biology, 63*(1), 49–61.

Burow, M., Losansky, A., Müller, R., Plock, A., Kliebenstein, D. J., & Wittstock, U. (2009). The genetic basis of constitutive and herbivore-induced ESP-independent nitrile formation in Arabidopsis. *Plant Physiology, 149*(1), 561–574.

Burow, M., Markert, J., Gershenzon, J., & Wittstock, U. (2006). Comparative biochemical characterization of nitrile-forming proteins from plants and insects that alter myrosinase-catalysed hydrolysis of glucosinolates. *FEBS Journal, 273*(11), 2432–2446.

Burow, M., Müller, R., Gershenzon, J., & Wittstock, U. (2006). Altered glucosinolate hydrolysis in genetically engineered *Arabidopsis thaliana* and its influence on the larval development of *Spodoptera littoralis*. *Journal of Chemical Ecology, 32*(11), 2333–2349.

Calmes, B., N'Guyen, G., Dumur, J., Brisach, C. A., Campion, C., Iacomi, B., … Simoneau, P. (2015). Glucosinolate-derived isothiocyanates impact mitochondrial function in fungal cells and elicit an oxidative stress response necessary for growth recovery. *Frontiers in Plant Science, 6*, 414.

Chaplin-Kramer, R., Kliebenstein, D. J., Chiem, A., Morrill, E., Mills, N. J., & Kremen, C. (2011). Chemically mediated tritrophic interactions: opposing effects of glucosinolates on a specialist herbivore and its predators. *Journal of Applied Ecology, 48*(4), 880–887.

Conn, E. E. (1980). Cyanogenic compounds. *Annual Review of Plant Physiology and Plant Molecular Biology, 31*, 433–451.

Cross, J. V., Rady, J. M., Foss, F. W., Jr., Lyons, C., Macdonald, T. L., & Templeton, D. J. (2009). Nutrient isothiocyanates covalently modify and inhibit the inflammatory cytokine macrophage migration inhibitory factor (MIF). *Biochemical Journal, 423*(3), 315–321.

David, W. A. L., & Gardiner, B. O. C. (1966). Mustard oil glucosides as feeding stimulants for *Pieris brassicae* larvae in a semi-synthetic diet. *Entomologia Experimentalis et Applicata, 9*(2), 247–255.

Dawson, G. W., Griffiths, D. C., Pickett, J. A., Wadhams, L. J., & Woodcock, C. M. (1987). Plant-derived synergists of alarm pheromone from turnip aphid, *Lipaphis (Hyadaphis) erysimi* (Homoptera, Aphididae). *Journal of Chemical Ecology, 13*(7), 1663–1671.

De Vos, M., & Jander, G. (2009). *Myzus persicae* (green peach aphid) salivary components induce defence responses in *Arabidopsis thaliana*. *Plant, Cell and Environment, 32*(11), 1548–1560.

De Vos, M., Kriksunov, K. L., & Jander, G. (2008). Indole-3-acetonitrile production from indole glucosinolates deters oviposition by *Pieris rapae*. *Plant Physiology, 146*(3), 916–926.

Deng, H., Huang, Y., Feng, Q., & Zheng, S. (2009). Two epsilon glutathione *S*-transferase cDNAs from the common cutworm, *Spodoptera litura*: characterization and developmental and induced expression by insecticides. *Journal of Insect Physiology, 55*(12), 1174–1183.

Després, L., David, J.-P., & Gallet, C. (2007). The evolutionary ecology of insect resistance to plant chemicals. *Trends in Ecology & Evolution, 22*(6), 298–307.

Discher, S., Burse, A., Tolzin-Banasch, K., Heinemann, S. H., Pasteels, J. M., & Boland, W. (2009). A versatile transport network for sequestering and excreting plant glycosides in leaf beetles provides an evolutionary flexible defense strategy. *ChemBioChem: A European Journal of Chemical Biology, 10*(13), 2223–2229.

Dussourd, D. E., & Eisner, T. (1987). Vein-cutting behavior: insect counterploy to the latex defense of plants. *Science, 237*(4817), 898–901.

Edger, P. P., Heidel-Fischer, H. M., Bekaert, M., Rota, J., Glöckner, G., Platts, A. E., … Wheat, C. W. (2015). The butterfly plant arms-race escalated by gene and genome duplications. *Proceedings of the National Academy of Sciences of the United States of America, 112*(27), 8362–8366.

Ehrlich, P. R., & Raven, P. H. (1964). Butterflies and plants: a study in coevolution. *Evolution: International Journal of Organic Evolution, 18*(4), 586–608.

Eklind, K. I., Morse, M. A., & Chung, F.-L. (1990). Distribution and metabolism of the natural anticarcinogen phenethyl isothiocyanate in A/J mice. *Carcinogenesis, 11*(11), 2033–2036.

Elbaz, M., Halon, E., Malka, O., Malitsky, S., Blum, E., Aharoni, A., & Morin, S. (2012). Asymmetric adaptation to indolic and aliphatic glucosinolates in the B and Q sibling species of *Bemisia tabaci* (Hemiptera: Aleyrodidae). *Molecular Ecology, 21*(18), 4533–4546.

Fahey, J. W., Zalcmann, A. T., & Talalay, P. (2001). The chemical diversity and distribution of glucosinolates and isothiocyanates among plants. *Phytochemistry, 56*(1), 5–51.

Falk, K. L., & Gershenzon, J. (2007). The desert locust, *Schistocerca gregaria*, detoxifies the glucosinolates of *Schouwia purpurea* by desulfation. *Journal of Chemical Ecology, 33*(8), 1542–1555.

Falk, K. L., Kästner, J., Bodenhausen, N., Schramm, K., Paetz, C., Vassão, D. G., … Meldau, S. (2014). The role of glucosinolates and the jasmonic acid pathway in resistance of *Arabidopsis thaliana* against molluscan herbivores. *Molecular Ecology, 23*(5), 1188–1203.

Fatouros, N. E., Broekgaarden, C., Bukovinszkine'Kiss, G., van Loon, J. J. A., Mumm, R., Huigens, M. E., … Hilker, M. (2008). Male-derived butterfly anti-aphrodisiac mediates induced indirect plant defense. *Proceedings of the National Academy of Sciences of the United States of America, 105*(29), 10033–10038.

Fatouros, N. E., Huigens, M. E., van Loon, J. J. A., Dicke, M., & Hilker, M. (2005). Chemical communication: butterfly anti-aphrodisiac lures parasitic wasps. *Nature, 433*(7027), 704.

Fraenkel, G. S. (1959). The raison d'être of secondary plant substances. *Science, 129*(3361), 1466–1470.

Francis, F., Gerkens, P., Harmel, N., Mazzucchelli, G., De Pauw, E., & Haubruge, E. (2006). Proteomics in *Myzus persicae*: effect of aphid host plant switch. *Insect Biochemistry and Molecular Biology, 36*(3), 219–227.

Francis, F., Haubruge, E., & Gaspar, C. (2000). Influence of host plants on specialist/generalist aphids and on the development of *Adalia bipunctata* (Coleoptera: Coccinellidae). *European Journal of Entomology, 97*(4), 481–485.
Francis, F., Lognay, G., Wathelet, J.-P., & Haubruge, E. (2001). Effects of allelochemicals from first (Brassicaceae) and second (*Myzus persicae* and *Brevicoryne brassicae*) trophic levels on *Adalia bipunctata*. *Journal of Chemical Ecology, 27*(2), 243–256.
Francis, F., Lognay, G., Wathelet, J.-P., & Haubruge, E. (2002). Characterisation of aphid myrosinase and degradation studies of glucosinolates. *Archives of Insect Biochemistry and Physiology, 50*(4), 173–182.
Francis, F., Vanhaelen, N., & Haubruge, E. (2005). Glutathione *S*-transferases in the adaptation to plant secondary metabolites in the *Myzus persicae* aphid. *Archives of Insect Biochemistry and Physiology, 58*(3), 166–174.
Frisch, T., Agerbirk, N., Davis, S., Cipollini, D., Olsen, C. E., Motawia, M. S., … Møller, B. L. (2014). Glucosinolate-related glucosides in *Alliaria petiolata*: sources of variation in the plant and different metabolism in an adapted specialist herbivore, *Pieris rapae*. *Journal of Chemical Ecology, 40*(10), 1063–1079.
Furlong, M. J., & Wright, D. J. (1994). Examination of stability of resistance and cross-resistance patterns to acylurea insect growth regulators in field populations of the diamondback moth, *Plutella xylostella*, from Malaysia. *Pesticide Science, 42*(4), 315–326.
Geu-Flores, F., Olsen, C. E., & Halkier, B. A. (2009). Towards engineering glucosinolates into non-cruciferous plants. *Planta, 229*(2), 261–270.
Ghaout, S., Louveaux, A., Mainguet, A. M., Deschamps, M., & Rahal, Y. (1991). What defense does *Schouwia purpurea* (Cruciferae) have against the desert locust? Secondary compounds and nutritive-value. *Journal of Chemical Ecology, 17*(8), 1499–1515.
Giamoustaris, A., & Mithen, R. (1995). The effect of modifying the glucosinolate content of leaves of oilseed rape (*Brassica napus* ssp. *oleifera*) on its interaction with specialist and generalist pests. *Annals of Applied Biology, 126*(2), 347–363.
Gleadow, R. M., & Woodrow, I. E. (2002). Constraints on effectiveness of cyanogenic glycosides in herbivore defense. *Journal of Chemical Ecology, 28*(7), 1301–1313.
Gloss, A. D., Vassão, D. G., Hailey, A. L., Nelson Dittrich, A. C., Schramm, K., Reichelt, M., … Whiteman, N. K. (2014). Evolution in an ancient detoxification pathway is coupled with a transition to herbivory in the Drosophilidae. *Molecular Biology and Evolution, 31*(9), 2441–2456.
Gols, R., Wagenaar, R., Bukovinszky, T., van Dam, N. M., Dicke, M., Bullock, J. M., & Harvey, J. A. (2008). Genetic variation in defense chemistry in wild cabbages affects herbivores and their endoparasitoids. *Ecology, 89*(6), 1616–1626.
Goodey, N. A., Florance, H. V., Smirnoff, N., & Hodgson, D. J. (2015). Aphids pick their poison: selective sequestration of plant chemicals affects host plant use in a specialist herbivore. *Journal of Chemical Ecology, 41*(10), 956–964.
Goyal, S., Lambert, C., Cluzet, S., Mérillon, J.M., & Ramawat, K.G. (2012). Secondary metabolites and plant defence. In H. M. T. Hokkanen (Series Ed.) & J. M. Mérillon, & K. G. Ramawat (Vol. Eds.), *Progress in biological control: Vol. 12. Plant defence: Biological control* (pp. 109–138). Netherlands: Springer.
Habig, W. H., Pabst, M. J., & Jakoby, W. B. (1974). Glutathione *S*-transferases – the first enzymatic step in mercapturic acid formation. *Journal of Biological Chemistry, 249*(22), 7130–7139.
Halkier, B. A., & Du, L. C. (1997). The biosynthesis of glucosinolates. *Trends in Plant Science, 2*(11), 425–431.
Halkier, B. A., & Gershenzon, J. (2006). Biology and biochemistry of glucosinolates. *Annual Review of Plant Biology, 57*, 303–333.

Hanschen, F. S., Brüggemann, N., Brodehl, A., Mewis, I., Schreiner, M., Rohn, S., & Kroh, L. W. (2012). Characterization of products from the reaction of glucosinolate-derived isothiocyanates with cysteine and lysine derivatives formed in either model systems or broccoli sprouts. *Journal of Agricultural and Food Chemistry, 60*(31), 7735–7745.

Hartmann, T. (1999). Chemical ecology of pyrrolizidine alkaloids. *Planta, 207*(4), 483–495.

Hartmann, T. (2007). From waste products to ecochemicals: fifty years research of plant secondary metabolism. *Phytochemistry, 68*(22–24), 2831–2846.

Harvey, J. A., Gols, R., Wagenaar, R., & Bezemer, T. M. (2007). Development of an insect herbivore and its pupal parasitoid reflect differences in direct plant defense. *Journal of Chemical Ecology, 33*(8), 1556–1569.

Heckel, D. G. (2014). Insect detoxification and sequestration strategies. In C. Voelckel, & G. Jander (Eds.), *Insect-plant interactions: Vol. 47. Annual plant reviews* (pp. 77–114). Chichester, UK: John Wiley & Sons, Ltd.

Herde, M., & Howe, G. A. (2014). Host plant-specific remodeling of midgut physiology in the generalist insect herbivore *Trichoplusia ni*. *Insect Biochemistry and Molecular Biology, 50*(0), 58–67.

Hicks, K. L. (1974). Mustard oil glucosides: feeding stimulants for adult cabbage flea beetles, *Phyllotreta cruciferae* (Coleoptera: Chrysomelidae). *Annals of the Entomological Society of America, 67*(2), 261–264.

Hilker, M., & Meiners, T. (2002). Induction of plant responses to oviposition and feeding by herbivorous arthropods: a comparison. *Entomologia Experimentalis et Applicata, 104*(1), 181–192.

Holzinger, F., Frick, C., & Wink, M. (1992). Molecular basis for the insensitivity of the monarch (*Danaus plexippus*) to cardiac glycosides. *FEBS Letters, 314*(3), 477–480.

Homolya, L., Váradi, A., & Sarkadi, B. (2003). Multidrug resistance-associated proteins: export pumps for conjugates with glutathione, glucuronate or sulfate. *BioFactors, 17*(1–4), 103–114.

Hopkins, R. J., van Dam, N. M., & van Loon, J. J. (2009). Role of glucosinolates in insect-plant relationships and multitrophic interactions. *Annual Review of Entomology, 54*, 57–83.

Hoy, C. W., Head, G. P., & Hall, F. R. (1998). Spatial heterogeneity and insect adaption to toxins. *Annual Review of Entomology, 43*, 571–594.

Hu, C., Eggler, A. L., Mesecar, A. D., & van Breemen, R. B. (2011). Modification of Keap1 cysteine residues by sulforaphane. *Chemical Research in Toxicology, 24*(4), 515–521.

Huang, X., & Renwick, J. A. A. (1993). Differential selection of host plants by two *Pieris* species: the role of oviposition stimulants and deterrents. *Entomologia Experimentalis et Applicata, 68*(1), 59–69.

Huang, Y., Xu, Z., Lin, X., Feng, Q., & Zheng, S. (2011). Structure and expression of glutathione *S*-transferase genes from the midgut of the common cutworm, *Spodoptera litura* (Noctuidae) and their response to xenobiotic compounds and bacteria. *Journal of Insect Physiology, 57*(7), 1033–1044.

Husebye, H., Arzt, S., Burmeister, W. P., Härtel, F. V., Brandt, A., Rossiter, J. T., & Bones, A. M. (2005). Crystal structure at 1.1 Å resolution of an insect myrosinase from *Brevicoryne brassicae* shows its close relationship to ß-glucosidases. *Insect Biochemistry and Molecular Biology, 35*(12), 1311–1320.

Husebye, H., Chadchawan, S., Winge, P., Thangstad, O. P., & Bones, A. M. (2002). Guard cell- and phloem idioblast-specific expression of thioglucoside glucohydrolase 1 (myrosinase) in Arabidopsis. *Plant Physiology, 128*(4), 1180–1188.

Iqbal, M., & Wright, D. J. (1997). Evaluation of resistance, cross-resistance and synergism of abamectin and teflubenzuron in a multi-resistant field population of *Plutella xylostella* (Lepidoptera: Plutellidae). *Bulletin of Entomological Research, 87*(5), 481–486.

Isidoro, N., Bartlet, E., Ziesmann, J., & Williams, I. H. (1998). Antennal contact chemosensilla in *Psylliodes chrysocephala* responding to cruciferous allelochemicals. *Physiological Entomology, 23*(2), 131–138.

Ivie, G. W., Bull, D. L., Beier, R. C., Pryor, N. W., & Oertli, E. H. (1983). Metabolic detoxification: mechanism of insect resistance to plant psoralens. *Science, 221*(4608), 374–376.

Jeffery, E. H., & Araya, M. (2009). Physiological effects of broccoli consumption. *Phytochemistry Reviews, 8*(1), 283–298.

Jeschke, V., Gershenzon, J., & Vassão, D. G. (2016). A mode of action of glucosinolate-derived isothiocyanates: detoxification depletes glutathione and cysteine levels with ramifications on protein metabolism in *Spodoptera littoralis*. *Insect Biochemistry and Molecular Biology, 71*, 37–48.

Jones, A. M. E., Bridges, M., Bones, A. M., Cole, R., & Rossiter, J. T. (2001). Purification and characterisation of a non-plant myrosinase from the cabbage aphid *Brevicoryne brassicae* (L.). *Insect Biochemistry and Molecular Biology, 31*(1), 1–5.

Kassahun, K., Davis, M., Hu, P., Martin, B., & Baillie, T. (1997). Biotransformation of the naturally occurring isothiocyanate sulforaphane in the rat: identification of phase I metabolites and glutathione conjugates. *Chemical Research in Toxicology, 10*, 1228–1233.

Kawakishi, S., & Kaneko, T. (1987). Interaction of proteins with allyl isothiocyanate. *Journal of Agricultural and Food Chemistry, 35*(1), 85–88.

Kawakishi, S., & Namiki, M. (1982). Oxidative cleavage of the disulfite bond of cysteine by allyl isothiocyanate. *Journal of Agricultural and Food Chemistry, 30*(3), 620–622.

Kazana, E., Pope, T. W., Tibbles, L., Bridges, M., Pickett, J. A., Bones, A. M., … Rossiter, J. T. (2007). The cabbage aphid: a walking mustard oil bomb. *Proceedings of the Royal Society B-Biological Sciences, 274*(1623), 2271–2277.

Kim, J. H., & Jander, G. (2007). *Myzus persicae* (green peach aphid) feeding on Arabidopsis induces the formation of a deterrent indole glucosinolate. *Plant Journal, 49*(6), 1008–1019.

Kim, J. H., Lee, B. W., Schroeder, F. C., & Jander, G. (2008). Identification of indole glucosinolate breakdown products with antifeedant effects on *Myzus persicae* (green peach aphid). *Plant Journal, 54*(6), 1015–1026.

Kissen, R., Rossiter, J. T., & Bones, A. M. (2009). The 'mustard oil bomb': not so easy to assemble?! Localization, expression and distribution of the components of the myrosinase enzyme system. *Phytochemistry Reviews, 8*(1), 69–86.

Kliebenstein, D., Pedersen, D., Barker, B., & Mitchell-Olds, T. (2002). Comparative analysis of quantitative trait loci controlling glucosinolates, myrosinase and insect resistance in *Arabidopsis thaliana*. *Genetics, 161*(1), 325–332.

Kliebenstein, D. J. (2004). Secondary metabolites and plant/environment interactions: a view through *Arabidopsis thaliana* tinged glasses. *Plant, Cell and Environment, 27*(6), 675–684.

Kliebenstein, D. J., Kroymann, J., Brown, P., Figuth, A., Pedersen, D., Gershenzon, J., & Mitchell-Olds, T. (2001). Genetic control of natural variation in *Arabidopsis glucosinolate* accumulation. *Plant Physiology, 126*, 811–825.

Kliebenstein, D. J., Kroymann, J., & Mitchell-Olds, T. (2005). The glucosinolate-myrosinase system in an ecological and evolutionary context. *Current Opinion in Plant Biology, 8*(3), 264–271.

Koroleva, O. A., Davies, A., Deeken, R., Thorpe, M. R., Tomos, A. D., & Hedrich, R. (2000). Identification of a new glucosinolate-rich cell type in Arabidopsis flower stalk. *Plant Physiology, 124*(2), 599–608.

Kos, M., Houshyani, B., Achhami, B. B., Wietsma, R., Gols, R., Weldegergis, B. T., … van Loon, J. J. A. (2012). Herbivore-mediated effects of glucosinolates on different natural enemies of a specialist aphid. *Journal of Chemical Ecology, 38*(1), 100–115.

Kos, M., Houshyani, B., Wietsma, R., Kabouw, P., Vet, L. E. M., van Loon, J. J. A., & Dicke, M. (2012). Effects of glucosinolates on a generalist and specialist leaf-chewing herbivore and an associated parasitoid. *Phytochemistry, 77*, 162–170.

Kos, M., Kabouw, P., Noordam, R., Hendriks, K., Vet, L. E. M., van Loon, J. J. A., & Dicke, M. (2011). Prey-mediated effects of glucosinolates on aphid predators. *Ecological Entomology, 36*(3), 377–388.

Kroymann, J., Donnerhacke, S., Schnabelrauch, D., & Mitchell-Olds, T. (2003). Evolutionary dynamics of an *Arabidopsis* insect resistance quantitative trait locus. *Proceedings of the National Academy of Sciences of the United States of America, 100*, 14587–14592.

Kuhn, J., Pettersson, E. M., Feld, B. K., Burse, A., Termonia, A., Pasteels, J. M., & Boland, W. (2004). Selective transport systems mediate sequestration of plant glucosides in leaf beetles: a molecular basis for adaptation and evolution. *Proceedings of the National Academy of Sciences of the United States of America, 101*(38), 13808–13813.

Lambrix, V., Reichelt, M., Mitchell-Olds, T., Kliebenstein, D. J., & Gershenzon, J. (2001). The Arabidopsis epithiospecifier protein promotes the hydrolysis of glucosinolates to nitriles and influences *Trichoplusia ni* herbivory. *Plant Cell, 13*(12), 2793–2807.

Lankau, R. A. (2007). Specialist and generalist herbivores exert opposing selection on a chemical defense. *New Phytologist, 175*(1), 176–184.

Li, Q., Eigenbrode, S. D., Stringham, G. R., & Thiagarajah, M. R. (2000). Feeding and growth *of Plutella xylostella* and *Spodoptera eridania* on *Brassica juncea* with varying glucosinolate concentrations and myrosinase activities. *Journal of Chemical Ecology, 26*(10), 2401–2419.

Lichtenstein, E. P., Strong, F. M., & Morgan, D. G. (1962). Naturally occurring insecticides: identification of 2-phenylethylisothiocyanate as an insecticide occurring naturally in edible part of turnips. *Journal of Agricultural and Food Chemistry, 10*(1), 30–33.

Lipka, U., Fuchs, R., & Lipka, V. (2008). *Arabidopsis* non-host resistance to powdery mildews. *Current Opinion in Plant Biology, 11*(4), 404–411.

Lipka, V., Dittgen, J., Bednarek, P., Bhat, R., Wiermer, M., Stein, M., … Schulze-Lefert, P. (2005). Pre- and postinvasion defenses both contribute to nonhost resistance in *Arabidopsis*. *Science, 310*(5751), 1180–1183.

Liu, S., Zhou, S., Tian, L., Guo, E., Luan, Y., Zhang, J., & Li, S. (2011). Genome-wide identification and characterization of ATP-binding cassette transporters in the silkworm, *Bombyx mori*. *BMC Genomics, 12*, 491.

Madsen, S. R., Kunert, G., Reichelt, M., Gershenzon, J., & Halkier, B. A. (2015). Feeding on leaves of the glucosinolate transporter mutant *gtr1gtr2* reduces fitness of *Myzus persicae*. *Journal of Chemical Ecology, 41*(11), 975–984.

Mainguet, A. M., Louveaux, A., El Sayed, G., & Rollin, P. (2000). Ability of a generalist insect, *Schistocerca gregaria*, to overcome thioglucoside defense in desert plants: tolerance or adaptation? *Entomologia Experimentalis et Applicata, 94*(3), 309–317.

Malka, O., Shekov, A., Reichelt, M., Gershenson, J., Vassão, D. G., & Morin, S. (2016). Glucosinolate desulfation by the phloem-feeding insect *Bemisia tabaci*. *Journal of Chemical Ecology, 42*(3), 230–235.

Marazzi, C., Patrian, B., & Städler, E. (2004). Secondary metabolites of the leaf surface affected by sulphur fertilisation and perceived by the diamondback moth. *Chemoecology, 14*(2), 81–86.

Marazzi, C., & Städler, E. (2004). *Arabidopsis thaliana* leaf-surface extracts are detected by the cabbage root fly (*Delia radicum*) and stimulate oviposition. *Physiological Entomology, 29*(2), 192–198.

Markovich, O., Kafle, D., Elbaz, M., Malitsky, S., Aharoni, A., Schwarzkopf, A., … Morin, S. (2013). *Arabidopsis thaliana* plants with different levels of aliphatic- and indolyl-glucosinolates affect host selection and performance of *Bemisia tabaci*. *Journal of Chemical Ecology, 39*(11–12), 1361–1372.

Matile, P. (1980). "The mustard oil bomb": compartmentation of the myrosinase system. *Biochemie und Physiologie der Pflanzen, 175*(8–9), 722–731.

Mewis, I., Appel, H. M., Hom, A., Raina, R., & Schultz, J. C. (2005). Major signaling pathways modulate Arabidopsis glucosinolate accumulation and response to both phloem-feeding and chewing insects. *Plant Physiology, 138*(2), 1149–1162.
Mewis, I., Tokuhisa, J. G., Schultz, J. C., Appel, H. M., Ulrichs, C., & Gershenzon, J. (2006). Gene expression and glucosinolate accumulation in *Arabidopsis thaliana* in response to generalist and specialist herbivores of different feeding guilds and the role of defense signaling pathways. *Phytochemistry, 67*(22), 2450–2462.
Mi, L., Di Pasqua, A. J., & Chung, F.-L. (2011). Proteins as binding targets of isothiocyanates in cancer prevention. *Carcinogenesis, 32*(10), 1405–1413.
Mithen, R., Bennett, R., & Marquez, J. (2010). Glucosinolate biochemical diversity and innovation in the Brassicales. *Phytochemistry, 71*(17–18), 2074–2086.
Mosleh Arany, A., de Jong, T. J., Kim, H. K., van Dam, N. M., Choi, Y. H., Verpoorte, R., & van der Meijden, E. (2008). Glucosinolates and other metabolites in the leaves of *Arabidopsis thaliana* from natural populations and their effects on a generalist and a specialist herbivore. *Chemoecology, 18*(2), 65–71.
Müller, C. (2009). Interactions between glucosinolate- and myrosinase-containing plants and the sawfly *Athalia rosae. Phytochemistry Reviews, 8*(1), 121–134.
Müller, C., Agerbirk, N., Olsen, C. E., Boevé, J.-L., Schaffner, U., & Brakefield, P. M. (2001). Sequestration of host plant glucosinolates in the defensive hemolymph of the sawfly *Athalia rosae. Journal of Chemical Ecology, 27*(12), 2505–2516.
Müller, C., & Brakefield, P. M. (2003). Analysis of a chemical defense in sawfly larvae: easy bleeding targets predatory wasps in late summer. *Journal of Chemical Ecology, 29*(12), 2683–2694.
Müller, C., & Sieling, N. (2006). Effects of glucosinolate and myrosinase levels in *Brassica juncea* on a glucosinolate-sequestering herbivore – and *vice versa. Chemoecology, 16*(4), 191–201.
Müller, C., van Loon, J., Ruschioni, S., De Nicola, G. R., Olsen, C. E., Iori, R., & Agerbirk, N. (2015). Taste detection of the non-volatile isothiocyanate moringin results in deterrence to glucosinolate-adapted insect larvae. *Phytochemistry, 118*, 139–148.
Müller, C., & Wittstock, U. (2005). Uptake and turn-over of glucosinolates sequestered in the sawfly *Athalia rosae. Insect Biochemistry and Molecular Biology, 35*(10), 1189–1198.
Müller, C., Zwaan, B. J., de Vos, H., & Brakefield, P. M. (2003). Chemical defence in a sawfly: genetic components of variation in relevant life-history traits. *Heredity, 90*(6), 468–475.
Müller, R., de Vos, M., Sun, J. Y., Sønderby, I. E., Halkier, B. A., Wittstock, U., & Jander, G. (2010). Differential effects of indole and aliphatic glucosinolates on lepidopteran herbivores. *Journal of Chemical Ecology, 36*(8), 905–913.
Mumm, R., Burow, M., Bukovinszkine'Kiss, G., Kazantzidou, E., Wittstock, U., Dicke, M., & Gershenzon, J. (2008). Formation of simple nitriles upon glucosinolate hydrolysis affects direct and indirect defense against the specialist herbivore, *Pieris rapae. Journal of Chemical Ecology, 34*(10), 1311–1321.
Najar-Rodriguez, A., Friedli, M., Klaiber, J., & Dorn, S. (2015). Aphid-deprivation from *Brassica* plants results in increased isothiocyanate release and parasitoid attraction. *Chemoecology, 25*(6), 303–311.
Nakamura, T., Kawai, Y., Kitamoto, N., Osawa, T., & Kato, Y. (2009). Covalent modification of lysine residues by allyl isothiocyanate in physiological conditions: plausible transformation of isothiocyanate from thiol to amine. *Chemical Research in Toxicology, 22*, 536–542.
Newton, E. L., Bullock, J. M., & Hodgson, D. J. (2009). Glucosinolate polymorphism in wild cabbage (*Brassica oleracea*) influences the structure of herbivore communities. *Oecologia, 160*(1), 63–76.

Nielsen, J. K., Hansen, M. L., Agerbirk, N., Petersen, B. L., & Halkier, B. A. (2001). Responses of the flea beetles *Phyllotreta nemorum* and *P. cruciferae* to metabolically engineered *Arabidopsis thaliana* with an altered glucosinolate profile. *Chemoecology, 11*(2), 75–83.

Nishida, R. (2002). Sequestration of defensive substances from plants by Lepidoptera. *Annual Review of Entomology, 47*, 57–92.

Ohara, Y., Nagasaka, K., & Ohsaki, N. (1993). Warning coloration in sawfly *Athalia rosae* larva and concealing coloration in butterfly *Pieris rapae* larva feeding on similar plants evolved through individual selection. *Researches on Population Ecology, 35*(2), 223–230.

Ohlen, M., Herfurth, A.-M., Kerbstadt, H., & Wittstock, U. (2016). Cyanide detoxification in an insect herbivore: molecular identification of β-cyanoalanine synthases from *Pieris rapae*. *Insect Biochemistry and Molecular Biology, 70*, 99–110.

Opitz, S. E. W., Jensen, S. R., & Müller, C. (2010). Sequestration of glucosinolates and iridoid glucosides in sawfly species of the genus *Athalia* and their role in defense against ants. *Journal of Chemical Ecology, 36*(2), 148–157.

Opitz, S. E. W., Mix, A., Winde, I. B., & Müller, C. (2011). Desulfation followed by sulfation: metabolism of benzylglucosinolate in *Athalia rosae* (Hymenoptera: Tenthredinidae). *ChemBioChem: A European Journal of Chemical Biology, 12*(8), 1252–1257.

Pedras, M. S. C., & Hossain, S. (2011). Interaction of cruciferous phytoanticipins with plant fungal pathogens: indole glucosinolates are not metabolized but the corresponding desulfo-derivatives and nitriles are. *Phytochemistry, 72*(18), 2308–2316.

Pentzold, S., Zagrobelny, M., Bjarnholt, N., Kroymann, J., Vogel, H., Olsen, C. E., … Bak, S. (2015). Metabolism, excretion and avoidance of cyanogenic glucosides in insects with different feeding specialisations. *Insect Biochemistry and Molecular Biology, 66*, 119–128.

Pentzold, S., Zagrobelny, M., Roelsgaard, P. S., Møller, B. L., & Bak, S. (2014). The multiple strategies of an insect herbivore to overcome plant cyanogenic glucoside defence. *PLoS One, 9*(3), e91337.

Pentzold, S., Zagrobelny, M., Rook, F., & Bak, S. (2014). How insects overcome two-component plant chemical defence: plant ß-glucosidases as the main target for herbivore adaptation. *Biological Reviews of the Cambridge Philosophical Society, 89*(3), 531–551.

Perkins, L. E., Cribb, B. W., Brewer, P. B., Hanan, J., Grant, M., de Torres, M., & Zalucki, M. P. (2013). Generalist insects behave in a jasmonate-dependent manner on their host plants, leaving induced areas quickly and staying longer on distant parts. *Proceedings of the Royal Society B-Biological Sciences, 280*(1756).

Petroski, R. J., & Kwolek, W. F. (1985). Interactions of a fungal thioglucoside glucohydrolase and cruciferous plant epithiospecifier protein to form 1-cyanoepithioalkanes: implications of an allosteric mechanism. *Phytochemistry, 24*(2), 213–216.

Pontoppidan, B., Ekbom, B., Eriksson, S., & Meijer, J. (2001). Purification and characterization of myrosinase from the cabbage aphid (*Brevicoryne brassicae*), a brassica herbivore. *European Journal of Biochemistry, 268*(4), 1041–1048.

Pratt, C., Pope, T. W., Powell, G., & Rossiter, J. T. (2008). Accumulation of glucosinolates by the cabbage aphid *Brevicoryne brassicae* as a defense against two coccinellid species. *Journal of Chemical Ecology, 34*(3), 323–329.

Proffit, M., Khallaf, M. A., Carrasco, D., Larsson, M. C., & Anderson, P. (2015). 'Do you remember the first time?' Host plant preference in a moth is modulated by experiences during larval feeding and adult mating. *Ecology Letters, 18*(4), 365–374.

Ramsey, J. S., Rider, D. S., Walsh, T. K., De Vos, M., Gordon, K. H. J., Ponnala, L., … Jander, G. (2010). Comparative analysis of detoxification enzymes in *Acyrthosiphon pisum* and *Myzus persicae*. *Insect Molecular Biology, 19*(Suppl. 2), 155–164.

Ramsey, J. S., Wilson, A. C. C., de Vos, M., Sun, Q., Tamborindeguy, C., Winfield, A., ... Jander, G. (2007). Genomic resources for *Myzus persicae*: EST sequencing, SNP identification, and microarray design. *BMC Genomics, 8*(423).
Ranson, H., Rossiter, L., Ortelli, F., Jensen, B., Wang, X., Roth, C. W., ... Hemingway, J. (2001). Identification of a novel class of insect glutathione *S*-transferases involved in resistance to DDT in the malaria vector *Anopheles gambiae*. *Biochemical Journal, 359*, 295–304.
Ratzka, A., Vogel, H., Kliebenstein, D. J., Mitchell-Olds, T., & Kroymann, J. (2002). Disarming the mustard oil bomb. *Proceedings of the National Academy of Sciences of the United States of America, 99*(17), 11223–11228.
Renwick, J. A. A., Radke, C. D., Sachdev-Gupta, K., & Städler, E. (1992). Leaf surface chemicals stimulating oviposition by *Pieris rapae* (Lepidoptera: Pieridae) on cabbage. *Chemoecology, 3*(1), 33–38.
Roessingh, P., Städler, E., Baur, R., Hurter, J., & Ramp, T. (1997). Tarsal chemoreceptors and oviposition behaviour of the cabbage root fly (*Delia radicum*) sensitive to fractions and new compounds of host-leaf surface extracts. *Physiological Entomology, 22*(2), 140–148.
Rothschild, M., & Schoonhoven, L. M. (1977). Assessment of egg load by *Pieris brassicae* (Lepidoptera: Pieridae). *Nature, 266*, 352–355.
Santolamazza-Carbone, S., Sotelo, T., Velasco, P., & Cartea, M. E. (2015). Antibiotic properties of the glucosinolates of *Brassica oleracea* var. *acephala* similarly affect generalist and specialist larvae of two lepidopteran pests. *Journal of Pest Science, 89*(1), 195–206.
Sarfraz, M., Dosdall, L. M., & Keddie, B. A. (2006). Diamondback moth-host plant interactions: implications for pest management. *Crop Protection, 25*(7), 625–639.
Sarosh, B. R., Wittstock, U., Halkier, B. A., & Ekbom, B. (2010). The influence of metabolically engineered glucosinolates profiles in *Arabidopsis thaliana* on *Plutella xylostella* preference and performance. *Chemoecology, 20*(1), 1–9.
Schlaeppi, K., Bodenhausen, N., Buchala, A., Mauch, F., & Reymond, P. (2008). The glutathione-deficient mutant *pad2-1* accumulates lower amounts of glucosinolates and is more susceptible to the insect herbivore *Spodoptera littoralis*, *Plant Journal, 55*(5), 774–786.
Schoonhoven, L. M., van Loon, J. J. A., & Dicke, M. (2005). *Insect-Plant Biology* (2nd ed.). Oxford University Press.
Schramm, K., Vassão, D. G., Reichelt, M., Gershenzon, J., & Wittstock, U. (2012). Metabolism of glucosinolate-derived isothiocyanates to glutathione conjugates in generalist lepidopteran herbivores. *Insect Biochemistry and Molecular Biology, 42*(3), 174–182.
Schweizer, F., Fernández-Calvo, P., Zander, M., Diez-Diaz, M., Fonseca, S., Glauser, G., ... Reymond, P. (2013). *Arabidopsis* basic helix-loop-helix transcription factors MYC2, MYC3, and MYC4 regulate glucosinolate biosynthesis, insect performance, and feeding behavior. *Plant Cell, 25*(8), 3117–3132.
Self, L. S., Guthrie, F. E., & Hodgson, E. (1964). Metabolism of nicotine by tobacco-feeding insects. *Nature, 204*(495), 300–301.
Seo, S. T., & Tang, C.-S. (1982). Hawaiian fruit flies (Diptera: Tephritidae): toxicity of benzyl isothiocyanate against eggs or 1st instars of three species. *Journal of Economic Entomology, 75*(6), 1132–1135.
Shikita, M., Fahey, J. W., Golden, T. R., Holtzclaw, W. D., & Talalay, P. (1999). An unusual case of 'uncompetitive activation' by ascorbic acid: purification and kinetic properties of a myrosinase from *Raphanus sativus* seedlings. *Biochemical Journal, 341*, 725–732.
Shroff, R., Schramm, K., Jeschke, V., Nemes, P., Vertes, A., Gershenzon, J., & Svatoš, A. (2015). Quantification of plant surface metabolites by matrix-assisted laser desorption-ionization mass spectrometry imaging: glucosinolates on *Arabidopsis thaliana* leaves. *Plant Journal, 81*, 961–972.

Shroff, R., Vergara, F., Muck, A., Svatoš, A., & Gershenzon, J. (2008). Nonuniform distribution of glucosinolates in *Arabidopsis thaliana* leaves has important consequences for plant defense. *Proceedings of the National Academy of Sciences of the United States of America, 105*(16), 6196–6201.

Siemens, D. H., & Mitchell-Olds, T. (1996). Glucosinolates and herbivory by specialists (Coleoptera: Chrysomelidae, Lepidoptera: Plutellidae): consequences of concentration and induced resistance. *Environmental Entomology, 25*(6), 1344–1353.

Smith, J. N. (1955). Comparative Detoxication. 4. Ethereal sulphate and glucoside conjugations in insects. *Biochemical Journal, 60*(1–4), 436–442.

Sønderby, I. E., Geu-Flores, F., & Halkier, B. A. (2010). Biosynthesis of glucosinolates – gene discovery and beyond. *Trends in Plant Science, 15*(5), 283–290.

Sønderby, I. E., Hansen, B. G., Bjarnholt, N., Ticconi, C., Halkier, B. A., & Kliebenstein, D. J. (2007). A systems biology approach identifies a R2R3 MYB gene subfamily with distinct and overlapping functions in regulation of aliphatic glucosinolates. *PLoS One, 2*(12), e1322.

Städler, E., Renwick, J. A. A., Radke, C. D., & Sachdev-Gupta, K. (1995). Tarsal contact chemoreceptor response to glucosinolates and cardenolides mediating oviposition in *Pieris rapae*. *Physiological Entomology, 20*(2), 175–187.

Stauber, E. J., Kuczka, P., van Ohlen, M., Vogt, B., Janowitz, T., Piotrowski, M., ... Wittstock, U. (2012). Turning the 'mustard oil bomb' into a 'cyanide bomb': aromatic glucosinoate metabolism in a specialist insect herbivore. *PLoS One, 7*(4), e35545.

Strauss, A. S., Peters, S., Boland, W., & Burse, A. (2013). ABC transporter functions as a pacemaker for sequestration of plant glucosides in leaf beetles. *eLife, 2*, e01096.

Sun, J. Y., Sønderby, I. E., Halkier, B. A., Jander, G., & de Vos, M. (2009). Non-volatile intact indole glucosinolates are host recognition cues for ovipositing *Plutella xylostella*. *Journal of Chemical Ecology, 35*(12), 1427–1436.

Terriere, L. C. (1984). Induction of detoxication enzymes in insects. *Annual Review of Entomology, 29*, 71–88.

Textor, S., & Gershenzon, J. (2009). Herbivore induction of the glucosinolate-myrosinase defense system: major trends, biochemical bases and ecological significance. *Phytochemistry Reviews, 8*(1), 149–170.

Thangstad, O. P., Gilde, B., Chadchawan, S., Seem, M., Husebye, H., Bradley, D., & Bones, A. M. (2004). Cell specific, cross-species expression of myrosinases in *Brassica napus, Arabidopsis thaliana* and *Nicotiana tabacum*. *Plant Molecular Biology, 54*(4), 597–611.

Thies, W. (1979). Detection and utilization of a glucosinolate sulfohydrolase in the edible snail, *Helix pomatia*. *Naturwissenschaften, 66*(7), 364–365.

Thöming, G., Larsson, M. C., Hansson, B. S., & Anderson, P. (2013). Comparison of plant preference hierarchies of male and female moths and the impact of larval rearing hosts. *Ecology, 94*(8), 1744–1752.

Tierens, K. F. M.-J., Thomma, B. P. H. J., Brouwer, M., Schmidt, J., Kistner, K., Porzel, A., ... Broekaert, W. F. (2001). Study of the role of antimicrobial glucosinolate-derived isothiocyanates in resistance of Arabidopsis to microbial pathogens. *Plant Physiology, 125*(4), 1688–1699.

Tjallingii, W. F., & Hogen Esch, T. (1993). Fine-structure of aphid stylet routes in plant-tissues in correlation with EPG signals. *Physiological Entomology, 18*(3), 317–328.

Traka, M., & Mithen, R. (2009). Glucosinolates, isothiocyanates and human health. *Phytochemistry Reviews, 8*(1), 269–282.

Van Geem, M., Harvey, J. A., Cortesero, A. M., Raaijmakers, C. E., & Gols, R. (2015). Interactions between a belowground herbivore and primary and secondary root metabolites in wild cabbage. *Journal of Chemical Ecology, 41*(8), 696–707.

Vanhaelen, N., Haubruge, E., Lognay, G., & Francis, F. (2001). Hoverfly glutathione *S*-transferases and effect of Brassicaceae secondary metabolites. *Pesticide Biochemistry and Physiology, 71*(3), 170–177.

Vergara, F., Svatoš, A., Schneider, B., Reichelt, M., Gershenzon, J., & Wittstock, U. (2006). Glycine conjugates in a lepidopteran insect herbivore – the metabolism of benzylglucosinolate in the cabbage white butterfly, *Pieris rapae*. *ChemBioChem: A European Journal of Chemical Biology, 7*, 1982–1989.

Wadleigh, R. W., & Yu, S. J. (1988). Detoxification of isothiocyanate allelochemicals by glutathione transferase in three lepidopterous species. *Journal of Chemical Ecology, 14*(4), 1279–1288.

Welte, C. U., de Graaf, R. M., van den Bosch, T. J. M., Op den Camp, H. J. M., van Dam, N. M., & Jetten, M. S. M. (2015). Plasmids from the gut microbiome of cabbage root fly larvae encode SaxA that catalyses the conversion of the plant toxin 2-phenylethyl isothiocyanate. *Environmental Microbiology, 18*(5), 1379–1390.

Wheat, C. W., Vogel, H., Wittstock, U., Braby, M. F., Underwood, D., & Mitchell-Olds, T. (2007). The genetic basis of a plant-insect coevolutionary key innovation. *Proceedings of the National Academy of Sciences of the United States of America, 104*(51), 20427–20431.

Winde, I., & Wittstock, U. (2011). Insect herbivore counteradaptations to the plant glucosinolate-myrosinase system. *Phytochemistry, 72*(13), 1566–1575.

Winde, I. B. (2011). *Entgiftung des Glucosinolat-Myrosinase-Systems durch generalistische Herbivoren der Lepidoptera* (Doctoral Thesis). Technical University Braunschweig.

Wink, M. (1988). Plant-breeding: importance of plant secondary metabolites for protection against pathogens and herbivores. *TAG. Theoretical and Applied Genetics. Theoretische und angewandte Genetik, 75*(2), 225–233.

Wittstock, U., Agerbirk, N., Stauber, E. J., Olsen, C. E., Hippler, M., Mitchell-Olds, T., … Vogel, H. (2004). Successful herbivore attack due to metabolic diversion of a plant chemical defense. *Proceedings of the National Academy of Sciences of the United States of America, 101*(14), 4859–4864.

Wittstock, U., & Burow, M. (2007). Tipping the scales – specifier proteins in glucosinolate hydrolysis. *IUBMB Life, 59*(12), 744–751.

Wittstock, U., & Burow, M. (2010). Glucosinolate breakdown in Arabidopsis: mechanism, regulation and biological significance. *Arabidopsis Book, 8*, e0134.

Wittstock, U., & Gershenzon, J. (2002). Constitutive plant toxins and their role in defense against herbivores and pathogens. *Current Opinion in Plant Biology, 5*(4), 300–307.

Wittstock, U., & Halkier, B. A. (2002). Glucosinolate research in the *Arabidopsis* era. *Trends in Plant Science, 7*(6), 263–270.

Wittstock, U., Kliebenstein, D. J., Lambrix, V., Reichelt, M., & Gershenzon, J. (2003). Glucosinolate hydrolysis and its impact on generalist and specialist insect herbivores. In J. T. Romeo (Ed.), *Integrative phytochemistry: From ethnobotany to molecular ecology: Vol. 37. Recent advances in phytochemisty* (pp. 101–125). Amsterdam: Elsevier.

Xiao, Z., Mi, L., Chung, F.-L., & Veenstra, T. D. (2012). Proteomic analysis of covalent modifications of tubulins by isothiocyanates. *Journal of Nutrition, 142*(7), 1377S–1381S.

Yang, R. S. H., & Wilkinson, C. F. (1973). Sulfotransferases and phosphotransferases in insects. *Comparative Biochemistry and Physiology, 46*(4B), 717–726.

You, M., Yue, Z., He, W., Yang, X., Yang, G., Xie, M., … Wang, J. (2013). A heterozygous moth genome provides insights into herbivory and detoxification. *Nature Genetics, 45*(2), 220–225.

Yu, S. J. (1984). Interactions of allelochemicals with detoxication enzymes of insecticide-susceptible and resistant fall armyworms. *Pesticide Biochemistry and Physiology, 22*, 60–68.

Yu, S. J. (1989). Purification and characterization of glutathione transferases from five phytophagous Lepidoptera. *Pesticide Biochemistry and Physiology, 35*, 97–105.

Yu, S. J. (1992). Plant-allelochemical-adapted glutathione transferases in Lepidoptera. In C. A. Mullin, & J. G. Scott (Eds.), *Molecular mechanisms of insecticide resistance: Vol. 505. ACS symposium series* (pp. 174–190).
Yu, S. J., & Hsu, E. L. (1993). Induction of detoxification enzymes in phytophagous insects: roles of insecticide synergists, larval age, and species. *Archives of Insect Biochemistry and Physiology, 24*, 21–32.
Zhao, Y., Hull, A. K., Gupta, N. R., Goss, K. A., Alonso, J., Ecker, J. R., … Celenza, J. L. (2002). Trp-dependent auxin biosynthesis in *Arabidopsis*: involvement of cytochrome P450s CYP79B2 and CYP79B3. *Genes & Development, 16*(23), 3100–3112.
Zou, X., Xu, Z., Zou, H., Liu, J., Chen, S., Feng, Q., & Zheng, S. (2016). Glutathione *S*-transferase SlGSTE1 in *Spodoptera litura* may be associated with feeding adaptation of host plants. *Insect Biochemistry and Molecular Biology, 70*, 32–43.

CHAPTER NINE

Health Benefits of Glucosinolates

M.H. Traka
Institute of Food Research, Norwich, United Kingdom
E-mail: maria.traka@ifr.ac.uk

Contents

Abstract

There is considerable evidence from observational population studies that diets rich in cruciferous vegetables are associated with improved health benefits, which include protection from cancer incidence and progression and from cardiovascular disease. These vegetables uniquely contain the sulphur-containing phytochemicals glucosinolates, which can be hydrolysed by plants or in the human gut to isothiocyanates. It is these isothiocyanates that have been linked to their bioactivity. By far the most commonly consumed and well-studied glucosinolates within our diets are those that derive from methionine, such as sulforaphane, erucin and allyl-glucosinolates, and aromatic glucosinolates, such as phenethyl- and benzyl-glucosinolate. A significant amount of cell and animal studies suggest a role for all these in interfering with key processes that drive carcinogenesis and metastasis, but also protect cells from redox imbalance that underlies the development of several chronic inflammatory diseases. In this review I will summarize the metabolic fate of the

Advances in Botanical Research, Volume 80
ISSN 0065-2296
http://dx.doi.org/10.1016/bs.abr.2016.06.004

common aliphatic and aromatic glucosinolates in humans, present the latest epidemiological evidence from metaanalyses studies and describe the mechanisms for their biological activity from cell and animal models, but also from the recent interventional studies in humans.

1. GLUCOSINOLATE STRUCTURE AND METABOLISM

Epidemiological evidence suggests that consumption of cruciferous vegetables is associated with reduced risk of developing disease, such as cancer and myocardial infarction. The majority of epidemiological studies do not distinguish between the types of cruciferous vegetables, but the most common worldwide include broccoli, cauliflower, cabbages, bok choy, kale, watercress, turnip and rocket. Although these vegetables are rich in a variety of nutrients, such as carotenoids, vitamins and minerals, it is the presence of glucosinolates that has been associated with their bioactivity. Glucosinolates are found in 16 dicotyledonous plant families, and at least 120 different structures have been identified so far (Fahey, Zalcmann, & Talalay, 2001). The glucosinolate molecule consists of a β-thioglucose moiety, a sulfonated oxime moiety and a variable side chain, derived from an amino acid. They are divided into three groups based on the amino acid from which they derive: aliphatic glucosinolates deriving mainly from methionine; aromatic from phenylalanine or tyrosine; and indolyl from tryptophan. The diversity in side-chain structures present in cruciferous vegetables belonging to the *Brassicaceae* family is achieved predominantly through side-chain elongation and secondary side-chain modifications (see, chapter 3 in this volume).

Glucosinolates that derive from chain-elongated forms of methionine are of particular importance due to their abundance (almost 50% of the known glucosinolates) and the biological activity of their degradation products. The degradation products of aromatic glucosinolates, such as phenylethyl- and benzyl-glucosinolate, and indole glucosinolates have also been associated with increased biological activity. Although glucosinolates themselves are considered biologically inactive molecules, upon mechanical damage, infection or pest attack, they are hydrolysed to give rise to a range of biologically active compounds (Bones & Rossiter, 1996). The hydrolysis is catalysed by the endogenous plant enzyme myrosinase and results in the cleavage of the thio-glucose bond to give rise to unstable thiohydroximate-O-sulfonate. This aglycone spontaneously rearranges to produce several products (Bones & Rossiter, 1996). Most frequently, it undergoes a Lossen rearrangement to

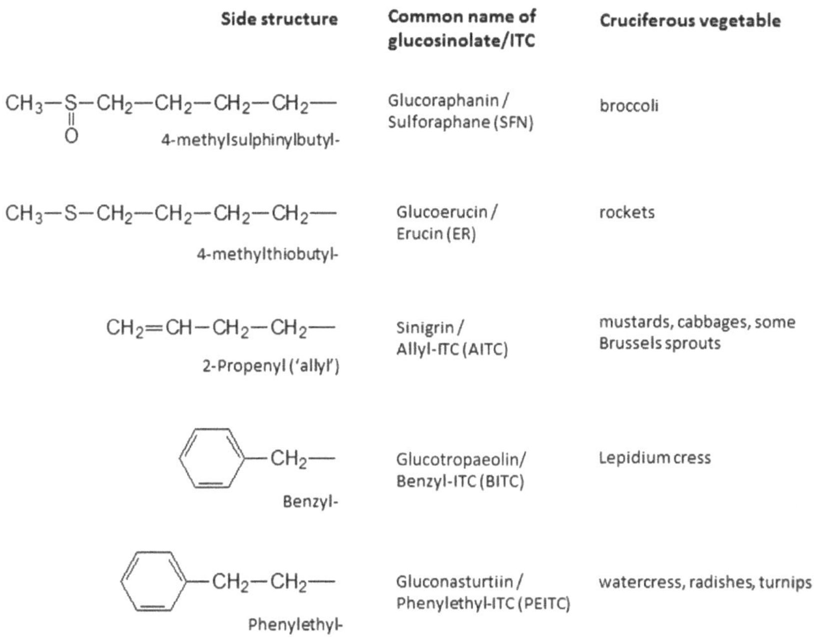

Figure 1 Side structures of the most commonly consumed isothiocyanates (ITCs).

produce an isothiocyanate (ITC) but can also produce thiocyanates, nitriles and elementary sulphur. It is the hydrolysis products and specifically ITCs that have been implicated in the bioactivity of cruciferous vegetables and in particular the protection against different types of cancer. The side structures of glucosinolates/ITCs with the most evidence for health benefits are shown in Fig. 1.

ITC production is favoured at neutral pH, whereas at lower pH nitriles are produced predominantly. Thiocyanates can potentially inhibit the uptake of iodine in the thyroid, and large amounts can have negative effects on the liver. However, only allyl-, benzyl- and 4-(methylthio)butylglucosinolates can degrade to thiocyanates (Bones & Rossiter, 1996). Finally, alkenyl glucosinolates give rise to epithioalkanes when a small labile protein known as epithiospecifier protein is present (Lambrix, Reichelt, Mitchell-Olds, Kliebenstein, & Gershenzon, 2001). Both myrosinase and the epithiospecifier protein are heat sensitive and are inactivated with cooking. However, myrosinase has a higher inactivation temperature threshold of 70°C compared to the epithiospecifier protein, which is inactivated at 60°C (Matusheski, Juvik, & Jeffery, 2004). This allows for an optimum cooking process that can maximize the production of ITCs and has been researched extensively (Bongoni, Verkerk, Steenbekkers, Dekker, & Stieger, 2014; Dosz & Jeffery, 2013; Rungapamestry, Duncan, Fuller, & Ratcliffe, 2006).

2. METABOLIC FATE IN HUMANS

Following ingestion of mildly cooked cruciferous vegetables, the glucosinolates present are hydrolysed by plant myrosinases in the mouth, and the resulting ITCs are rapidly absorbed in the upper gastrointestinal tract and subsequently metabolized. When myrosinases in the plant tissue are deactivated, e.g., by excess cooking, then enzymes with myrosinase activity produced by intestinal bacteria present in the human distal gut are capable of hydrolysing the ingested glucosinolates (Luang-In et al., 2014; Rouzaud, Rabot, Ratcliffe, & Duncan, 2003). Human gut bacterial metabolism is poorly understood and highly variable amongst individuals, with some being high and others low metabolizers of glucosinolates into ITCs (Li, Hullar, Beresford, & Lampe, 2011).

Once ITCs are produced they are taken up by enterocytes through a process of passive diffusion (Fig. 2). Conjugation with glutathione occurs spontaneously and is further promoted by glutathione-*S*-transferases (GSTs), which include, amongst others, Mu (GSTM1), considered the major determinant of ITC metabolism, Pi (GSTP1) and Theta (GSTT1) (Hayes, Flanagan, & Jowsey, 2005). The glutathione conjugate is then rapidly exported extracellularly with the help of multidrug resistance—associated protein-1 (MRP1), MRP2 and P-glycoprotein-1 (Pgp-1) (Payen, Courtois, Loewert, Guillouzo, & Fardel, 2001; Zhang & Callaway, 2002). Following export to the extracellular matrix, ITC metabolites are transported to the liver through the hepatic portal vein, where they are metabolized via the mercapturic acid pathway, a major pathway for elimination of many xenobiotics. They undergo enzymatic modifications including cleavage of glutamine, which yields cysteine-glycine-conjugates, cleavage of glycine, yielding cysteine-conjugates and finally acetylation to produce N-acetyl-cysteine (NAC)-conjugates that are excreted in urine (Brusewitz et al., 1977). Free ITCs and their metabolites are also taken up by hepatocytes and passed on to blood systemic circulation where they can reach peripheral tissues, such as breast and prostate, albeit at significantly lower concentrations than those found in the liver. The metabolic fate of

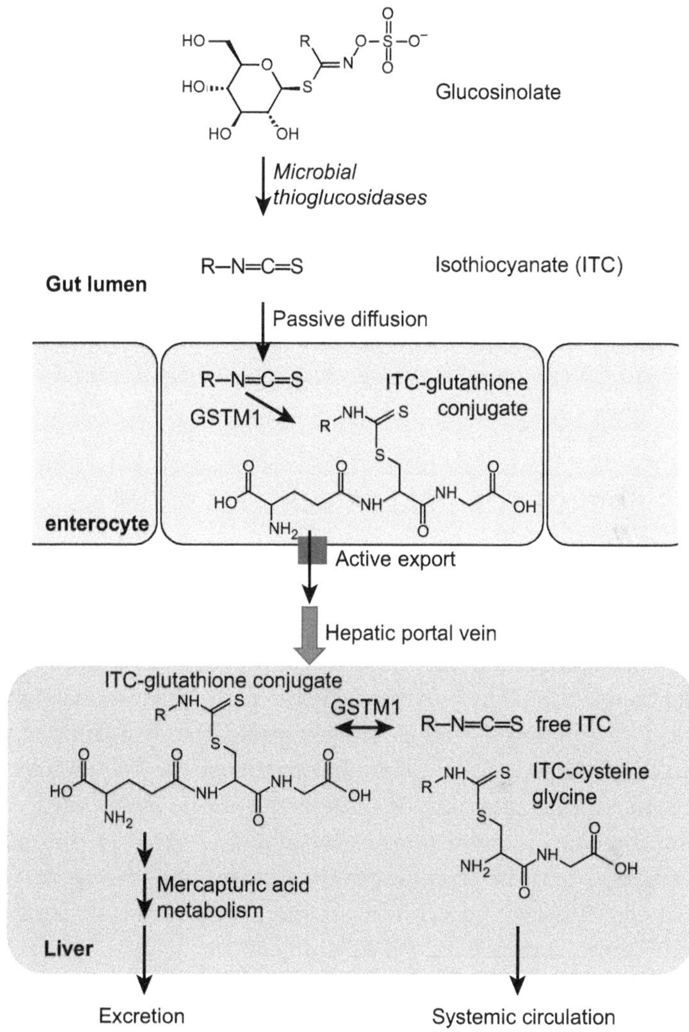

Figure 2 Metabolic fate of ingested glucosinolates in humans.

glucosinolates in humans has been well documented following ingestion of broccoli, but this would be similar for other glucosinolates. Broccoli contains 4-methyl-sulfinyl-butyl glucosinolate, also known as glucoraphanin, which is hydrolysed to produce sulforaphane. The major metabolites found in urine are the sulforaphane-NAC- and sulforaphane-cysteine-conjugates, whereas in plasma free sulforaphane and the sulforaphane-cysteine-glycine-conjugate are found (Gasper et al., 2005; Saha et al., 2012). Presence of circulating free sulforaphane, at levels up to 45% of those ingested, is likely due to GSTM1 activity on the sulforaphane-glutathione conjugate, as GSTM1 is capable of catalysing both the conjugation and the dissociation of glutathione to ITCs, although the decomposition of the glutathione conjugates is very slow in comparison to the rate of conjugate formation (Gasper et al., 2005; Meyer, Crease, & Ketterer, 1995; Zhang, Kolm, Mannervik, & Talalay, 1995).

Following ingestion of glucosinolates from cruciferous vegetables the levels of ITCs and their metabolites that are circulating and reaching peripheral tissues are in the region of low μM (Table 1). These levels are likely to be higher in the liver where ITCs enter through the hepatic portal vein to be further metabolized.

3. EVIDENCE FOR HEALTH BENEFITS FROM EPIDEMIOLOGY

The initial indication for the health benefits of consumption of cruciferous vegetables comes from observational epidemiological studies that draw associations between disease risk and consumption of a particular food group in selected populations. The two main types of such observational studies are 'cohort' and 'case–control' studies. Cohort studies follow a group of people with defined characteristics to determine incidence of, or mortality from, some specific disease and can be prospective or retrospective, whereas case–control studies identify subjects by outcome status at the outset of the investigation and are predominantly retrospective (Song & Chung, 2010). In comparison to cohort studies, case–control studies are quick and require comparatively fewer subjects but have the potential for recall bias because individuals with cancer may associate their malignancy with a previous 'bad diet' and thus overreport foods considered less healthy.

There is a plethora of epidemiological evidence that suggests that consumption of cruciferous vegetables, which are uniquely rich in glucosinolates, is associated with a significantly lower risk of myocardial infarction (Cornelis, El-Sohemy, & Campos, 2007) and developing cancer at different sites, such as the lung (London et al., 2000; Spitz et al., 2000; Wang, Giovanucci, et al., 2004; Zhao et al., 2001), stomach (Hansson et al., 1993), colorectal (Lin et al., 1998; Seow et al., 2002), breast (Ambrosone et al., 2004; Fowke et al., 2003), bladder (Zhao et al., 2007) and prostate (Cohen, Kristal, & Stanford, 2000; Giovannucci, Rimm, Liu, Stampfer, & Willett, 2003; Kirsh et al., 2007). Diets that are rich in cruciferous vegetables

Table 1 Plasma Concentration of Isothiocyanate (ITC) Metabolites in Humans Following Glucosinolate-Rich Cruciferous Vegetable Interventions

Intervention Arms	Glucosinolate/ITC Content of Diet	Study Design	Plasma Concentration of ITC Metabolites	References
Broccoli-Related Interventions				
Fresh broccoli sprouts (FBS)	200 μmol sulforaphane (SFN)	N = 10 parallel	0.7 μM total SFN metabolites (C_{max})	Atwell et al. (2015)
Myrosinase-treated broccoli sprout extract (BSE)	200 μmol SFN		1.9 μM total SFN metabolites (C_{max})	
100 g lightly cooked fresh broccoli soup	4.16 mg SFN (23.5 μmol)	N = 18 crossover	0.21 μM total SFN metabolites (C_{max})	Saha et al. (2012)
100 g lightly cooked frozen broccoli soup	18.6 mg glucosinolates (42.5 μmol)		0.02 μM total SFN metabolites (C_{max})	
42 g of 5-day-old intact fresh broccoli sprouts	70 μmol SFN	N = 4 crossover	1.53 μM total ITC metabolites (C_{max})	Cramer, Teran-Garcia, and Jeffery (2012)
2 g glucoraphanin-rich powder	120 μmol SFN (after myrosinase)		0.37 μM total ITC metabolites (C_{max})	
Combination of the above	Not measured		2.86 μM total ITC metabolites (C_{max})	
Broccoli sprouts	150 μmol glucoraphanin and 71 μmol glucoerucin	N = 12 crossover	~3 μM total SF and ER metabolites (C_{max})	Clarke et al. (2011)

(Continued)

Table 1 Plasma Concentration of Isothiocyanate (ITC) Metabolites in Humans Following Glucosinolate-Rich Cruciferous Vegetable Interventions—cont'd

Intervention Arms	Glucosinolate/ITC Content of Diet	Study Design	Plasma Concentration of ITC Metabolites	References
Supplement BroccoMax	121 μmol glucoraphanin and 40 μmol glucoerucin		~0.5 μM total SF and ER metabolites (C_{max})	
2 g air-dried broccoli sprouts	70 μmol SFN	N = 4 crossover	0.9 μM total ITC metabolites (C_{max})	Cramer and Jeffery (2011)
2 g glucoraphanin-rich powder	120 μmol SFN (after myrosinase)		No change from control	
Combination of the above	Not measured		2.1 μM total ITC metabolites (C_{max})	
200 g of crushed microwave-cooked broccoli	61.4 μmol glucoraphanin, 5.4 μmol glucoiberin and 101.6 μmol glucobrassicins	N = 8 crossover	31 nM total ITC metabolites (C_{max})	Vermeulen, Klopping-Ketelaars, van den Berg, and Vaes (2008)
200 g of crushed raw broccoli	9.92 μmol SFN		103 nM total ITC metabolites (C_{max})	
Standard broccoli	~101 μM SFN metabolites	N = 16 crossover	2.4 μM total SFN metabolites (C_{max})	Gasper et al. (2005)
Super broccoli	~344 μM SFN metabolites		7.4 μM total SFN metabolites (C_{max})	

Myrosinase-treated extract of 3-day-old broccoli sprouts	200 μmol of ITCs (largely SFN with lesser amounts of iberin and erucin)	N = 4	Range from 1.6 to 2.2 μM total ITC metabolites (C_{max})	Ye et al. (2002)
200 g steamed broccoli	200 μmol glucosinolates (measured as ITC equivalents)	N = 12 crossover study	No change from baseline	Conaway et al. (2000)
200 g fresh broccoli	216 μmol glucosinolates (measured as ITC equivalents)		Range from 0.8 to 1.7 μM total ITC metabolites (C_{max})	
Nonbroccoli Interventions				
PEITC capsules	20 mg	N = 2		
10 g freeze-dried nasturtium dissolved in 50 mL water	Not measured	N = 4	BITC−CysGly 2.61 μM (C_{max}) BITC−NAC 1.25 μM (C_{max}) BITC−Cys 0.57 μM (C_{max})	Platz et al. (2013)
100 g watercress	25 mg PEITC	N = 4	928.5 nM PEITC (C_{max})	Ji and Morris (2003)

such as broccoli have also been associated with a reduction in progression from localized to more aggressive forms of prostate cancer (Richman, Carroll, & Chan, 2012), which suggests that glucosinolates may also be beneficial once initiation of disease has occurred.

A more robust systematic approach to identifying the best evidence for chemoprotection comes from metaanalysis of observational studies in epidemiology, where a collection of epidemiological studies are analysed collectively. Such studies report strong inverse associations between consumption of cruciferous vegetables and incidence of gastric, colorectal, bladder, prostate, ovarian, breast, endometrial and renal cancer (Table 2). In the few such studies that looked at the types of cruciferous vegetables that were being consumed, high broccoli intake is identified as being strongly associated with reduced risk. Interestingly, US-based studies compared to Europe-based ones tend to demonstrate a higher association, which is likely correlated with the fact that broccoli is the prevalent crucifer consumed in the United States. Additionally, the strongest evidence comes from case–control studies rather than prospective cohort studies, which is likely due to the recall bias mentioned previously. The focus should now be in undertaking well-designed human interventional trials that will provide experimental evidence for such associations.

4. DIET–GENE INTERACTIONS AND THE ROLE OF GLUTATHIONE-*S*-TRANSFERASE GENOTYPES

Several epidemiological studies suggest that the protective effects of crucifer consumption are modulated by polymorphisms within GSTs. Polymorphisms have been described in many genes of this family (Hayes et al., 2005; Hayes & Strange, 2000), including null mutations in GSTM1 and GSTT1, resulting in the absence of a functional gene product. The frequency of the homozygous null genotype of GSTM1 varies between 39 and 63%, while homozygous null frequencies of GSTT1 are 10–21% for Caucasians, but can be as high as 64% within some Asian populations (Cotton, Sharp, Little, & Brockton, 2000).

It is interesting to note that US-based epidemiological studies have concluded that GSTM1-positive individuals gain greater cancer protection from either broccoli consumption or total cruciferous vegetable consumption than GSTM1 nulls (Joseph et al., 2004; Spitz et al., 2000; Wang, Giovanucci, et al., 2004). This was also observed in a human interventional study in which there was a significant reduction in colorectal adenomas

Table 2 Metaanalysis of Observational Studies Reporting Associations Between Cruciferous Vegetable Consumption and Cancer Risk

Cancer Type	Types of Studies Assessed	Relative Risk (95% Confidence Interval) for Crucifer Intake	Further Subgroup Analyses	References
Gastric cancer	16 case–control and 6 prospective studies	0.81 (0.75–0.88)	• Case–control studies showed significant reduction, but prospective studies showed reduction of borderline statistical significance	Wu, Yang, Wang, Han, and Xiang (2013)
Colorectal cancer	24 case–control (24,275 cases) and 11 prospective studies	0.82 (0.75–0.90)	• RR 0.82 (95% CI 0.65–1.02) specifically for high broccoli intake from 3 cohort studies and 3 case–control studies	Wu, Yang, Vogtmann, et al. (2013)
Bladder cancer	7 cohort and case–control (2437 cases) studies	0.85 (0.69–1.06)		Vieira et al. (2015)
Bladder cancer	12 case–control (6496 cases) and cohort studies	0.84 (0.77–0.91)	• Significant reduction in studies conducted in the USA (RR 0.80, 95% CI 0.73–0.88), but not in studies in Europe (RR 0.99, 95% CI 0.81–1.20)	Yao et al. (2014)
Bladder cancer	5 cohort and 5 case–control (5772 cases) studies	0.80 (0.69–0.92)	• Significant reduction in case–control studies (RR 0.78, 95% CI, 0.67–0.89), but not in cohort studies (RR 0.86, 95% CI, 0.61–1.11) • Significant reduction in US-based studies (RR 0.73, 95% CI 0.57–0.89)	Liu, Mao, Lin, Zhou, and Xie (2013)

(Continued)

Table 2 Metaanalysis of Observational Studies Reporting Associations Between Cruciferous Vegetable Consumption and Cancer Risk—cont'd

Cancer Type	Types of Studies Assessed	Relative Risk (95% Confidence Interval) for Crucifer Intake	Further Subgroup Analyses	References
			but not in Europe (RR 0.88, 95% CI 0.75–1.02)	
Prostate cancer	7 cohort and 6 population-based case–control studies	0.90 (0.85–0.96)	• Significant reduction in case–control studies (RR 0.79, 95% CI, 0.69–0.89), but not in cohort studies (RR 0.95, 95% CI, 0.88–1.02)	Liu, Mao, Cao, and Xie (2012)
Ovarian cancer	6 case–control (4306 cases) and 5 cohort studies	0.84 (0.75–0.94)	• Significant reduction in case–control studies (RR 0.84, 95% CI, 0.75–0.94), but not in cohort studies (RR 1.00, 95% CI, 0.85–1.11)	Han, Li, and Yu (2014)
Ovarian cancer	4 case–control and 4 cohort studies	0.89 (0.81–0.99)	• RR 0.78 (95% CI 0.58–1.06) specifically for broccoli consumption from 2 studies	Hu, Hu, Hu, and Zheng (2015)
Breast cancer	11 case–control (18,673 cases) and 2 cohort studies	0.85 (0.77–0.94)		Liu and Lv (2013)
Endometrial cancer	1 cohort study and 16 case–control studies	Odds ratio: 0.79; 95% CI: 0.69–0.90 per 100 g/day		Bandera, Kushi, Moore, Gifkins, and McCullough (2007)
Renal cell carcinoma	6 cohorts and 6 case–control (5773 cases) studies	0.81 (0.72–0.91)	• No significant reduction amongst studies conducted in Europe (RR 0.87, 95% CI 0.71–1.07)	Zhao and Zhao (2013)

only in GSTM1-positive individuals who consumed a single portion of broccoli per week, although both GSTM1 positives and nulls had reduced incidence of adenomas with higher levels of broccoli consumption (Lin et al., 1998). Conversely, studies based in Asia conclude that GSTM1 nulls and GSTT1 nulls may gain greater protection than GSTM1 and GSTT1 positives from cruciferous vegetable consumption (which would comprise mainly Chinese cabbage as opposed to broccoli) estimated either by food frequency questionnaires (Zhao et al., 2001) or by quantification of ITCs in urine (Fowke et al., 2003; London et al., 2000), a biomarker of crucifer intake (Chung, Jiao, Getahun, & Yu, 1998; Seow et al., 1998).

Polymorphisms also occur at the GSTT1 locus, with about 20% of Caucasians and up to 60% of people of Asiatic descent having a homozygous deletion. GSTT1 genotype has been associated with modulating the reduction in cancer risk through cruciferous vegetable consumption in some studies, but not in others (Lin et al., 2002; Moore et al., 2007; Spitz et al., 2000). Although GSTT1 has not been extensively studied, there is some evidence that a combination of GSTM1 and GSTT1 genotype might influence cancer risk following brassica consumption (Seow et al., 2002). Consumption of cruciferous vegetables was also associated with a lower risk of myocardial infarction amongst those individuals with a functional GSTT1 allele (Cornelis et al., 2007).

Based on the predominant function of GSTs to conjugate ITCs for excretion, there is an assumption that a GSTM1- or GSTT1-null genotype would result in a reduced rate of metabolism or excretion of ITC. However, human interventional studies with watercress report there is no difference in the ITC excretion rate between positive and null genotypes (Dyba et al., 2010). Similar studies with broccoli have shown that GSTM1-null individuals excreted a greater proportion of ingested sulforaphane via mercapturic acid metabolism than GSTM1-positive individuals (Gasper et al., 2005; Steck, Gammon, Hebert, Wall, & Zeisel, 2007). One possible explanation is that within the low glutathione environment of the plasma, GSTM1 activity results in the deconjugation of the ITC and glutathione following export from epithelial cells. The contrasting enzymology of GSTs with ITCs with different side-chain structures (Zhang et al., 1995) may account for the apparent contradictory diet–GST interactions from epidemiology studies. However, considerably more research is required to clarify the interactions between GST polymorphisms, ITC metabolism and the health benefits of crucifer consumption.

In contrast to GSTM1 and GSTT1, to date, no statistically significant association has been found between GSTP1 polymorphisms and cruciferous vegetable intake at least in breast cancer from studies based in Shanghai (Fowke et al., 2003) and the United States (Steck, Gaudet, et al., 2007).

Recently, in a first of its kind intervention study looking at genomewide associations by SNP analysis a novel diet—gene interaction was identified. Individuals who received a high-glucoraphanin broccoli intervention modulated their metabolic profiles differently depending on polymorphisms within the poly-A polymerase gamma (PAPOLG) gene (Armah et al., 2013). PAPOLG has not been previously linked to diet or metabolic regulation, and its importance in mediating the effects of glucosinolates remains to be explored. However, this approach demonstrates that other yet undiscovered diet—gene interactions may also be regulating the biological response to glucosinolates.

5. EVIDENCE FOR HEALTH BENEFITS FROM HUMAN INTERVENTIONAL STUDIES

In contrast to observational studies that are providing associations between glucosinolate-rich diets and prevention of disease, experimental studies in humans have the potential to provide a direct link for the health benefits of increasing glucosinolates in the diet. However, in the case of cancer prevention such human intervention studies with defined clinical endpoints would require a significant number of participants to obtain appropriate power, and their duration would be prohibiting. Instead human intervention studies can be used to directly assess the biological response, predominantly in blood but also in relatively accessible tissues such as prostate, following a diet rich in glucosinolates. As such studies are beginning to emerge we now have direct experimental evidence that glucosinolate-rich diets are increasing the detoxification of carcinogens, are modulating key metabolic pathways and are altering signalling pathways that drive disease.

In two studies that were based in the Qidong region of China, an area documented to have high incidence of hepatocellular carcinoma and high exposure to airborne pollutants; glucosinolate-rich interventions of broccoli sprouts significantly increased the levels of carcinogen excretion, especially in GSTT1-positive individuals, suggesting that the intervention increased the Phase 2 detoxification capacity of these individuals (Egner et al., 2014; Kensler et al., 2012). Similarly, a watercress intervention increased enzyme activity of the Phase 2 enzymes glutathione peroxidase and superoxide dismutase in red blood cells of individuals with GSTM1-null genotype (Hofmann et al., 2009).

In addition to activation of xenobiotic metabolism glucosinolate-rich diets have been effective in modulating metabolic parameters. A 12-week intervention with a novel high-glucoraphanin broccoli significantly reduced cholesterol levels in individuals at risk of cardiovascular disease (Armah et al., 2015). Reduction in cholesterol levels was also seen following a short 4-day mustard dietary intervention (Lamy, Garcia-Kaufer, Prinzhorn, & Mersch-Sundermann, 2012). The reduction of cholesterol is likely related to modulation of lipid metabolism and oxidative phosphorylation as evidenced by high-throughput metabolite profiling of plasma from volunteers who received the high-glucoraphanin broccoli (Armah et al., 2013).

Finally, in the only study that assessed biopsy tissue from men at significant risk of developing prostate cancer, a high-glucoraphanin diet for 12 months significantly altered major transcriptional pathways, such as transforming growth factor β (TGFβ) and insulin signalling, that are key in driving carcinogenesis (Traka et al., 2008).

6. MECHANISMS OF BIOACTIVITY FROM ANIMAL AND CELL MODELS

The health benefits of glucosinolates have largely been attributed to the biological activity of ITCs. There is a wealth of studies that have investigated the chemopreventive activity of sulforaphane, derived from glucoraphanin found in broccoli, but other ITCs also have similar activity. What has become clear from such studies is that the chemopreventive properties of glucosinolates/ITCs do not involve a single mechanism of action but multiple mechanisms, which probably interact together to reduce the risk of carcinogenesis. These include modulation of xenobiotic metabolism (inhibition of Phase I and induction of Phase II metabolism enzymes), regulation of oxidative stress, induction of apoptosis and cell cycle arrest, antiinflammatory properties and inhibition of angiogenesis (Fig. 3). It is important to note that the vast majority of these come from cell and animal models, where the concentrations used, typically more than 15 μM ITCs, are far in excess of what could be achieved by normal dietary consumption of glucosinolates (Table 1). Also, the experimental design adopted in these studies, where cells are exposed to ITCs for 24 h before being assessed, is not representative of what is happening in vivo, where the gastrointestinal

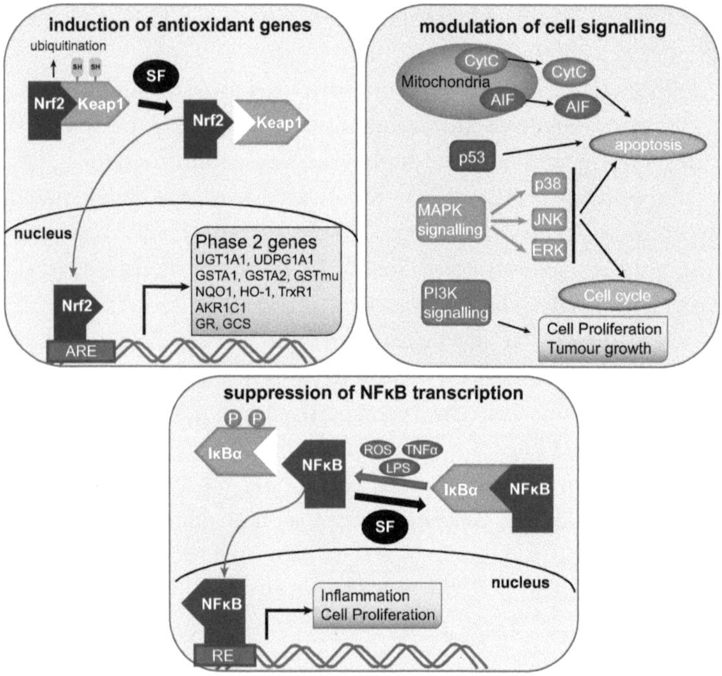

Figure 3 Main mechanisms of action of isothiocyanates from cell and animal models.

tract and the liver are more likely to be exposed for a short period of time to high concentrations, whereas peripheral tissues will be exposed for longer periods to lower concentrations. Nevertheless, they have been extremely valuable in understanding what is the putative mechanistic basis for the health benefits of glucosinolates and serve as the basis for identifying such mechanisms in humans.

6.1 Regulation of Xenobiotic Metabolism

Enzymes involved in the metabolism of carcinogens include those that activate carcinogens (Phase I), consisting of members of the cytochrome P450 (CYP) family, and Phase II enzymes that detoxify activated carcinogens. During Phase I metabolism chemical molecules, including dietary and environmental carcinogens, are converted into highly reactive intermediates that can potentially be harmful by binding to critical macromolecules such as DNA, RNA and protein. There is substantial evidence indicating that the hydrolysis products of glucosinolates are involved in inhibiting Phase I and inducing Phase II enzymes thereby facilitating protection.

6.2 Modulation of Phase I Enzymes

In vitro incubation of rat liver microsomes with phenylethyl-ITC (PEITC) showed a dose-dependent inhibition of CYP1A1, CYP1A2 and CYP2B1 activity (Thapliyal & Maru, 2001). Similarly, in intact human and rat hepatocytes sulforaphane was shown to inhibit enzyme activity of several members of the CYP family (Barcelo, Gardiner, Gescher, & Chipman, 1996; Juge, Mithen, & Traka, 2007; Maheo et al., 1997). Sulforaphane (10 and 50 μM) but not PEITC reduced expression of genes involved in aflatoxin B(1) bioactivation (CYP3A4 and CYP1A2), which greatly decreased aflatoxin–DNA adduct levels in primary cultures of human hepatocytes (Gross-Steinmeyer et al., 2010). Consistently with in vitro models, experiments undertaken with precision-cut liver slices from human donors showed that 24 h incubation with PEITC suppressed the levels of CYP1A2 apoprotein levels (Konsue & Ioannides, 2010). Interestingly, when the intact glucosinolates glucoerucin and glucoraphanin were incubated with precision-cut rat liver slices, both elevated CYP1 activity and increased the apoprotein levels of microsomal CYP1A1, CYP1A2 and CYP1B1, and the addition of myrosinase abolished these effects (Abdull Razis, Bagatta, De Nicola, Iori, & Ioannides, 2010). Also, in animal feeding experiments it seems that glucosinolates are effective in modulating Phase I metabolic enzymes. When rats were administered orally or intraperitoneally a glucosinolate-rich extract of Tuscan black cabbage, several CYP members were reduced significantly (Melega et al., 2013).

6.3 Modulation of Phase II Enzymes

During Phase II metabolism, the activated Phase I metabolites are conjugated in the liver with glutathione (GSH), sulphate, glycine or glucuronic acid to form more soluble or polar metabolites that are easily excreted by the kidneys. Phase II enzymes that have been linked to ITCs include GST, UDP-glucuronosyl transferase (UGT), glutamate-cysteine ligase catalytic (GCLC) and modifier (GCLM) subunits that catalyse the rate-limiting step in glutathione biosynthesis, thioredoxin (TXN) and thioredoxin reductase (TXNRD), aldo-keto reductase (AKR), NAD(P)H:quinone oxidoreductase (NQO1), superoxide dismutase (SOD1) and haeme oxygenase 1 (HO-1). Activity of all these Phase II enzymes was induced in cell and animal models following treatment with several ITCs including PEITC and sulforaphane (reviewed in Cheung & Kong, 2010; Dinkova-Kostova, 2013; Juge et al., 2007). This increase has been observed at both the transcriptional and enzyme activity level in model cell lines of different origin (Basten, Bao, & Williamson, 2002; Brooks, Paton, & Vidanes, 2001; Jiang, Chen, Yang, Hebbar, & Kong, 2003; Matusheski & Jeffery, 2001; Rose, Faulkner, Williamson, & Mithen, 2000). The use of whole-genome transcriptional profiling has provided a detailed picture for the specific targets of different ITCs, which all converge to the significant activation of Phase II metabolism by all ITCs studied (Agyeman et al., 2012; Chambers et al., 2009; Hu et al., 2006; Traka et al., 2005).

One of the key regulators of xenobiotic metabolism is the transcription factor nuclear factor (erythroid 2)-like 2 (nrf2), which induces expression of a range of detoxification and antioxidant genes (Hayes, McMahon, Chowdhry, & Dinkova-Kostova, 2010; Thimmulappa et al., 2002). Under normal homeostatic conditions nrf2 forms a dimer with Kelch-like ECH-associated protein 1 (Keap1), which targets nrf2 for proteosomal degradation by ubiquitination (Baird & Dinkova-Kostova, 2011). Certain thiol-reactive inducers, such as ITCs derived from cruciferous vegetables, react with specific cysteine residues on Keap1 causing detrimental conformational changes to Keap1 and accumulation of nrf2 in the cytoplasm (Baird, Swift, Lleres, & Dinkova-Kostova, 2014). Consequently, accumulated nrf2 translocates to the nucleus, where it binds to the antioxidant response elements (AREs) present in promoter regions of several Phase II antioxidant genes and activates their transcription (Greco, Shafer, & Fiskum, 2011; Kobayashi & Yamamoto, 2005; Thimmulappa et al., 2002).

6.4 Regulation of Oxidative Stress

Oxidative stress, resulting in excessive levels of free radicals and reactive oxygen species (ROS), is involved in the pathogenesis of chronic disease. The prooxidant capacity of ITCs associated with ROS production has been linked to their chemopreventive properties. One example is PEITC, which selectively induces cell death in cancerous cells. Cancerous cells exhibit increased basal levels of ROS compared to noncancerous cells, thus exposure to PEITC would increase ROS levels above a 'tipping point' only in cancerous cells (Trachootham et al., 2006). Certain ITCs, such as erucin, enhanced the cytotoxic effects of the arsenic trioxide (ATO), therapeutically used in the treatment of acute promyelocytic leukaemia, through a ROS-dependent mechanism in leukaemic cells (Doudican, Bowling, & Orlow, 2010). Depletion of intracellular glutathione is tightly linked with the direct prooxidant effect of ITCs. This will itself trigger an nrf2-mediated response from the cell to replenish glutathione stores through transcriptional activation of Phase II enzymes, such as GCLC and GCLM, responsible for glutathione biosynthesis. The increased nrf2-mediated antioxidant capacity of the cell will then protect the cells from exposure to carcinogens. Conversely, glutathione depletion and detrimental ROS production in the cancerous cell will target it for cell death, thus achieving protection from carcinogenesis.

6.5 Metabolic Regulation

Recently, there is some evidence to suggest a role of glucosinolates in affecting metabolic parameters, such as lipid and glucose metabolism. Sulforaphane was shown to prevent lipid accumulation in the liver of mice and activated AMPK pathway in adipose tissue (Choi et al., 2014), but the molecular mechanisms were not investigated. Supplementation of diets with sulforaphane also prevented lipid accumulation in cardiac tissue in mice with T2DM (Zhang, Wang, et al., 2014). In mouse embryonic fibroblasts sulforaphane increased glucose uptake, induced nrf2-mediated transcription (Heiss, Schachner, Zimmermann, & Dirsch, 2013) and prevented lipid accumulation and triglyceride content (Xu, Kulkarni, Donepudi, More, & Slitt, 2012). Sulforaphane also attenuated the activity of PKC in response to high glucose stimulation of HMEC-1 endothelial cells and increased activity of the pentose phosphate pathway enzyme transketolase (Xue et al., 2008).

There is increasing appreciation of the interplay between redox status and energy metabolism. In mitochondria, energy production through the tricarboxylic acid (TCA) cycle and oxidative phosphorylation is a major producer of ROS, through the action of alpha-ketoglutarate dehydrogenase (α-KGDH), a key enzyme in TCA cycle (Tretter & Adam-Vizi, 2005). Paradoxically, TCA cycle is also inhibited by ROS and oxidative stress, which suggests that redox modulates energy production. Nrf2-mediated transcription and subsequent redox modulation by several ITCs has the potential to affect central metabolism (Fig. 4). Additionally, several nrf2-regulated Phase 2 enzymes require the supply of NADPH as a cofactor for their enzymatic action. This is provided by the pentose phosphate pathway, which is redirecting glucose away from glycolysis that produces pyruvate, thus interfering with key metabolic pathways that are driving carcinogenesis (Hayes & Dinkova-Kostova, 2014).

6.6 Induction of Apoptosis

Apoptosis, or programmed cell death, is a highly regulated process that plays important roles in the development and maintenance of homeostasis and in the elimination of cells that are damaged or no longer necessary for the

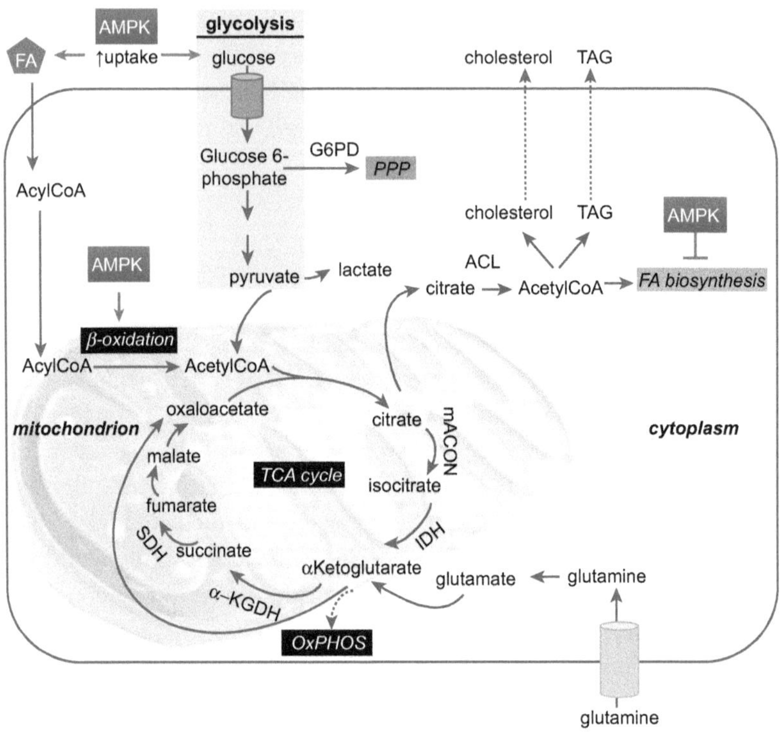

Figure 4 Metabolic pathways that are regulated by glucosinolate-rich diets.

organism. Inappropriate regulation of apoptosis may cause serious disorders, such as neural degeneration, autoimmune diseases and cancers. The classical hallmarks of apoptosis such as chromatin condensation, translocation of phosphatidylserine across the plasma membrane and DNA fragmentation have been shown to be induced by ITCs in a variety of colon (Gamet-Payrastre et al., 2000), prostate (Choi et al., 2006; Singh, Xiao, Lew, Dhir, & Singh, 2004), medulloblastoma (Gingras et al., 2004) and mammary cell lines (Jackson & Singletary, 2004a). The mechanisms by which isothiocyanates induce apoptosis have been extensively explored with cell models, and several key processes within the apoptotic pathways have been shown to be affected by ITCs, such as induction of caspases (Gingras et al., 2004; Xu & Thornalley, 2000; Yu, Mandlekar, Harvey, Ucker, & Kong, 1998), disruption of mitochondrial integrity (Singh et al., 2005; Tang & Zhang, 2005) through induction of the proapoptotic members of the Bcl-2 family and release of mitochondrial proteins cytochrome c, Smac/DIABLO and AIF (Choi & Singh, 2005; Gamet-Payrastre et al., 2000; Karmakar, Weinberg,

Banik, Patel, & Ray, 2006; Singh et al., 2005). ITCs may also induce apoptosis via the JNK/MAPK signalling pathway (Hu, Kim, Chen, Hebbar, & Kong, 2003; Xiao & Singh, 2002; Xu et al., 2006).

6.7 Cell Cycle Arrest

ITCs are also capable of arresting the cell cycle at different stages of its progression. This arrest has been documented in colon, prostate, breast, bladder and T-cells and resulted in both G0/G1 and G2/M phase block depending on the cell type. Additionally, S-phase block as a result of ITC treatment has been reported in human UM-UC-3 bladder cells (Tang & Zhang, 2004). Several genes have been involved in cell cycle arrest by ITCs. Consistent with a block in the G0/G1 phase of the cell cycle ITCs downregulated expression of the Cyclin D1 protein in prostate and colon cells (Chiao et al., 2002; Shen, Xu, Chen, Hebbar, & Kong, 2006), reduced levels of its associated kinase cdk4 in DU-145 prostate cells (Wang, Liu, et al., 2004) and reduced phosphorylation of the Rb tumour suppressor protein that activates the transition from G(1) phase to S phase (Wang, Liu, et al., 2004). Significant reduction in the levels of Cyclin D3 and slight reduction in Cyclin D2 and associated CDK4 and Cdk6 were also reported in non-transformed T lymphocytes (Fimognari et al., 2002). ITCs also induce cell cycle arrest at the G2/M phase by regulating expression of Cyclin B1. In human colon and breast cells 15 μM ITCs for up to 48 h increased Cyclin B1 protein expression (Gamet-Payrastre et al., 2000; Jackson & Singletary, 2004b). In HT29 cells this G2/M phase arrest was achieved by maintaining the cdc2 kinase in its active dephosphorylated form and was associated with phosphorylation/activation of the Rb protein (Parnaud et al., 2004). In contrast, ITC treatment on prostate PC-3 and bladder UM-UC-3 cells resulted in reduced expression of Cyclin B1 (Singh, Herman-Antosiewicz, et al., 2004; Tang & Zhang, 2004). The G2/M phase arrest in PC-3 cells activated the DNA damage checkpoint pathway through activation of checkpoint kinase 2 (Chk2) and subsequent phosphorylation of cell division cycle 25C (Singh, Herman-Antosiewicz, et al., 2004). Finally, S-phase arrest reported in UM-UC-3 cells in response to ITC treatment was also associated with reduced levels of Cyclin A (Tang & Zhang, 2004).

6.8 Inhibition of Angiogenesis and Metastasis

Angiogenesis is a prerequisite for the growth of solid tumours and metastasis. Without new blood vessel formation leading to the development of intratumoral capillary networks, tumour progression is severely limited. Several studies suggest that ITCs can interfere with all essential steps of neovascularization from proangiogenic signalling and basement membrane integrity to endothelial cell proliferation, migration and tube formation. ITCs have been shown to suppress proliferation and migration of cultured human umbilical vein endothelial cells (HUVECs), an in vitro system for the study of angiogenesis (Asakage et al., 2006; Thejass & Kuttan, 2007; Xiao & Singh, 2007). These were directly linked to downregulation of vascular endothelial growth factor (VEGF) and proinflammatory cytokines and suppression of matrix metalloproteinase (MMP) production, which are key drivers of neovascularisation and angiogenesis.

The inhibitory metastatic potential of ITCs has been studied in vitro using wound healing and cell invasion assays (Lai et al., 2014; Wolf & Claudio, 2014; Zhang, Hao, et al., 2014). The antimetastatic effects are closely linked to cancer initiation prevention by ITCs and, in particular, are mediated through modulation of MAPK signalling, TGFβ/Smad2 signalling and apoptosis. Evidence from animal models also supports a role for ITCs in protection from metastasis. More recently, diets containing benzyl ITC (BITC) were shown to suppress mouse mammary tumour metastasis in the lung and liver as a result of high-fat diet (Kim et al., 2015). Similarly, gavage with BITC suppressed pulmonary tumour nodules and the total pulmonary metastatic volume in a xenograft model of murine mammary metastasis (Kim, Hong, Eom, Lee, & Park, 2011). Oral gavage of sulforaphane was also effective in reducing pulmonary metastasis incidence and multiplicity in TRAMP mice, a model of prostate cancer progression associated with immune modulation (Singh et al., 2009).

6.9 Other Biological Activities

Glucosinolates/ITCs have also demonstrated biological activities that are indirectly linked with cancer prevention amongst other things. These include (1) antiinflammatory properties through regulation of nuclear factor-kappa B (NF-κB) and its downstream signalling and inhibition of TNF-α and lipopolysaccharide (LPS)-stimulated inflammatory response (Dey, Ribnicky, Kurmukov, & Raskin, 2006; Folkard et al., 2014; Heiss, Herhaus, Klimo, Bartsch, & Gerhauser, 2001; Karmakar et al., 2006; Rose, Won, Ong, & Whiteman, 2005; Wierinckx et al., 2005; Xu, Shen, Chen, Gelinas, & Kong, 2005), (2) downregulation of androgen receptor signalling in prostate cancer prevention through suppression of AR transcription and protein levels (Gibbs, Schwartzman, Deng, & Alumkal, 2009; Kim & Singh, 2009; Wang, Liu, & Chiao, 2006), (3) epigenetic modification (Chen et al., 2016; Qu et al., 2015; Rajendran et al., 2015) and (4) antibacterial properties against *Helicobacter pylori*, a known cause for gastritis and peptic ulcers, but also associated with enhanced risk of gastric cancer (Haristoy, Fahey, Scholtus, & Lozniewski, 2005; Moon, Kim, Ahn, & Shibamoto, 2010; Yanaka et al., 2009).

7. CONCLUSION

There is now a strong body of evidence from epidemiological studies that increasing glucosinolate intake from cruciferous vegetables reduces the risk of developing several chronic diseases, in particular, cancer. This is likely due to the multifaceted biological response that their degradation products, isothiocyanates, are eliciting as documented in a variety of cell and animal models. Although the mechanisms of action from such studies may partly explain the observational evidence, there is a need to obtain experimental evidence for such mechanisms directly in vivo by undertaking short- and long-term human intervention studies. In addition, we must begin to adopt a systems approach to the study of glucosinolate bioactivity. This will be greatly facilitated by the use of whole-genome profiling and high-throughput analytical methods, which have the potential to significantly impact the field by providing an unbiased snapshot of the biological pathways involved. Such technologies are already providing fascinating results by identifying novel links between glucosinolate-rich diets and regulation of metabolism.

REFERENCES

Abdull Razis, A. F., Bagatta, M., De Nicola, G. R., Iori, R., & Ioannides, C. (2010). Intact glucosinolates modulate hepatic cytochrome P450 and phase II conjugation activities and may contribute directly to the chemopreventive activity of cruciferous vegetables. *Toxicology, 277*, 74–85.

Agyeman, A. S., Chaerkady, R., Shaw, P. G., Davidson, N. E., Visvanathan, K., Pandey, A., & Kensler, T. W. (2012). Transcriptomic and proteomic profiling of KEAP1 disrupted and sulforaphane-treated human breast epithelial cells reveals common expression profiles. *Breast Cancer Research and Treatment, 132*, 175–187.

Ambrosone, C. B., McCann, S. E., Freudenheim, J. L., Marshall, J. R., Zhang, Y., & Shields, P. G. (2004). Breast cancer risk in premenopausal women is inversely associated with consumption of broccoli, a source of isothiocyanates, but is not modified by GST genotype. *Journal of Nutrition, 134*, 1134–1138.

Armah, C. N., Derdemezis, C., Traka, M. H., Dainty, J. R., Doleman, J. F., Saha, S., ... Mithen, R. F. (2015). Diet rich in high glucoraphanin broccoli reduces plasma LDL cholesterol: evidence from randomised controlled trials. *Molecular Nutrition & Food Research, 59*, 918–926.

Armah, C. N., Traka, M. H., Dainty, J. R., Defernez, M., Janssens, A., Leung, W., ... Mithen, R. F. (2013). A diet rich in high-glucoraphanin broccoli interacts with genotype to reduce discordance in plasma metabolite profiles by modulating mitochondrial function. *American Journal of Clinical Nutrition, 98*, 712–722.

Asakage, M., Tsuno, N. H., Kitayama, J., Tsuchiya, T., Yoneyama, S., Yamada, J., ... Nagawa, H. (2006). Sulforaphane induces inhibition of human umbilical vein endothelial cells proliferation by apoptosis. *Angiogenesis, 9*, 83–91.

Atwell, L. L., Hsu, A., Wong, C. P., Stevens, J. F., Bella, D., Yu, T. W., ... Ho, E. (2015). Absorption and chemopreventive targets of sulforaphane in humans following consumption of broccoli sprouts or a myrosinase-treated broccoli sprout extract. *Molecular Nutrition & Food Research, 59*, 424–433.

Baird, L., & Dinkova-Kostova, A. T. (2011). The cytoprotective role of the Keap1-Nrf2 pathway. *Archives of Toxicology, 85*, 241–272.

Baird, L., Swift, S., Lleres, D., & Dinkova-Kostova, A. T. (2014). Monitoring Keap1-Nrf2 interactions in single live cells. *Biotechnology Advances, 32*, 1133–1144.

Bandera, E. V., Kushi, L. H., Moore, D. F., Gifkins, D. M., & McCullough, M. L. (2007). Fruits and vegetables and endometrial cancer risk: a systematic literature review and meta-analysis. *Nutrition and Cancer, 58*, 6–21.

Barcelo, S., Gardiner, J. M., Gescher, A., & Chipman, J. K. (1996). CYP2E1-mediated mechanism of anti-genotoxicity of the broccoli constituent sulforaphane. *Carcinogenesis, 17*, 277–282.

Basten, G. P., Bao, Y., & Williamson, G. (2002). Sulforaphane and its glutathione conjugate but not sulforaphane nitrile induce UDP-glucuronosyl transferase (UGT1A1) and glutathione transferase (GSTA1) in cultured cells. *Carcinogenesis, 23*, 1399–1404.

Bones, A. M., & Rossiter, J. T. (1996). The myrosinase-glucosinolate system, its organisation and biochemistry. *Physiologia Plantarum, 97*, 194–208.

Bongoni, R., Verkerk, R., Steenbekkers, B., Dekker, M., & Stieger, M. (2014). Evaluation of different cooking conditions on broccoli (*Brassica oleracea* var. *italica*) to improve the nutritional value and consumer acceptance. *Plant Foods for Human Nutrition, 69*, 228–234.

Brooks, J. D., Paton, V. G., & Vidanes, G. (2001). Potent induction of phase 2 enzymes in human prostate cells by sulforaphane. *Cancer Epidemiology, Biomarkers & Prevention: A Publication of the American Association for Cancer Research, Cosponsored by the American Society of Preventive Oncology, 10*, 949–954.

Brusewitz, G., Cameron, B. D., Chasseaud, L. F., Gorler, K., Hawkins, D. R., Koch, H., & Mennicke, W. H. (1977). The metabolism of benzyl isothiocyanate and its cysteine conjugate. *Biochemical Journal, 162*, 99–107.

Chambers, K. F., Bacon, J. R., Kemsley, E. K., Mills, R. D., Ball, R. Y., Mithen, R. F., & Traka, M. H. (2009). Gene expression profile of primary prostate epithelial and stromal cells in response to sulforaphane or iberin exposure. *Prostate, 69*, 1411–1421.

Chen, Y., Cang, S., Han, L., Liu, C., Yang, P., Solangi, Z., ... Chiao, J. W. (2016). Establishment of prostate cancer spheres from a prostate cancer cell line after phenethyl isothiocyanate treatment and discovery of androgen-dependent reversible differentiation between sphere and neuroendocrine cells. *Oncotarget* [Epub ahead of print].

Cheung, K. L., & Kong, A. N. (2010). Molecular targets of dietary phenethyl isothiocyanate and sulforaphane for cancer chemoprevention. *AAPS Journal, 12*, 87–97.

Chiao, J. W., Chung, F. L., Kancherla, R., Ahmed, T., Mittelman, A., & Conaway, C. C. (2002). Sulforaphane and its metabolite mediate growth arrest and apoptosis in human prostate cancer cells. *International Journal of Oncology, 20*, 631–636.

Choi, K. M., Lee, Y. S., Kim, W., Kim, S. J., Shin, K. O., Yu, J. Y., ... Yoo, H. S. (2014). Sulforaphane attenuates obesity by inhibiting adipogenesis and activating the AMPK pathway in obese mice. *Journal of Nutritional Biochemistry, 25*, 201–207.

Choi, S., Lew, K. L., Xiao, H., Herman-Antosiewicz, A., Xiao, D., Brown, C. K., & Singh, S. V. (2006). D,L-Sulforaphane-induced cell death in human prostate cancer cells is regulated by inhibitor of apoptosis family proteins and Apaf-1. *Carcinogenesis, 28*, 151–162.

Choi, S., & Singh, S. V. (2005). Bax and Bak are required for apoptosis induction by sulforaphane, a cruciferous vegetable-derived cancer chemopreventive agent. *Cancer Research, 65*, 2035–2043.

Chung, F. L., Jiao, D., Getahun, S. M., & Yu, M. C. (1998). A urinary biomarker for uptake of dietary isothiocyanates in humans. *Cancer Epidemiology, Biomarkers & Prevention: A Publication of the American Association for Cancer Research, Cosponsored By the American Society of Preventive Oncology, 7*, 103–108.

Clarke, J. D., Hsu, A., Riedl, K., Bella, D., Schwartz, S. J., Stevens, J. F., & Ho, E. (2011). Bioavailability and inter-conversion of sulforaphane and erucin in human subjects consuming broccoli sprouts or broccoli supplement in a cross-over study design. *Pharmacological Research: The Official Journal of the Italian Pharmacological Society, 64*, 456–463.

Cohen, J. H., Kristal, A. R., & Stanford, J. L. (2000). Fruit and vegetable intakes and prostate cancer risk. *Journal of the National Cancer Institute, 92*, 61–68.

Conaway, C. C., Getahun, S. M., Liebes, L. L., Pusateri, D. J., Topham, D. K., Botero-Omary, M., & Chung, F. L. (2000). Disposition of glucosinolates and sulforaphane in humans after ingestion of steamed and fresh broccoli. *Nutrition and Cancer, 38*, 168–178.

Cornelis, M. C., El-Sohemy, A., & Campos, H. (2007). GSTT1 genotype modifies the association between cruciferous vegetable intake and the risk of myocardial infarction. *American Journal of Clinical Nutrition, 86*, 752–758.

Cotton, S. C., Sharp, L., Little, J., & Brockton, N. (2000). Glutathione S-transferase polymorphisms and colorectal cancer: a HuGE review. *American Journal of Epidemiology, 151*, 7–32.

Cramer, J. M., & Jeffery, E. H. (2011). Sulforaphane absorption and excretion following ingestion of a semi-purified broccoli powder rich in glucoraphanin and broccoli sprouts in healthy men. *Nutrition and Cancer, 63*, 196–201.

Cramer, J. M., Teran-Garcia, M., & Jeffery, E. H. (2012). Enhancing sulforaphane absorption and excretion in healthy men through the combined consumption of fresh broccoli sprouts and a glucoraphanin-rich powder. *British Journal of Nutrition, 107*, 1333–1338.

Dey, M., Ribnicky, D., Kurmukov, A. G., & Raskin, I. (2006). In vitro and in vivo anti-inflammatory activity of a seed preparation containing phenethylisothiocyanate. *Journal of Pharmacology and Experimental Therapeutics, 317*, 326–333.

Dinkova-Kostova, A. T. (2013). Chemoprotection against cancer by isothiocyanates: a focus on the animal models and the protective mechanisms. *Topics in Current Chemistry, 329*, 179–201.

Dosz, E. B., & Jeffery, E. H. (2013). Modifying the processing and handling of frozen broccoli for increased sulforaphane formation. *Journal of Food Science, 78*, H1459–H1463.

Doudican, N. A., Bowling, B., & Orlow, S. J. (2010). Enhancement of arsenic trioxide cytotoxicity by dietary isothiocyanates in human leukemic cells via a reactive oxygen species-dependent mechanism. *Leukemia Research, 34*, 229–234.

Dyba, M., Wang, A., Noone, A. M., Goerlitz, D., Shields, P., Zheng, Y. L., … Chung, F. L. (2010). Metabolism of isothiocyanates in individuals with positive and null GSTT1 and M1 genotypes after drinking watercress juice. *Clinical Nutrition: Official Journal of the European Society of Parenteral and Enteral Nutrition, 29*, 813–818.

Egner, P. A., Chen, J. G., Zarth, A. T., Ng, D. K., Wang, J. B., Kensler, K. H., … Kensler, T. W. (2014). Rapid and sustainable detoxication of airborne pollutants by broccoli sprout beverage: results of a randomized clinical trial in China. *Cancer Prevention Research (Philadelphia), 7*, 813–823.

Fahey, J. W., Zalcmann, A. T., & Talalay, P. (2001). The chemical diversity and distribution of glucosinolates and isothiocyanates among plants. *Phytochemistry, 56*, 5–51.

Fimognari, C., Nusse, M., Berti, F., Iori, R., Cantelli-Forti, G., & Hrelia, P. (2002). Cyclin D3 and p53 mediate sulforaphane-induced cell cycle delay and apoptosis in non-transformed human T lymphocytes. *Cellular and Molecular Life Sciences: CMLS, 59*, 2004–2012.

Folkard, D. L., Melchini, A., Traka, M. H., Al-Bakheit, A., Saha, S., Mulholland, F., ... Mithen, R. F. (2014). Suppression of LPS-induced transcription and cytokine secretion by the dietary isothiocyanate sulforaphane. *Molecular Nutrition & Food Research, 58*, 2286–2296.

Fowke, J. H., Chung, F. L., Jin, F., Qi, D., Cai, Q., Conaway, C., ... Zheng, W. (2003). Urinary isothiocyanate levels, brassica, and human breast cancer. *Cancer Research, 63*, 3980–3986.

Gamet-Payrastre, L., Li, P., Lumeau, S., Cassar, G., Dupont, M. A., Chevolleau, S., ... Terce, F. (2000). Sulforaphane, a naturally occurring isothiocyanate, induces cell cycle arrest and apoptosis in HT29 human colon cancer cells. *Cancer Research, 60*, 1426–1433.

Gasper, A. V., Al-Janobi, A., Smith, J. A., Bacon, J. R., Fortun, P., Atherton, C., ... Mithen, R. F. (2005). Glutathione S-transferase M1 polymorphism and metabolism of sulforaphane from standard and high-glucosinolate broccoli. *American Journal of Clinical Nutrition, 82*, 1283–1291.

Gibbs, A., Schwartzman, J., Deng, V., & Alumkal, J. (2009). Sulforaphane destabilizes the androgen receptor in prostate cancer cells by inactivating histone deacetylase 6. *Proceedings of the National Academy of Sciences of the United States of America, 106*, 16663–16668.

Gingras, D., Gendron, M., Boivin, D., Moghrabi, A., Theoret, Y., & Beliveau, R. (2004). Induction of medulloblastoma cell apoptosis by sulforaphane, a dietary anticarcinogen from Brassica vegetables. *Cancer Letters, 203*, 35–43.

Giovannucci, E., Rimm, E. B., Liu, Y., Stampfer, M. J., & Willett, W. C. (2003). A prospective study of cruciferous vegetables and prostate cancer. *Cancer Epidemiology, Biomarkers & Prevention: A Publication of the American Association for Cancer Research, Cosponsored by the American Society of Preventive Oncology, 12*, 1403–1409.

Greco, T., Shafer, J., & Fiskum, G. (2011). Sulforaphane inhibits mitochondrial permeability transition and oxidative stress. *Free Radical Biology & Medicine, 51*, 2164–2171.

Gross-Steinmeyer, K., Stapleton, P. L., Tracy, J. H., Bammler, T. K., Strom, S. C., & Eaton, D. L. (2010). Sulforaphane- and phenethyl isothiocyanate-induced inhibition of aflatoxin B1-mediated genotoxicity in human hepatocytes: role of GSTM1 genotype and CYP3A4 gene expression. *Toxicological Sciences: An Official Journal of the Society of Toxicology, 116*, 422–432.

Han, B., Li, X., & Yu, T. (2014). Cruciferous vegetables consumption and the risk of ovarian cancer: a meta-analysis of observational studies. *Diagnostic Pathology [Electronic Resource], 9*, 7.

Hansson, L. E., Nyren, O., Bergstrom, R., Wolk, A., Lindgren, A., Baron, J., & Adami, H. O. (1993). Diet and risk of gastric cancer. A population-based case-control study in Sweden. *International Journal of Cancer, 55*, 181–189.

Haristoy, X., Fahey, J. W., Scholtus, I., & Lozniewski, A. (2005). Evaluation of the antimicrobial effects of several isothiocyanates on *Helicobacter pylori*. *Planta Medica, 71*, 326–330.

Hayes, J. D., & Dinkova-Kostova, A. T. (2014). The Nrf2 regulatory network provides an interface between redox and intermediary metabolism. *Trends in Biochemical Sciences, 39*, 199–218.

Hayes, J. D., Flanagan, J. U., & Jowsey, I. R. (2005). Glutathione transferases. *Annual Review of Pharmacology and Toxicology, 45*, 51–88.

Hayes, J. D., McMahon, M., Chowdhry, S., & Dinkova-Kostova, A. T. (2010). Cancer chemoprevention mechanisms mediated through the Keap1-Nrf2 pathway. *Antioxidants & Redox Signaling, 13*, 1713–1748.

Hayes, J. D., & Strange, R. C. (2000). Glutathione S-transferase polymorphisms and their biological consequences. *Pharmacology, 61*, 154–166.

Heiss, E., Herhaus, C., Klimo, K., Bartsch, H., & Gerhauser, C. (2001). Nuclear factor kappa B is a molecular target for sulforaphane-mediated anti-inflammatory mechanisms. *Journal of Biological Chemistry, 276*, 32008–32015.

Heiss, E. H., Schachner, D., Zimmermann, K., & Dirsch, V. M. (2013). Glucose availability is a decisive factor for Nrf2-mediated gene expression. *Redox Biology, 1*, 359–365.

Hofmann, T., Kuhnert, A., Schubert, A., Gill, C., Rowland, I. R., Pool-Zobel, B. L., & Glei, M. (2009). Modulation of detoxification enzymes by watercress: in vitro and in vivo investigations in human peripheral blood cells. *European Journal of Nutrition, 48*, 483–491.

Hu, J., Hu, Y., Hu, Y., & Zheng, S. (2015). Intake of cruciferous vegetables is associated with reduced risk of ovarian cancer: a meta-analysis. *Asia Pacific Journal of Clinical Nutrition, 24*, 101–109.

Hu, R., Kim, B. R., Chen, C., Hebbar, V., & Kong, A. N. (2003). The roles of JNK and apoptotic signaling pathways in PEITC-mediated responses in human HT-29 colon adenocarcinoma cells. *Carcinogenesis, 24*, 1361–1367.

Hu, R., Xu, C., Shen, G., Jain, M. R., Khor, T. O., Gopalkrishnan, A., … Kong, A. N. (2006). Identification of Nrf2-regulated genes induced by chemopreventive isothiocyanate PEITC by oligonucleotide microarray. *Life Sciences, 79*, 1944–1955.

Jackson, S. J., & Singletary, K. W. (2004a). Sulforaphane: a naturally occurring mammary carcinoma mitotic inhibitor, which disrupts tubulin polymerization. *Carcinogenesis, 25*, 219–227.

Jackson, S. J., & Singletary, K. W. (2004b). Sulforaphane inhibits human MCF-7 mammary cancer cell mitotic progression and tubulin polymerization. *Journal of Nutrition, 134*, 2229–2236.

Ji, Y., & Morris, M. E. (2003). Determination of phenethyl isothiocyanate in human plasma and urine by ammonia derivatization and liquid chromatography-tandem mass spectrometry. *Analytical Biochemistry, 323*, 39–47.

Jiang, Z. Q., Chen, C., Yang, B., Hebbar, V., & Kong, A. N. (2003). Differential responses from seven mammalian cell lines to the treatments of detoxifying enzyme inducers. *Life Sciences, 72*, 2243–2253.

Joseph, M. A., Moysich, K. B., Freudenheim, J. L., Shields, P. G., Bowman, E. D., Zhang, Y., … Ambrosone, C. B. (2004). Cruciferous vegetables, genetic polymorphisms in glutathione S-transferases M1 and T1, and prostate cancer risk. *Nutrition and Cancer, 50*, 206–213.

Juge, N., Mithen, R. F., & Traka, M. (2007). Molecular basis for chemoprevention by sulforaphane: a comprehensive review. *Cellular and Molecular Life Sciences: CMLS, 64*, 1105–1127.

Karmakar, S., Weinberg, M. S., Banik, N. L., Patel, S. J., & Ray, S. K. (2006). Activation of multiple molecular mechanisms for apoptosis in human malignant glioblastoma T98G and U87MG cells treated with sulforaphane. *Neuroscience, 141*, 1265–1280.

Kensler, T. W., Ng, D., Carmella, S. G., Chen, M., Jacobson, L. P., Munoz, A., … Hecht, S. S. (2012). Modulation of the metabolism of airborne pollutants by glucoraphanin-rich and sulforaphane-rich broccoli sprout beverages in Qidong, China. *Carcinogenesis, 33*, 101–107.

Kim, E. J., Hong, J. E., Eom, S. J., Lee, J. Y., & Park, J. H. (2011). Oral administration of benzyl-isothiocyanate inhibits solid tumor growth and lung metastasis of 4T1 murine mammary carcinoma cells in BALB/c mice. *Breast Cancer Research and Treatment, 130*, 61–71.

Kim, M., Cho, H. J., Kwon, G. T., Kang, Y. H., Kwon, S. H., Her, S., … Park, J. H. (2015). Benzyl isothiocyanate suppresses high-fat diet-stimulated mammary tumor progression via the alteration of tumor microenvironments in obesity-resistant BALB/c mice. *Molecular Carcinogenesis, 54*, 72–82.

Kim, S. H., & Singh, S. V. (2009). D,L-Sulforaphane causes transcriptional repression of androgen receptor in human prostate cancer cells. *Molecular Cancer Therapeutics, 8*, 1946–1954.

Kirsh, V. A., Peters, U., Mayne, S. T., Subar, A. F., Chatterjee, N., Johnson, C. C., … Prostate, Lung, Colorectal and Ovarian Cancer Screening Trial. (2007). Prospective study of fruit and vegetable intake and risk of prostate cancer. *Journal of the National Cancer Institute, 99*, 1200–1209.

Kliebenstein, D. J., & Cacho, N. I. (2016). Nonlinear selection and a blend of convergent, divergent and parallel evolution shapes: natural variation in glucosinolates. In S. Kopriva (Ed.), *Glucosinolates: Vol. 80* (in this volume).

Kobayashi, M., & Yamamoto, M. (2005). Molecular mechanisms activating the Nrf2-Keap1 pathway of antioxidant gene regulation. *Antioxidants & Redox Signaling*, 7, 385–394.

Konsue, N., & Ioannides, C. (2010). Modulation of carcinogen-metabolising cytochromes P450 in human liver by the chemopreventive phytochemical phenethyl isothiocyanate, a constituent of cruciferous vegetables. *Toxicology, 268*, 184–190.

Lai, K. C., Lu, C. C., Tang, Y. J., Chiang, J. H., Kuo, D. H., Chen, F. A., … Yang, J. S. (2014). Allyl isothiocyanate inhibits cell metastasis through suppression of the MAPK pathways in epidermal growth factor stimulated HT29 human colorectal adenocarcinoma cells. *Oncology Reports, 31*, 189–196.

Lambrix, V., Reichelt, M., Mitchell-Olds, T., Kliebenstein, D. J., & Gershenzon, J. (2001). The Arabidopsis epithiospecifier protein promotes the hydrolysis of glucosinolates to nitriles and influences *Trichoplusia ni* herbivory. *Plant Cell, 13*, 2793–2807.

Lamy, E., Garcia-Kaufer, M., Prinzhorn, J., & Mersch-Sundermann, V. (2012). Antigenotoxic action of isothiocyanate-containing mustard as determined by two cancer biomarkers in a human intervention trial. *European Journal of Cancer Prevention: The Official Journal of the European Cancer Prevention Organisation (ecp), 21*, 400–406.

Li, F., Hullar, M. A., Beresford, S. A., & Lampe, J. W. (2011). Variation of glucoraphanin metabolism in vivo and ex vivo by human gut bacteria. *British Journal of Nutrition, 106*, 408–416.

Lin, H. J., Probst-Hensch, N. M., Louie, A. D., Kau, I. H., Witte, J. S., Ingles, S. A., … Haile, R. W. (1998). Glutathione transferase null genotype, broccoli, and lower prevalence of colorectal adenomas. *Cancer Epidemiology, Biomarkers & Prevention: A Publication of the American Association for Cancer Research, Cosponsored by the American Society of Preventive Oncology*, 7, 647–652.

Lin, H. J., Zhou, H. Y., Dai, A. H., Huang, H. F., Lin, J. H., Frankl, H. D., … Haile, R. W. (2002). Glutathione transferase GSTT1, broccoli, and prevalence of colorectal adenomas. *Pharmacogenetics, 12*, 175–179.

Liu, B., Mao, Q., Cao, M., & Xie, L. (2012). Cruciferous vegetables intake and risk of prostate cancer: a meta-analysis. *International Journal of Urology: Official Journal of the Japanese Urological Association, 19*, 134–141.

Liu, B., Mao, Q., Lin, Y., Zhou, F., & Xie, L. (2013). The association of cruciferous vegetables intake and risk of bladder cancer: a meta-analysis. *World Journal of Urology, 31*, 127–133.

Liu, X., & Lv, K. (2013). Cruciferous vegetables intake is inversely associated with risk of breast cancer: a meta-analysis. *Breast, 22*, 309–313.

London, S. J., Yuan, J. M., Chung, F. L., Gao, Y. T., Coetzee, G. A., Ross, R. K., & Yu, M. C. (2000). Isothiocyanates, glutathione S-transferase M1 and T1 polymorphisms, and lung-cancer risk: a prospective study of men in Shanghai, China. *Lancet, 356*, 724–729.

Luang-In, V., Narbad, A., Nueno-Palop, C., Mithen, R., Bennett, M., & Rossiter, J. T. (2014). The metabolism of methylsulfinylalkyl- and methylthioalkyl-glucosinolates by a selection of human gut bacteria. *Molecular Nutrition & Food Research, 58*, 875–883.

Maheo, K., Morel, F., Langouet, S., Kramer, H., Le Ferrec, E., Ketterer, B., & Guillouzo, A. (1997). Inhibition of cytochromes P-450 and induction of glutathione S-transferases by sulforaphane in primary human and rat hepatocytes. *Cancer Research, 57*, 3649–3652.

Matusheski, N. V., & Jeffery, E. H. (2001). Comparison of the bioactivity of two glucoraphanin hydrolysis products found in broccoli, sulforaphane and sulforaphane nitrile. *Journal of Agricultural and Food Chemistry, 49*, 5743–5749.

Matusheski, N. V., Juvik, J. A., & Jeffery, E. H. (2004). Heating decreases epithiospecifier protein activity and increases sulforaphane formation in broccoli. *Phytochemistry, 65*, 1273–1281.

Melega, S., Canistro, D., Pagnotta, E., Iori, R., Sapone, A., & Paolini, M. (2013). Effect of sprout extract from Tuscan black cabbage on xenobiotic-metabolizing and antioxidant enzymes in rat liver. *Mutation Research, 751*, 45–51.

Meyer, D. J., Crease, D. J., & Ketterer, B. (1995). Forward and reverse catalysis and product sequestration by human glutathione S-transferases in the reaction of GSH with dietary aralkyl isothiocyanates. *The Biochemical Journal, 306*, 565–569.

Moon, J. K., Kim, J. R., Ahn, Y. J., & Shibamoto, T. (2010). Analysis and anti-Helicobacter activity of sulforaphane and related compounds present in broccoli (*Brassica oleracea* L.) sprouts. *Journal of Agricultural and Food Chemistry, 58*, 6672–6677.

Moore, L., Brennan, P., Karami, S., Hung, R., Hsu, C., Boffetta, P., ... Chow, W. H. (2007). Glutathione S-transferase polymorphisms, cruciferous vegetable intake, and cancer risk in the Central and Eastern European Kidney Cancer study. *Carcinogenesis, 28*(9), 1960–1964.

Parnaud, G., Li, P., Cassar, G., Rouimi, P., Tulliez, J., Combaret, L., & Gamet-Payrastre, L. (2004). Mechanism of sulforaphane-induced cell cycle arrest and apoptosis in human colon Cancer cells. *Nutrition and Cancer, 48*, 198–206.

Payen, L., Courtois, A., Loewert, M., Guillouzo, A., & Fardel, O. (2001). Reactive oxygen species-related induction of multidrug resistance-associated protein 2 expression in primary hepatocytes exposed to sulforaphane. *Biochemical and Biophysical Research Communications, 282*, 257–263.

Platz, S., Kuhn, C., Schiess, S., Schreiner, M., Mewis, I., Kemper, M., ... Rohn, S. (2013). Determination of benzyl isothiocyanate metabolites in human plasma and urine by LC-ESI-MS/MS after ingestion of nasturtium (*Tropaeolum majus* L.). *Analytical and Bioanalytical Chemistry, 405*, 7427–7436.

Qu, X., Proll, M., Neuhoff, C., Zhang, R., Cinar, M. U., Hossain, M. M., ... Uddin, M. J. (2015). Sulforaphane epigenetically regulates innate immune responses of porcine monocyte-derived dendritic cells induced with lipopolysaccharide. *PLoS One, 10*, e0121574.

Rajendran, P., Dashwood, W. M., Li, L., Kang, Y., Kim, E., Johnson, G., ... Dashwood, R. H. (2015). Nrf2 status affects tumor growth, HDAC3 gene promoter associations, and the response to sulforaphane in the colon. *Clinical Epigenetics, 7*, 102.

Richman, E. L., Carroll, P. R., & Chan, J. M. (2012). Vegetable and fruit intake after diagnosis and risk of prostate cancer progression. *International Journal of Cancer, 131*, 201–210.

Rose, P., Faulkner, K., Williamson, G., & Mithen, R. (2000). 7-Methylsulfinylheptyl and 8-methylsulfinyloctyl isothiocyanates from watercress are potent inducers of phase II enzymes. *Carcinogenesis, 21*, 1983–1988.

Rose, P., Won, Y. K., Ong, C. N., & Whiteman, M. (2005). Beta-phenylethyl and 8-methylsulphinyloctyl isothiocyanates, constituents of watercress, suppress LPS induced production of nitric oxide and prostaglandin E2 in RAW 264.7 macrophages. *Nitric Oxide: Biology and Chemistry, 12*, 237–243.

Rouzaud, G., Rabot, S., Ratcliffe, B., & Duncan, A. J. (2003). Influence of plant and bacterial myrosinase activity on the metabolic fate of glucosinolates in gnotobiotic rats. *British Journal of Nutrition, 90*, 395–404.

Rungapamestry, V., Duncan, A. J., Fuller, Z., & Ratcliffe, B. (2006). Changes in glucosinolate concentrations, myrosinase activity, and production of metabolites of glucosinolates in cabbage (*Brassica oleracea* Var. *capitata*) cooked for different durations. *Journal of Agricultural and Food Chemistry, 54*, 7628–7634.

Saha, S., Hollands, W., Teucher, B., Needs, P. W., Narbad, A., Ortori, C. A., … Kroon, P. A. (2012). Isothiocyanate concentrations and interconversion of sulforaphane to erucin in human subjects after consumption of commercial frozen broccoli compared to fresh broccoli. *Molecular Nutrition & Food Research, 56*, 1906–1916.

Seow, A., Shi, C. Y., Chung, F. L., Jiao, D., Hankin, J. H., Lee, H. P., … Yu, M. C. (1998). Urinary total isothiocyanate (ITC) in a population-based sample of middle-aged and older Chinese in Singapore: relationship with dietary total ITC and glutathione S-transferase M1/T1/P1 genotypes. *Cancer Epidemiology, Biomarkers & Prevention: A Publication of the American Association for Cancer Research, Cosponsored by the American Society of Preventive Oncology, 7*, 775–781.

Seow, A., Yuan, J. M., Sun, C. L., Van Den Berg, D., Lee, H. P., & Yu, M. C. (2002). Dietary isothiocyanates, glutathione S-transferase polymorphisms and colorectal cancer risk in the Singapore Chinese Health Study. *Carcinogenesis, 23*, 2055–2061.

Shen, G., Xu, C., Chen, C., Hebbar, V., & Kong, A. N. (2006). p53-independent G1 cell cycle arrest of human colon carcinoma cells HT-29 by sulforaphane is associated with induction of p21CIP1 and inhibition of expression of cyclin D1. *Cancer Chemotherapy and Pharmacology, 57*, 317–327.

Singh, A. V., Xiao, D., Lew, K. L., Dhir, R., & Singh, S. V. (2004). Sulforaphane induces caspase-mediated apoptosis in cultured PC-3 human prostate cancer cells and retards growth of PC-3 xenografts in vivo. *Carcinogenesis, 25*, 83–90.

Singh, S. V., Herman-Antosiewicz, A., Singh, A. V., Lew, K. L., Srivastava, S. K., Kamath, R., … Baskaran, R. (2004). Sulforaphane-induced G2/M phase cell cycle arrest involves checkpoint kinase 2-mediated phosphorylation of cell division cycle 25C. *Journal of Biological Chemistry, 279*, 25813–25822.

Singh, S. V., Srivastava, S. K., Choi, S., Lew, K. L., Antosiewicz, J., Xiao, D., … Antosiewicz, A. (2005). Sulforaphane-induced cell death in human prostate cancer cells is initiated by reactive oxygen species. *Journal of Biological Chemistry, 280*, 19911–19924.

Singh, S. V., Warin, R., Xiao, D., Powolny, A. A., Stan, S. D., Arlotti, J. A., … Chambers, W. H. (2009). Sulforaphane inhibits prostate carcinogenesis and pulmonary metastasis in TRAMP mice in association with increased cytotoxicity of natural killer cells. *Cancer Research, 69*, 2117–2125.

Song, J. W., & Chung, K. C. (2010). Observational studies: cohort and case-control studies. *Plastic and Reconstructive Surgery, 126*, 2234–2242.

Spitz, M. R., Duphorne, C. M., Detry, M. A., Pillow, P. C., Amos, C. I., Lei, L., … Wu, X. (2000). Dietary intake of isothiocyanates: evidence of a joint effect with glutathione S-transferase polymorphisms in lung cancer risk. *Cancer Epidemiology Biomarkers & Prevention, 9*, 1017–1020.

Steck, S. E., Gammon, M. D., Hebert, J. R., Wall, D. E., & Zeisel, S. H. (2007). GSTM1, GSTT1, GSTP1, and GSTA1 polymorphisms and urinary isothiocyanate metabolites following broccoli consumption in humans. *Journal of Nutrition, 137*, 904–909.

Steck, S. E., Gaudet, M. M., Britton, J. A., Teitelbaum, S. L., Terry, M. B., Neugut, A. I., … Gammon, M. D. (2007). Interactions among GSTM1, GSTT1 and

GSTP1 polymorphisms, cruciferous vegetable intake and breast cancer risk. *Carcinogenesis, 28*, 1954–1959.

Tang, L., & Zhang, Y. (2004). Dietary isothiocyanates inhibit the growth of human bladder carcinoma cells. *Journal of Nutrition, 134*, 2004–2010.

Tang, L., & Zhang, Y. (2005). Mitochondria are the primary target in isothiocyanate-induced apoptosis in human bladder cancer cells. *Molecular Cancer Therapeutics, 4*, 1250–1259.

Thapliyal, R., & Maru, G. B. (2001). Inhibition of cytochrome P450 isozymes by curcumins in vitro and in vivo. *Food and Chemical Toxicology: An International Journal Published for the British Industrial Biological Research Association, 39*, 541–547.

Thejass, P., & Kuttan, G. (2007). Inhibition of endothelial cell differentiation and proinflammatory cytokine production during angiogenesis by allyl isothiocyanate and phenyl isothiocyanate. *Integrative Cancer Therapies, 6*, 389–399.

Thimmulappa, R. K., Mai, K. H., Srisuma, S., Kensler, T. W., Yamamoto, M., & Biswal, S. (2002). Identification of Nrf2-regulated genes induced by the chemopreventive agent sulforaphane by oligonucleotide microarray. *Cancer Research, 62*, 5196–5203.

Trachootham, D., Zhou, Y., Zhang, H., Demizu, Y., Chen, Z., Pelicano, H., ... Huang, P. (2006). Selective killing of oncogenically transformed cells through a ROS-mediated mechanism by beta-phenylethyl isothiocyanate. *Cancer Cell, 10*, 241–252.

Traka, M., Gasper, A. V., Melchini, A., Bacon, J. R., Needs, P. W., Frost, V., ... Mithen, R. F. (2008). Broccoli consumption interacts with GSTM1 to perturb oncogenic signalling pathways in the prostate. *PLoS One, 3*, e2568.

Traka, M., Gasper, A. V., Smith, J. A., Hawkey, C. J., Bao, Y., & Mithen, R. F. (2005). Transcriptome analysis of human colon Caco-2 cells exposed to sulforaphane. *Journal of Nutrition, 135*, 1865–1872.

Tretter, L., & Adam-Vizi, V. (2005). Alpha-ketoglutarate dehydrogenase: a target and generator of oxidative stress. Philosophical transactions of the Royal Society of London Series B. *Biological Sciences, 360*(1464), 2335–2345.

Vermeulen, M., Klopping-Ketelaars, I. W., van den Berg, R., & Vaes, W. H. (2008). Bioavailability and kinetics of sulforaphane in humans after consumption of cooked versus raw broccoli. *Journal of Agricultural and Food Chemistry, 56*, 10505–10509.

Vieira, A. R., Vingeliene, S., Chan, D. S., Aune, D., Abar, L., Navarro Rosenblatt, D., ... Norat, T. (2015). Fruits, vegetables, and bladder cancer risk: a systematic review and meta-analysis. *Cancer Medicine [Electronic Resource], 4*, 136–146.

Wang, L., Liu, D., Ahmed, T., Chung, F. L., Conaway, C., & Chiao, J. W. (2004). Targeting cell cycle machinery as a molecular mechanism of sulforaphane in prostate cancer prevention. *International Journal of Oncology, 24*, 187–192.

Wang, L. G., Liu, X. M., & Chiao, J. W. (2006). Repression of androgen receptor in prostate cancer cells by phenethyl isothiocyanate. *Carcinogenesis, 27*, 2124–2132.

Wang, L. I., Giovannucci, E. L., Hunter, D., Neuberg, D., Su, L., & Christiani, D. C. (2004). Dietary intake of Cruciferous vegetables, Glutathione S-transferase (GST) polymorphisms and lung cancer risk in a Caucasian population. *Cancer Causes & Control: CCC, 15*, 977–985.

Wierinckx, A., Breve, J., Mercier, D., Schultzberg, M., Drukarch, B., & Van Dam, A. M. (2005). Detoxication enzyme inducers modify cytokine production in rat mixed glial cells. *Journal of Neuroimmunology, 166*, 132–143.

Wolf, M. A., & Claudio, P. P. (2014). Benzyl isothiocyanate inhibits HNSCC cell migration and invasion, and sensitizes HNSCC cells to cisplatin. *Nutrition and Cancer, 66*, 285–294.

Wu, Q. J., Yang, Y., Vogtmann, E., Wang, J., Han, L. H., Li, H. L., & Xiang, Y. B. (2013). Cruciferous vegetables intake and the risk of colorectal cancer: a meta-analysis of observational studies. *Annals of Oncology, 24*, 1079–1087.

Wu, Q. J., Yang, Y., Wang, J., Han, L. H., & Xiang, Y. B. (2013). Cruciferous vegetable consumption and gastric cancer risk: a meta-analysis of epidemiological studies. *Cancer Science, 104*, 1067–1073.
Xiao, D., & Singh, S. V. (2002). Phenethyl isothiocyanate-induced apoptosis in p53-deficient PC-3 human prostate cancer cell line is mediated by extracellular signal-regulated kinases. *Cancer Research, 62*, 3615–3619.
Xiao, D., & Singh, S. V. (2007). Phenethyl isothiocyanate inhibits angiogenesis in vitro and ex vivo. *Cancer Research, 67*, 2239–2246.
Xu, C., Shen, G., Chen, C., Gelinas, C., & Kong, A. N. (2005). Suppression of NF-kappaB and NF-kappaB-regulated gene expression by sulforaphane and PEITC through IkappaBalpha, IKK pathway in human prostate cancer PC-3 cells. *Oncogene, 24*, 4486–4495.
Xu, C., Shen, G., Yuan, X., Kim, J. H., Gopalkrishnan, A., Keum, Y. S., ... Kong, A. N. (2006). ERK and JNK signaling pathways are involved in the regulation of activator protein 1 and cell death elicited by three isothiocyanates in human prostate cancer PC-3 cells. *Carcinogenesis, 27*, 437–445.
Xu, J., Kulkarni, S. R., Donepudi, A. C., More, V. R., & Slitt, A. L. (2012). Enhanced Nrf2 activity worsens insulin resistance, impairs lipid accumulation in adipose tissue, and increases hepatic steatosis in leptin-deficient mice. *Diabetes, 61*, 3208–3218.
Xu, K., & Thornalley, P. J. (2000). Studies on the mechanism of the inhibition of human leukaemia cell growth by dietary isothiocyanates and their cysteine adducts in vitro. *Biochemical Pharmacology, 60*, 221–231.
Xue, M., Qian, Q., Adaikalakoteswari, A., Rabbani, N., Babaei-Jadidi, R., & Thornalley, P. J. (2008). Activation of NF-E2-related factor-2 reverses biochemical dysfunction of endothelial cells induced by hyperglycemia linked to vascular disease. *Diabetes, 57*, 2809–2817.
Yanaka, A., Fahey, J. W., Fukumoto, A., Nakayama, M., Inoue, S., Zhang, S., ... Yamamoto, M. (2009). Dietary sulforaphane-rich broccoli sprouts reduce colonization and attenuate gastritis in *Helicobacter pylori*-infected mice and humans. *Cancer Prevention Research (Philadelphia), 2*, 353–360.
Yao, B., Yan, Y., Ye, X., Fang, H., Xu, H., Liu, Y., ... Zhao, Y. (2014). Intake of fruit and vegetables and risk of bladder cancer: a dose-response meta-analysis of observational studies. *Cancer Causes & Control: CCC, 25*, 1645–1658.
Ye, L., Dinkova-Kostova, A. T., Wade, K. L., Zhang, Y., Shapiro, T. A., & Talalay, P. (2002). Quantitative determination of dithiocarbamates in human plasma, serum, erythrocytes and urine: pharmacokinetics of broccoli sprout isothiocyanates in humans. *Clinica Chimica Acta; International Journal of Clinical Chemistry, 316*, 43–53.
Yu, R., Mandlekar, S., Harvey, K. J., Ucker, D. S., & Kong, A. N. (1998). Chemopreventive isothiocyanates induce apoptosis and caspase-3-like protease activity. *Cancer Research, 58*, 402–408.
Zhang, L., Hao, Q., Bao, L., Liu, W., Fu, X., Chen, Y., ... Wu, H. (2014). Phenethyl isothiocyanate suppresses cervical carcinoma metastasis potential and its molecular mechanism. *Molecular Medicine Reports, 10*, 2675–2680.
Zhang, Y., & Callaway, E. C. (2002). High cellular accumulation of sulphoraphane, a dietary anticarcinogen, is followed by rapid transporter-mediated export as a glutathione conjugate. *Biochemical Journal, 364*, 301–307.
Zhang, Y., Kolm, R. H., Mannervik, B., & Talalay, P. (1995). Reversible conjugation of isothiocyanates with glutathione catalyzed by human glutathione transferases. *Biochemical and Biophysical Research Communications, 206*, 748–755.
Zhang, Z., Wang, S., Zhou, S., Yan, X., Wang, Y., Chen, J., ... Cai, L. (2014). Sulforaphane prevents the development of cardiomyopathy in type 2 diabetic mice probably by

reversing oxidative stress-induced inhibition of LKB1/AMPK pathway. *Journal of Molecular and Cellular Cardiology, 77*, 42–52.

Zhao, B., Seow, A., Lee, E. J., Poh, W. T., Teh, M., Eng, P., et al. (2001). Dietary isothiocyanates, glutathione S-transferase -M1, -T1 polymorphisms and lung cancer risk among Chinese women in Singapore. *Cancer Epidemiology, Biomarkers & Prevention: A Publication of the American Association for Cancer Research, Cosponsored By the American Society of Preventive Oncology, 10*, 1063–1067.

Zhao, H., Lin, J., Grossman, H. B., Hernandez, L. M., Dinney, C. P., & Wu, X. (2007). Dietary isothiocyanates, GSTM1, GSTT1, NAT2 polymorphisms and bladder cancer risk. *International Journal of Cancer, 120*, 2208–2213.

Zhao, J., & Zhao, L. (2013). Cruciferous vegetables intake is associated with lower risk of renal cell carcinoma: evidence from a meta-analysis of observational studies. *PLoS One, 8*, e75732.

CHAPTER TEN

Glucosinolates – The Agricultural Story

E. Schnug[1], S. Haneklaus
Julius Kuehn-Institut (JKI), Braunschweig, Germany
[1]Corresponding author: E-mail: ewald.schnug@julius-kuehn.de

Contents

Abstract

Glucosinolates are the only plant biochemicals which caused unforgettable turmoil in agricultural science and practical farming. In the middle of the 1980s, new oilseed rape varieties with reduced seed glucosinolate content were bred and released in Northern Europe. A low glucosinolate content is the result of a genetically impaired sulphur metabolism. The new varieties converged contemporaneously with an environment that was characterised by depleting atmospheric sulphur inputs in the aftermath of environmental policy and fertiliser practice. Glucosinolates became a buzzword for a broad community of breeders, traders, farmers and scientists, and most importantly decision makers in agricultural policies. Issues covered a wide range of subjects from genetics over plant physiology to economics. This chapter reflects milestones of the perhaps most exciting 20 years of agricultural science in the new era: the rise of the double low oilseed rape varieties.

1. INTRODUCTION

There are numerous stories to tell about glucosinolates, these sulphur-containing molecules consisting of an aglucone bound to a glucose body.

Advances in Botanical Research, Volume 80
ISSN 0065-2296
http://dx.doi.org/10.1016/bs.abr.2016.07.003

Glucosinolates are characteristic compounds of at least 15 dicotyledonous taxa. Out of these, *Brassicaceae* are the most important agricultural crops (Schnug, 1990). There are surveys summarising their significance for botanical systematics (Heaney, Fenwick, Mithen, & Lewis, 1987; Rodman, 1991; Sebanek et al. 2012), food and health (Johnson, 2009; Kumar et al., 2015; Traka & Mithen, 2009), plant resistance against pests and diseases (Gershenzon, 1984; Rosa, Heaney, Fenwick, & Portas, 1997; Schnug, 1987) and ecological interactions and communication of plants with other organisms (Bialy, Oleszek, Lewis, & Fenwick, 1990; Haneklaus, Bloem, & Schnug, 2006; Jabran & Farooq, 2013). The agronomic, in particular, the impact of sulphur fertilisation on the glucosinolate content of different crop plants, is summarised by Schnug and Haneklaus (1998) and Haneklaus, Bloem, Schnug, De Kok, and Stulen (2006). Such reviews provide a proper basis for information on the fundamental and biochemical aspects of glucosinolates which are not the subject of this section. The objective of this chapter is to outline the historic impact glucosinolates had on agricultural science and farming in light of the vast expansion of the oilseed rape cropping area in Northern Europe during the last quarter of the 20th century.

2. OILSEED RAPE – FROM INDUSTRIAL TO FOOD PLANT

What makes glucosinolates exceptionally remarkable is the fact that they created a furore in agriculture like no other biochemical substance ever before or after. The agricultural story of glucosinolates began in the 1970s when the then European Economic Community fostered a strategy to make Europe less reliant on international imports of plant fat. The strategy involved two steps: firstly, the up-to-then common rapeseed varieties with high erucic acid content were substituted by those without erucic acid (so-called single low, or '0' varieties). The oil of these new varieties was suitable for food production, while before it could only be used for technical purposes. Since 1975, all oilseed rape varieties grown in West Germany have been of the 0-type. The switch to the single low varieties led consequently to a considerable enlargement of the cropping area of oilseed rape: the peak of oilseed rape cultivation was in the late 1980s to early 1990s. In some northern regions of Germany, oilseed rape was grown on more than one-third of the entire agricultural land (Fig. 1).

Figure 1 Oilseed rape dominates the landscape in Northern Germany at the start of the 1990s.

Despite this breakthrough in breeding, the seeds of 0-oilseed rape varieties still contained another metabolite which hampered the full economic exploitation of seeds, that is to say glucosinolates which are commonly named mustard oils. Glucosinolates remain in the extracted rapeseed meal which is the solid residuum after crushing the oil. Although the meal has a high protein content that is rich in essential amino acids, it was rated as being unsuitable for feeding because goitrogenic[1] effects were attributed to glucosinolates (Berdanier, 2002). In addition, their off-taste limited the intake by animals (Brak & Henkel, 1978; EFSA, 2008). In the mid of the 1970s, a feverish screening of oilseed rape cultivars for low glucosinolate concentrations in seeds took place in the international breeding community. The single variety with seed glucosinolate concentrations significantly lower than the rest of the variety population was the Polish Landrace *Bronowski* which genes provided the basis for a new type of oilseed rape cultivars, the so-called double low or 00-varieties (Byczynska, 1972; Cleemput, 2011, p. 104; Krzymanski, 1970; 1972; Litynski, 1964). These were free of erucic acid and showed low seed glucosinolate concentrations. At the time when the first 00-varieties were

[1] Glucosinolates decrease iodide uptake by competitive inhibition resulting in a reduced thyroxine and triiodothyronine secretion by the gland which again stimulates the growing of the organ.

grown on production fields the typical total glucosinolate content of air-dry (8% water, 42% fat) 0-seeds was about 100 μmol/g, that of 00-seeds lower than 20 μmol/g. Worth to mention that by legal definition (European Commission, 1990), the term total glucosinolate content addresses the sum of 4 allyl- and 2 indole-glucosinolates: glucobrassicanapine, glucobrassicine, gluconapine, napoleiferine, progoitrine and 4-hydroxiglucobrassicine. Vegetative tissue contains distinctly lower glucosinolate concentrations which amount to roughly 5–10% of the glucosinolate concentration in seeds (Schnug, 1990).

Already in 1974, the first 00-varieties of oilseed rape were released by Canadian breeders (Canola Council of Canada, 2016) coining it *Canola* and by this addressing both quality features: the oil being free of erucic acid and the meal being low in glucosinolates. In addition, the name *Canola* avoided the unfortunate association of the word 'rape' in the English language. All *Canola* varieties released by the Canadian breeders were spring-grown cultivars. After the 00-feature had been successfully implemented in winter cultivars, 00-varieties dominated the landscapes in Northern Europe from 1986 onwards. To promote an area-wide fast switch, the European Community recompensed farmers with substantial financial subsidies. A premium was paid for 00-seeds which propelled the price up to 10 times above the world market price for 0-seeds during peak times. The limit value for the glucosinolate content of seeds was first set to 35 μmol/g, later and finally to 20 μmol/g (air-dry seeds) to receive the premium. And exactly here the first big problem of the big project emerged: 0- and 00-seeds are identical in optical appearance and thousand grain weight. The only difference between both varieties is the 'invisible' glucosinolate content. *Brassica* species are masters in propagation and survival: a proper crop of oilseed rape can easily yield up to 5 t/ha seeds, and it is not uncommon that during combining 10% of the potential harvest yield is lost to the soil and germinating seeds appear as volunteers in the following crops. Keeping in mind that for sowing usually less than 10 kg/ha seeds are required, a newly introduced variety with novel quality features such as a low glucosinolate content faces a vast number of competitive seeds which are already in the soil and which have a survival time of more than 30 years. This means that a contamination of the harvest with just 6% of volunteers from old 0-varieties with an assumed glucosinolate content of 100 μmol/g was sufficient to ruin the hopes for a premium based on the 20 μmol/g limit value if the freshly sown 00-variety had a glucosinolate content of 15 μmol/g.

3. CONSEQUENCES OF AN ABRUPT SWITCH OF OILSEED RAPE VARIETIES IN AGRICULTURAL PRODUCTION

The second large variety switch in oilseed rape cropping started throughout Germany and Northern Europe in 1986. Farmers, seed traders and oil mills had to complete the legerdemain to distinguish between seed batches above or below the officially set intervention level of 20 μmol/g before premium payments were approved. The analytical methods used in the hunt for low glucosinolate-containing varieties were rather primitive (Fig. 2).

The hammer test (Fig. 2) was merely useful for a rough and fast screening of varieties during the breeding process, but did not deliver precise glucosinolate concentrations. The idea behind this test was the detection of enzymatically released glucose which led to more sophisticated analytical quick checks employing test sets developed for diabetes control (Lein, 1970). They all failed finally because it proved to be not possible to control factors affecting the enzymatic release of glucose. Accurate methods were gas chromatography to quantify the allyl-glucosinolate content and later high-performance liquid chromatography (HPLC) which determined indole- and allyl-glucosinolates (Quinsac, 2005). Both procedures were time consuming, tedious and full of pitfalls. Reason is the multitude of

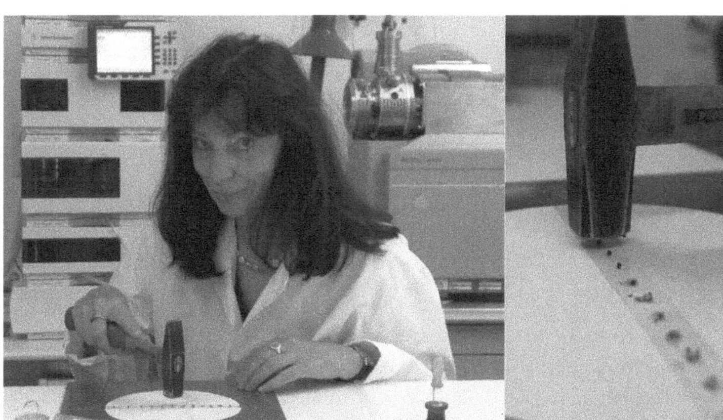

Figure 2 The 'hammer test' (Lein, 1970; McGregor & Downey, 1975): seeds were crushed between filter paper with a hammer, lightly moistened and the glucose released by the enzyme myrosinase detected by means of glucose-sensitive papers employed for diabetes diagnosis.

analytical steps involved in the procedure, the instability of the enzymatic conversions during preparation and last but not least the fact that the individual glucosinolates undergo a number of transformation processes during seed maturation, but it was not possible to quantify such intermediate metabolites in the chromatograms.

All these methods were completely inoperative, when it came to separating seed batches during harvest at the point of acceptance either at country manufacturers or oil mills (Schnug & Haneklaus, 1988a). The maximum time for a Cinderella's[2] decision was less than 30 min, a target none of the methods known at this time was able to perform with an acceptable accuracy. The big breakthrough came in 1986 with the discovery that the total glucosinolate content of oilseed rape correlates extremely closely with its total sulphur content and that the total sulphur content of the seeds could be determined accurately and highly precise by means of wavelength X-ray fluorescence spectroscopy (X-RF; Fig. 3) (Schnug & Haneklaus, 1988b, 1990). With just three steps, the total glucosinolate content of a seed batch could be determined in less than 2 min and with an accuracy and precision 10 times better than the HPLC reference methods (Schnug et al., 2000;

Figure 3 Laboratory (left) and modified field scale (right) X-RF spectrometer for the determination of the total glucosinolate content in seeds of oilseed rape employed in 1986.

[2] Cinderella: The good ones into the pot; the bad ones into the crop.

Schnug & Kallweit, 1987). Never before a method has been applied for the determination of an organic compound which is exclusively employed in purely inorganic, elemental analysis.

A great advantage of determining the total glucosinolate via the total sulphur concentration is the fact that this method is not affected by structural changes of glucosinolates during seed maturation so that the method permits also the forecast of the final glucosinolate content at harvest already weeks before the physical harvest takes place (Schnug & Haneklaus, 1988c).

The European Commission adopted X-RF together with HPLC as reference methods for the glucosinolate determination in oilseed rape (ISO/CD, 1991). Three standard rapeseed reference materials were provided by the BCR (which was at this time the EU official office for standardisation) in which the total glucosinolate and the total sulphur concentration were certified (Wagstaffe, Boenke, Schnug, & Lindsey, 1992). Chromatographic methods determine the total glucosinolate content as the sum of six individual glucosinolates (European Commission, 1990). In this context, it is still an enigma that it was possible to certify the total glucosinolate content of the reference specimens, but not the concentration of individual glucosinolates (Wathelet, Marlier, & Biston, 1987; Wathelet, Marlier, Severin, 1989). From this the conclusion can be drawn that the relationship of individual glucosinolates within seeds of oilseed rape is constant over a wide range of concentrations and independent of environmental and genetic factors. This further implies that most studies investigating the impact of genetic and environmental factors on the variability of individual glucosinolates need to be considered as random. The constant ratio between individual aliphatic glucosinolates which is independent of the total glucosinolate content allows emulating the concentration of individual glucosinolates from the total glucosinolate content on basis of the total sulphur content (Schnug & Haneklaus, 2016). As indole-glucosinolates represent a constant background value of the total glucosinolate content, their estimated concentration is added to the calculated sum of aliphatic glucosinolates to obtain an emulated total glucosinolate content (Schnug & Haneklaus, 2016).

In addition to being fast and reliable, the X-RF method can be used to analyse the glucosinolate content in single seeds (Schnug, Murray, & Haneklaus, 1993), and finally to backtrack the glucosinolate content in processed oilseed rape meals to the original seed batch (Haneklaus & Schnug, 1993). Albeit the high costs for the machinery at the beginning of the 1990s, approximately 100 laboratories worldwide were employing X-RF for glucosinolate analysis in oilseed rapeseeds.

4. KNOCKOUT OF DOUBLE LOW OILSEED RAPE VARIETIES BY SEVERE SULPHUR DEFICIENCY

The deterioration of seed quality by volunteers and the lack of adequate analytical methods were pressing problems and threat to the new crop. Although the essentiality of sulphur for higher plants was discovered by *Wilhelm Knop* and *Julius Sachs* already in 1865 and cruciferous crops known to be plants with an elevated sulphur demand because of the sulphur bound in glucosinolates, the general opinion of plant breeders was at that time that the glucosinolate content is primarily a maternally (genetic) inherited feature of a variety (Magrath & Mithen, 1993).

The 1980s were the prime time for breeding new double low varieties, a decade when in its first half sulphur depositions from industrial emissions fully satisfied the sulphur demand of oilseed rape. Sulphur had the same value like the proverbial Manna to the Israelites during their exodus from Egypt. Sulphur fertilisation was no issue at all, and sulphur-containing fertilisers had been substituted downright by sulphur-free products such as ammoniumnitrate, triplephosphate and high-grade compound (NPK) fertilisers (Schnug & Haneklaus, 1994). The situation changed completely when clean air acts came into force in the beginning of the 1980s in Germany, during the second half of the 1980s in most Western European countries, and in Eastern countries after Perestroika caused a policy reform and breakdown of sulphur emissions (Haneklaus, Bloem, & Schnug, 2008). Within only one decade, atmospheric sulphur emissions and consequently sulphur loads to soils dropped by 90% (Haneklaus et al., 2008).

It was a red-letter day for plant sulphur research, when first studies from field surveys in remote areas of Northern Germany revealed in 1981 that at this time the majority of field-grown oilseed rape was suffering from sulphur deficiency. Severe macroscopic symptoms became visible on oilseed rape production fields 3 years later, 30 km north of Kiel in Schleswig-Holstein (Fig. 4).

The instantaneous response of one of the leading pioneers, Prof. Dr h.c. mult. Gerhard Röbbelen in double low oilseed rape breeding at the annual meeting of the agricultural faculty of Kiel University was proverbially '...sulphur is a forgettable issue for the future double low varieties, they don't need it anymore in quantities like the single lows' (Anonymous, 2012). Arduous research work in the following years proved that exactly the opposite was true: double and single low oilseed rape varieties respond with the same changes in glucosinolate concentrations in both vegetative

Figure 4 First appearance of macroscopic sulphur deficiency symptoms in a field-grown oilseed rape crop west of Kiel in Schleswig-Holstein (N 54.434237, E 9.872343) in 1984. [0-variety: *Jet Neuf*, growth stage 31 (BBCH), 0.29% total S (d.m.) in fully differentiated leaves]

and generative plant parts in relation to the external sulphur supply. The most unexpected perception, however, was that severe sulphur deficiency reduced growth of double low oilseed rape varieties stronger than that of single lows (Fig. 5, Table 1).

Figure 5 Growth performance of three winter 00-oilseed rape varieties compared to a single low variety at low sulphur supply (from left to right: JET-NEUF (0-variety), LINDORA, LIROPA, ELENA (00-varieties). [0.38% total S (d.m.) in younger leaves; no statistically significant difference between the varieties]. *Data extracted from Schnug, E. (1988).* Quantitative und qualitative Aspekte der Diagnose und Therapie der Schwefelversorgung von Raps (Brassica napus L.) unter besonderer Berücksichtigung glucosinolatarmer Sorten. Habilitationsschrift, *(D.sc. thesis), Kiel: Agrarwiss. Fakultät der Christian-Albrechts-Universität.*

Table 1 Total Sulphur Concentrations in Younger, Fully Developed Leaves and Seed Yield of Oilseed Rape in Relation to Graded Rates of Sulphur (Sand Culture in Greenhouse, Mean of Three Replicates for Four 0-Varieties (*Jet-Neuf, Korina, Lirakotta, Mirander*) and Eight 00-Varieties (*Rubin, Ceres, Elena, Lindora, Liropa, Wiebke, Glumander, Arabella*)

S Supply (mg/pot)	Total S in Younger Leaves (%)	Relative Seed Yield[a] (%)	
		0-Variety	00-Variety
20	0.09	1	1
40	0.19	22	3
60	0.26	52	33
80	0.33	71	65
100	0.38	76	80
120	0.43	86	88
140	0.54	91	100
160	0.58	94	96
180	0.63	98	99
200	0.68	100	100
$LSD_{5\%}$	0.04		

[a]100% seed yield (g/pot) = 14.6 (0-varieties) and 13.7 (00-varieties).
Data extracted from Schnug, E. (1988). *Quantitative und qualitative Aspekte der Diagnose und Therapie der Schwefelversorgung von Raps (Brassica napus L.) unter besonderer Berücksichtigung glucosinolatarmer Sorten. Habilitationsschrift*, (D.sc. thesis), Kiel: Agrarwiss. Fakultät der Christian-Albrechts-Universität.

Breeding of 00-varieties took place during the same time when atmospheric S depositions peaked. 00-varieties which were bred under high sulphur input conditions and later released into a sulphur depleted agro-environment showed symptoms of severe sulphur deficiency which became rapidly the most widespread nutritional disorder in agricultural production.

5. GENETIC CHANGES IN GLUCOSINOLATE METABOLISM MODIFY SULPHUR UTILISATION, NATURAL PLANT HEALTH AND WILDLIFE INTERACTIONS

0- and 00-varieties show the same growth development if sulphur is sufficiently supplied (Table 1). Only if sulphur is insufficiently supplied, 00-varieties react with a stronger loss in dry matter and yield production (Table 2, Fig. 6, Schnug, 1988). Sulphur fertilisation will satisfy the nutrient demand of the crop, but will also increase the glucosinolate content. This has been a very early finding (Schnug, 1988).

Table 2 Internal Sulphur Balance of Field-Grown 0- (*Mirander*) and 00- (*Glumander*) Oilseed Rape (About 4 t/ha Seed Yield)

Growth Stage		Leaves	Stems[a]	Pods		Seeds		Total[a]
	BBCH[b]			0-cv	00-cv	0-cv	00-cv	
Total Sulphur Concentration (%)								
Before winter	19	0.69	0.39					0.49
Start of vegetation	30	0.65	0.32					0.39
Start of flowering	61	0.63	0.33	0.60	0.51			0.44
Mid flowering	65	0.52	0.30	0.27	0.64	0.86	0.45	0.40
End flowering	69	0.40	0.29	0.21	0.67	0.87	0.46	0.44
10 d after flowering	71	0.37	0.29	0.19	0.66	0.85	0.46	0.45
Harvest	89	0.37	0.28	0.19	0.65	0.85	0.46	0.48
$LSD_{5\%}$		0.05	0.04	0.03	0.05	0.05	0.04	0.05
Sulphur Uptake (kg/ha)								
Before winter	19	3.45	3.9					7.4
Start of vegetation	30	1.63	2.7					4.3
Start of flowering	61	6.62	9.5	2.1	1.6			18.0
Mid flowering	65	4.94	13.2	2.4	5.8	7.7	4.1	28.1
End flowering	69	3.00	14.2	7.7	24.5	31.8	16.8	57.6
10 d after flowering	71	1.11	13.9	8.3	27.9	35.9	19.4	61.7
Harvest	89	1.00	12.3	8.0	26.3	34.4	18.6	56.9
$LSD_{5\%}$		0.31	1.7	0.9	1.1	1.9	1.0	3.9

[a]Differences between single (0) and double low (00) cultivars (cv) were statistically not significant. Thus the mean values are reported for both varieties.
[b]Stauss et al. (1994) and Meier et al. (2009).
Data extracted from Schnug, E. (1988). *Quantitative und qualitative Aspekte der Diagnose und Therapie der Schwefelversorgung von Raps (Brassica napus L.) unter besonderer Berücksichtigung glucosinolatarmer Sorten. Habilitationsschrift*, (D.sc. thesis), Kiel: Agrarwiss. Fakultät der Christian-Albrechts-Universität.

According to Schnug (1988), an increase of the total sulphur content (d.m.) in younger, fully differentiated leaves of oilseed rape at the beginning of stem elongation by 0.1 % requires 50 kg/ha S under the growth conditions of Northern German moraine soils. This amount of sulphur secures seed yield but increases the total glucosinolate content on an average by 8.3 µmol/g in 0-varieties and by 1.6 µmol/g in 00-varieties of oilseed rape, respectively. These findings implied for farmers that they had to adjust the sulphur fertiliser rate to match exactly the demand as any additional sulphur supply yields an undesired increase of the glucosinolate content so that the premium payment for 00-quality was endangered.

The extract of all the research work on glucosinolates has been transferred into a small BASIC programme which has been attached to this

Figure 6 Single low volunteer (0-variety) plant without symptoms of sulphur deficiency (right) grown in a 00-variety field (left and background to the right) expressing severe symptoms of sulphur deficiency. The total sulphur content in younger leaves of both plants was 0.30% S.

chapter (see Supplementary data). It is based on data about the total sulphur concentration in the dry matter of younger fully differentiated leaves of oilseed rape at growth stage 30 (BBCH) which is the most suitable and best proven criteria for diagnosing the plant sulphur status. All other diagnostic criteria discussed over the last 50 years as, for instance, the sulphate content, the N:S ratio and the sulphate:malate ratio are clearly less predictable, more complicated and biased in other factors than the sulphur nutritional status (Schnug & Haneklaus, 1998). Other input data are about the genetic quality in terms of the glucosinolate content of the seeds to be sown and the contamination of the soil with volunteers. Based on this information, the program calculates under *ceteris paribus* conditions the amount of sulphur fertiliser (kg/ha) which is required to achieve maximum yield, the expected glucosinolate content of the seeds at harvest and the utilisation coefficient for sulphur in the fertilisers employed. An executable form of the program is available via the author's account in ResearchGate (www.researchgate.net).

The reason for the higher sensitivity of double low varieties towards sulphur deficiency is quite obviously related to differences in their glucosinolate metabolism (Schnug, 1990).

Table 2 shows the shifts in the internal sulphur balance of 0- and 00-oilseed rape varieties. The total amount of sulphur that is assimilated is the same in both variety types. However, that amount of sulphur which is allocated to glucosinolates in seeds of the 0-variety remains in the pod walls of the 00-variety. This underlines the perception that glucosinolates are synthesised in the pod walls and only structural complete glucosinolate molecules are transported from there into the seeds. The data of Bloem, Haneklaus, and Schnug (2007) confirm these findings.

This hypothesis finds approval by the fact that the total sulphur concentration of a seed can be attributed nearly entirely to sulphur in proteins and sulphur in fully developed glucosinolates. This is likewise the reason for the successful operation of the X-RF method. The question about the chemical nature of the compounds left over from the altered glucosinolate metabolism of 00-varieties is still open. The studies of Glendening and Poulton (1988) showed that the big difference in the biochemical glucosinolate pathway of 0- and 00-varieties lies in the last step, where a sulphotransferase adds the sulphate moiety to the molecule.

Under conditions of sulphur starvation, intact glucosinolates are recycled into the primary sulphur metabolism (Borgen, 2002; Cleemput, 2011; Nikiforova et al., 2005; Rednovnikovic, Gliveti, Delonga, & Vorkapic-Fura, 2008; Schnug, 1990). This mechanism is triggered through an increase of myrosinase activity which again is regulated by the ascorbic acid concentration in the vacuoles (Bones & Rossiter, 1996; Schnug, 1988). 0-varieties show a better growth performance under sulphur stress than 00-varieties (Fig. 5, Table 1). Under the same sulphur stress, growth of 00-varieties decreases together with their genetic glucosinolate level (Fig. 7).

Figure 7 Sulphur deficiency symptoms in two 00-oilseed rape varieties (BBCH 37) with different genetic glucosinolate potential grown in Northern Germany (N 54.358146, E 10.480757). [The 00-variety *Liporta* (12 μmol/g) in front of the fields shows more severe sulphur deficiency symptoms than the 00-variety *Lirabon* (20 μmol/g) in the back. The total sulphur content of younger, fully developed leaves was similar in both crops with 0.29% S (d.m.)]

Figure 8 *Brassica campestris* lines with different genetic glucosinolate level grown at the research station of Agriculture Canada in Saskatoon (N 52.133989, E 106.635159) in 1988. [mid-row: line with genetically 0 μmol/g seed glucosinolate content; left and right row: lines with genetically 1–10 μmol/g seed glucosinolate level; for details see Schnug (1988)]

The link of an intact glucosinolate metabolism to plant vigour is underlined as field-grown glucosinolate knockout genotypes reveal; plants nearly died off (Fig. 8). Glucosinolates are storage products in oilseed rape plants, which significance for the total sulphur balance changes during plant development: in vegetative tissue of young plants, less than 10% of the total sulphur is bound in glucosinolates, a fact which some authors inspired to negate the significance of glucosinolates as a vital sulphur storage (Aghajanzadeh, Hawkesford, & De Kok, 2014; Fieldsend & Milford, 1994; Zukalova, Vasak, Nerad, & Stranc, 2002). At maturity, however, the majority of the sulphur taken up by a 0-variety is allocated to seed glucosinolates. In 00-varieties, less sulphur is bound in seed glucosinolates, but the missing fraction accumulates in so far unspecified glucosinolate precursors in pod walls (Table 2). Glucosinolates stored in seeds are not a sulphur source under conditions of sulphur starvation and will not contribute to the primary sulphur metabolism. Only the degradation of glucosinolates in vegetative tissues can deliver sulphate for metabolisation.

Eventually it is not only the significance of the amount of sulphur itself that is recycled from glucosinolates in vegetative tissues which affects the vigour of 00-varieties under sulphur stress, but also the downregulation of glucosinolate synthesis which causes an upregulation of genes controlling sulphur uptake and assimilation (Falk, Tokuhisa, & Gershenzon, 2007).

These results clearly stress the need to compare single and double low oilseed rape varieties to disclose genetic differences in the diversified recycling of glucosinolates. The use of double low varieties with different glucosinolate content or drawing on different *Brassica* species is of no avail as the genetic system that regulates glucosinolate metabolism can only be found in single and double low oilseed rape varieties. Any metabolic changes in glucosinolate synthesis will be manifested at the start of the phylogeny.

A link to signalling via plant hormones within these processes seems unlikely. Neither an enhanced synthesis of auxin as a consequence of the degradation of indole-glucosinolates and hereby strengthened tryptophan formation (Kutz et al., 2002) nor cytokinins seem to be involved in the regulation of sulphate uptake (Dan, Yang, & Zheng, 2007; Maruyama-Nakashita, Nakamura, Yamaya, & Takahashi, 2004). It can be concluded at this point that more comparative studies on sulphur nutrition in 0- and 00-oilseed rape varieties are required for a complemented understanding of this unique physiological phenomenon. Yet another open question is the significance and regulatory mechanisms of glucosinolate transport processes within a plant in the dark. So far research focuses on the biosynthesis of seed constituents in pod walls, neglecting the overall role of stems and leaves. Great potential for exploring this innovative field of research will have the use of synthetic α-glucosinolates and xenobiotic aglucones such as glucofaline[3] (Rollin & Schnug, 2000).

A characteristic floral symptom of severe sulphur deficiency has been identified correctly by oilseed rape breeders in the 1980s, but its causal background misinterpreted. The phenomenon of white flowering was attributed to a genetic instability caused by inbreed of 'Canadian blood'. But what was falsely evaluated as genetic inheritance was in effect caused by sulphur deficiency (Fig. 9 Schnug & Haneklaus, 1998; Schnug & Haneklaus, 2005). White flowering is caused by the formation of colourless anthocyanins in petals because of an increasing accumulation of glucose. The reason is that protein synthesis is hampered by sulphur deficiency (Schnug & Haneklaus, 2005).

The cuss Canadian blood can be traced back to the fact that white flowering was common in oilseed rape crops grown in Canada. Canada was at that time already a notorious sulphur deficiency area. During the same

[3] The name was an acknowledgement to the former Federal Agricultural Research Center (FAL) in Braunschweig, Germany (1948–2008).

Figure 9 Symptoms of severe sulphur deficiency in flowers of oilseed rape affecting plant growth early during plant development (left) and shortly before full bloom (right).

period a rapeseed multiplication crop lost its official approval as a pure variety in Europe when white flowers were found during official inspections.

When sulphur deficiency hampers plant development early during vegetative growth, the colour of flowers changes uniformly from bright yellow to pale yellow/white. Only a part of petals and flowers alters the colour when sulphur deficiency affects plant growth shortly before and during flowering (Fig. 9). In relation to the duration of sulphur deficiency, not only the colour but also the size and shape of oilseed rape flowers are affected. A detailed description of sulphur deficiency symptoms during flowering is provided by Schnug and Haneklaus (2005).

The new 00-varieties showed a higher susceptibility against pests and diseases (Fig. 10). The term sulphur-induced resistance (SIR) denotes the

Figure 10 Influence of soil-applied sulphur fertilisation (100 kg/ha S) on visual manifestation of infection with *Pyrenopeziza brassicae* in southern Denmark. *Photo: C.A. Pedersen, Skejby.*

reinforcement of the natural resistance of plants against fungal pathogens through triggering the stimulation of metabolic processes involving sulphur by targeted soil-applied fertiliser strategies. Meanwhile, major pathways, mechanisms and metabolites involved in SIR have been identified and strategies developed to transform this knowledge into practical applications on production fields. Main findings are summarised in Bloem, Haneklaus, Salac, Wickenhäuser, and Schnug (2007) and Haneklaus, Bloem, and Schnug (2007, 2009).

The newly introduced oilseed rape varieties bore not only hidden biochemical secrets for scientists but also surprises for the public. Like many large scale changes in the human environment, the switch to 00-varieties stirred its fair share of xenophobia. The uproar started with some dead hares found in late autumn by hunters in the middle of Germany (Anonymous, 1986) and the rather simple-minded explanation bellowed that the animals were killed by devouring the newly introduced 00-oilseed rape cultivars which were supposed to be tastier because of an also assumed lower concentrations of less off-tasting glucosinolates (Reidt, 1988). The hysteria spread quickly in Europe. In Switzerland it triggered an ad hoc moratorium for 00-oilseed rape (Häberli, 1995). Some 5 years later, the hoax was over as intensive research showed that under the same environmental conditions 0- and 00-varieties did not unfold any detrimental effect on roe deer. Thus the reasoning of a preferable taste of 00-varieties could be ruled out. Rather changes in the frequency of oilseed rape in the environment reduced the biodiversity of the food supply. Meanwhile 00-varieties of oilseed rape are not considered to be a higher potential risk for wildlife health than 0-varieties (Askew, 1990).

Honeybees are attracted by scent, colour and form of the honey-bearing plants. Though the sulphur nutritional status affects all of these parameters, and isothiocyanates and other sulphur-containing volatiles are present in significant amounts in the headspace of oilseed rape flowers, it is another circumstance that is responsible for a significantly reduced number of visiting honeybees in sulphur-deficient oilseed rape during flowering (Schnug & Haneklaus, 2005; Haneklaus, Brauer, Bloem, & Schnug, 2005). Oilseed rape crops visited by bees show earlier petal fall, probably because they set flowers earlier, resulting in a more uniform pod ripening and ease of harvest. The reflective pattern of flowers provides visitors with clues as to the age of the flowers and presence of food rewards (Kevan & Baker, 1983). During senescence of rapeseed flowers, which begins immediately after pollination, the yellow petal colour vanishes and the petals shrink quickly before falling to the ground. A pollinated and fading rapeseed flower is therefore similar to an unpollinated sulphur deficient one and thus less attractive to honeybees.

6. EPILOGUE

Research on glucosinolates in agriculture was a scientific bushfire during a rather short period of only 15 years with a peak in 1990. In the 1970s still known only to a handful of scientists, at the end of the 1980s glucosinolates were in every pheasant's vocabulary. At the same time, sulphur excessively delivered by atmospheric pollution was regarded as the yellow peril of environment killing forests and damaging human health. Twenty years later, serious sulphur fertilisation is vital to combat deficiency. These special features made the last 20 years of the bygone millennium a unique scenario for plant sulphur research in all its facets.

The introduction of double low oilseed rape varieties stirred a unique combination of developments in science and agricultural production on a scale which is unlikely to recap in the foreseeable future. Because of its significant economic impact the subject stimulated science from different disciplines as, for instance, genetics, analytical chemistry and plant nutrition. But as fast as the issue rose as fast it faded away from the stages of science, simply because the agricultural policy of the European Community changed from a subsidy based on quality to an archaic area-based support system of oilseed rape growers. Interest in glucosinolates as a quality factor vanished quickly.

Unfortunately not many scientists have been aware of the great opportunities which rapeseed with its unique coexistence of varieties with a natural 'knockout' gene for glucosinolates offered. As a consequence, only a limited number of comparative metabolic studies with 0- and 00-varieties have been conducted. This leaves still a number of challenging questions to understand the complex nature of glucosinolate biosynthesis and its link to sulphur metabolism in the dark. Examples are the biochemical nature of intermediates of the genetically impeded glucosinolate metabolism in double low oilseed rape varieties and their significance with view to the sulphur balance of the crop because substantially higher amounts of organically bound sulphur remain with the harvest residues in the soil. The use of synthetic alpha-glucosinolates as natural tracers is another innovative assignment for investigating regulatory source/sink mechanisms in oilseed rape plants. Although without a promise for easy funding, the topic offers an exciting playground for future generations of researchers.

SUPPLEMENTARY DATA

Supplementary data related to this article can be found at http://dx.doi.org/10.1016/bs.abr.2016.07.003

REFERENCES

Aghajanzadeh, T., Hawkesford, M., & De Kok, L. J. (2014). The significance of glucosinolates for sulfur storage in *Brassicaceae* seedlings. *Frontiers in Plant Science, 5* (Article No 704).

Anonymous. (1986). *Süßer Tod. Der Spiegel 52, 22.12.1986.* http://www.spiegel.de/spiegel/print/d-13522441.html.

Anonymous. (2012). *50 Jahre Rapszüchtung.* Rapsmagazin der UFOP, 20-25 http://www.ufop.de/files/2613/3879/2280/UFOP_0995_RAPSMAG_WEB.indd.pdf.

Askew, M. (1990). *Rapeseed and intoxication of wildlife. Commission of the European Communities Directorate-General — Telecommunications, Information Industries and Innovation.* Scientific and Technical Communication Unit, L-2920 Luxembourg; CEC EUR 11771 EN.

Berdanier, C. D. (2002). Food constituents. In D. Berdanier, E. B. Feldman, W. P. Flatt, & A. T. St. Jeor (Eds.), *Handbook of nutrition and food* (pp. 3—98). Boca Raton: CRC Press.

Bialy, Z., Oleszek, W., Lewis, J., & Fenwick, G. R. (1990). *Allelopathic potential of glucosinolates (mustard oil glycosides) and their degradation products against wheat.*

Bloem, E., Haneklaus, S., Salac, I., Wickenhäuser, P., & Schnug, E. (2007). Facts and fiction about sulfur metabolism in relation to plant-pathogen interactions. *Plant Biology, 9*, 596—607.

Bloem, E., Haneklaus, S., & Schnug, E. (2007). Changes in the sulphur and glucosinolate content of developing pods and seeds of oilseed rape (*Brassica napus* L.) in relation to different cultivars. *Landbauforschung Völkenrode, 57*, 297—306.

Bones, A. M., & Rossiter, J. T. (1996). The myrosinase-glucosinolate system, its organization and biochemistry. *Physiologia Plantarum, 97*, 194—208.

Borgen, B. H. (2002). *Functional analysis of plant idioblasts (myrosin cells) and their role in defense, development and growth.* Trondheim University (Ph.D. thesis).

Brak, B., & Henkel, H. (1978). Untersuchungen ueber den Einsatz glucosinolatarmer Rapssaaten bzw. Extraktionsschrote als Futtermittel fuer monogastrische Tiere. *Fette, Seife, Anstrichmittel, 80*, 104—105.

Byczynska, B. (1972). The variation in contents of thioglycosides in seeds of the genus *Brassica. Biuletyn Instytutu Hodowli i Aklimatyzacji Roslin, 5/6*, 217—220.

Canola Council of Canada. (2016). *The history of canola.* http://www.canolacouncil.org/oil-and-meal/what-is-canola/the-history-of-canola/.

Cleemput, S. (2011). *Breeding for a reduced glucosinolate green mass of rapeseed to improve its suitability for biogas production* (Ph.D. thesis). Göttingen University.

Commission Regulation (EEC) No 1864/90. (1990). *Annex VIII. Official Journal of the European Communities, L 170* (pp. 27—34).

Dan, H., Yang, G., & Zheng, Z.-L. (2007). A negative role for auxin in sulphate deficiency response in *Arabidopsis thaliana. Plant Molecular Biology, 63*, 221—235.

EFSA. (2008). Glucosinolates as undesirable substances in animal feed. *The EFSA Journal, 590*, 1—76.

Falk, K. L., Tokuhisa, J. G., & Gershenzon, J. (2007). The effect of sulfur nutrition on plant glucosinolate content: physiology and molecular mechanisms. *Plant Biology, 9*, 573–581.

Fieldsend, J., & Milford, G. F. J. (1994). Changes in glucosinolates during crop development in single- and double-low genotypes of winter oilseed rape (*Brassica napus*): I. Production and distribution in vegetative tissues and developing pods during development and potential role in the recycling of sulphur within the crop. *Annals of Applied Biology, 124*, 531–542.

Gershenzon, J. (1984). Changes in the levels of plant secondary metabolites under water and nutrient stress. In B. N. Timmermann, C. Steelink, & A. Loewus (Eds.), *Plant adaptations to stress* (pp. 273–321). New York: Plenum Press.

Glendening, T. M., & Poulton, J. E. (1988). Glucosinolate biosynthesis – sulfatation of desulfobenzylglucosinolate by cell-free extracts of cress (*Lepidium sativum* L.) seedlings. *Plant Physiology, 86*, 319–321.

Häberli, R. (1995). *00-Raps bedingte, klinisch-pathologische Veränderungen bei freilebendem Rehwild in der Schweiz* (Med.-vet. Ph.D. thesis, Bern, Switzerland).

Haneklaus, S., Bloem, E., & Schnug, E. (2006). Sulphur interactions in crop ecosystems. In M. J. Hawkesford, & L. J. De Kok (Eds.), *Sulfur in plants – An ecological perspective* (pp. 17–58). Dordrecht: Springer.

Haneklaus, S., Bloem, E., & Schnug, E. (2007). Sulfur and plant disease. In L. E. Datnoff, W. H. Elmer, & D. M. Huber (Eds.), *Mineral nutrition and plant disease* (pp. 101–118). Minnesota, MN: APS Press. St. Paul.

Haneklaus, S., Bloem, E., & Schnug, E. (2008). History of sulfur deficiency in crops. In J. Jez (Ed.), *Sulfur: A missing link between soils, crops, and nutrition. B. Crops and sulfur* (pp. 45–58). CSSA-ASA-SSSA.

Haneklaus, S., Bloem, E., & Schnug, E. (2009). Plant disease control by nutrient management: sulphur. In D. Walters (Ed.), *Disease control in crops: Biological and environmentally friendly approaches* (pp. 221–236). Oxford: Blackwell.

Haneklaus, S., Bloem, E., Schnug, E., De Kok, L., & Stulen, I. (2006). Sulphur. In A. V. Barker, & D. J. Pilbeam (Eds.), *Handbook of plant nutrition* (pp. 183–238). Boca Raton: CRC Press.

Haneklaus, S., Brauer, A., Bloem, E., & Schnug, E. (2005). Relationship between sulphur deficiency in oilseed rape (*Brassica napus* L.) and its attractiveness for honey bees. *Landbauforschung Völkenrode Sonderheft, 283*, 37–43.

Haneklaus, S., & Schnug, E. (1993). Determination of total and destroyed glucosinolates in processed oilseed rape meal. *Landbauforschung Völkenrode, 43*, 73–76.

Heaney, R. K., Fenwick, G. R., Mithen, R. F., & Lewis, B. G. (1987). Glucosinolates of wild and cultivated *Brassica* species. *Phytochemistry, 26*, 1969–1973.

ISO/CD. (1991). *9167.2-Rapeseed – Determination of total glucosinolates – Part 2*. X–Ray Fluorescence Spectrometry (XRF).

Jabran, K., & Farooq, M. (2013). Implications of potential allelopathic crops in agricultural systems. In Z. A. Cheema, & M. Farooq (Eds.), *Allelopathy – Current trends and future applications* (pp. 349–385). Heidelberg: Springer Verlag.

Johnson, I. T. (2009). Glucosinolates: bioavailability and importance to health. *International Journal for Vitamin and Nutrition Research, 72*, 26–31.

Kevan, P. G., & Baker, H. G. (1983). Insects as flower visitors and pollinators. *Annual Review of Entomology, 28*, 407–453.

Krzymanski, J. (1970). Inheritance of thioglucoside content by rapeseed (*Brassica napus*). In *International conference on oil rape, Paris, 26–30 May 1970* (pp. 212–218).

Krzymanski, J. (1972). Current problems and directions in swede-rape breeding. *Biuletyn Instytutu Hodowli i Aklimatyzacji Roslin, 5/6*, 205–209.

Kumar, G., Tuli, H. S., Mittal, S., Shandilya, J. K., Tiwari, A., & Sandhu, S. S. (2015). Isothiocyanates: a class of bioactive metabolites with chemopreventive potential. *Tumour Biology, 36*, 4005—4016.

Kutz, A., Müller, A., Hennig, P., Kaiser, W. M., Piotrowski, M., & Weiler, E. W. (2002). A role for nitrilase 3 in the regulation of root morphology in sulphur-starving *Arabidopsis thaliana. The Plant Journal, 30*, 95—106.

Lein, A. (1970). Quantitative Bestimmungsmethoden fuer Samenglucosinolate in *Brassica*-Arten und ihre Anwendung in der Züchtung von glucosinolatarmen Raps. *Zeitschrift Pflanzenzuechtung, 63*, 137—154.

Litynski, A. (1964). The development of breeding oil crops in Poland in the light of local investigations. *Biuletyn Instytutu Hodowli i Aklimatyzacji Roslin, 5-6*, 97—112.

Magrath, R., & Mithen, R. (1993). Maternal effects on the expression of individual aliphatic glucosinolates in seeds and seedlings of *Brassica napus. Plant Breeding, 111*, 249—252.

Maruyama-Nakashita, A., Nakamura, Y., Yamaya, T., & Takahashi, H. (2004). A novel regulatory pathway of sulfate uptake in Arabidopsis roots: implication of CRE1/WOL/AHK4-mediated cytokinins-dependent regulation. *The Plant Journal, 38*, 779—789.

McGregor, D. I., & Downey, R. K. (1975). A rapid and simple assay for identifying low glucosinolates rapeseed. *Canadian Journal of Plant Science, 55*, 191—196.

Meier, U., Bleiholder, H., Buhr, L., Feller, C., Hack, H., Heß, M., ... Zwerger, P. (2009). The BBCH system to coding the phenological growth stages of plants — history and publications. *Journal of Cultivated Plants, 61*, 41—52.

Nikiforova, V. J., Kopka, J., Tolstikov, V., Fiehn, O., Hopkins, L., Hawkesford, M. J., ... Hoefgen, R. (2005). Systems rebalancing of metabolism in response to sulfur deprivation, as revealed by metabolome analysis of *Arabidopsis* plants. *Plant Physiology, 138*, 304—318.

Quinsac, A. (2005). Progress in standardisation for glucosinolate analysis. GCIRC Bulletin. In *Technical meeting, Winnipeg, Canada, June, 28th, 2005 1/8.*

Rednovnikovic, I., Gliveti, T., Delonga, K., & Vorkapic-Fura, J. (2008). Glucosinolates and their potential role in plant. *Periodicim Biologorum, 110*, 297—309.

Reidt, L. (1988). *Rätselhaftes Hasensterben.* Die Zeit. online http://www.zeit.de/1988/41/raetselhaftes-hasensterben.

Rodman, J. E. (1991). A taxonomic analysis of glucosinolate-producing plants, part 1: phenetics. *Systematic Botany, 16*, 598—618.

Rollin, P., & Schnug, E. (2000). Synthetic glucosinolates as tracers for the investigation of the secondary sulfur metabolism. In *The role of sulfur metabolism in plant adaptation to stress; physiological and environmental aspects; January 7—9, 2000, Lisbon, Portugal* (p. 14).

Rosa, E. A. S., Heaney, R. K., Fenwick, G. R., & Portas, C. A. M. (1997). Glucosinolates in crop plants. *Horticultural Reviews, 19*, 99—215.

Schnug, E. (1987). Relations between sulfur-supply and glucosinolate content of 0- and 00 oilseed rape. In *Proceedings of 7th international rapeseed congress, Poznan, Poland* (pp. 682—686).

Schnug, E. (1988). *Quantitative und qualitative Aspekte der Diagnose und Therapie der Schwefelversorgung von Raps (*Brassica napus *L.) unter besonderer Berücksichtigung glucosinolatarmer Sorten. Habilitationsschrift* (D.sc. thesis). Kiel: Agrarwiss. Fakultät der Christian-Albrechts-Universität.

Schnug, E. (1990). Glucosinolates — fundamental, environmental and agricultural aspects. In H. Rennenberg, C. Brunold, L. J. De Kok, & I. Stulen (Eds.), *Sulfur nutrition and sulfur assimilation in higher plants* (pp. 97—106). The Hague, The Netherlands: SPB Academic Publishing bv.

Schnug, E., & Haneklaus, S. (1988a). Röntgenfluoreszenzanalytische Bestimmung des Gesamtglucosinolat-gehaltes in Rapssamen und Rapsextraktionsschrot. *VDLUFA-Schriftenreihe 23, Kongreßband, 1987*, 883—898.

Schnug, E., & Haneklaus, S. (1988b). The sulphur concentration as a standard for the total glucosinolate content of rapeseed and meal and its determination by X-ray fluorescence spectroscopy (X-RF method). *Journal of the Science of Food and Agriculture, 45*, 243–254.
Schnug, E., & Haneklaus, S. (1988c). Vor-Erntebestimmung des Gesamtglucosinolatgehaltes von Rapssaat mittels Röntgenfluoreszenzanalyse. *Raps, 6*, 146–148.
Schnug, E., & Haneklaus, S. (1990). Quantitative glucosinolate analysis in *Brassica* seeds by X-ray fluorescence spectroscopy. *Phytochemical Analysis, 1*, 40–43.
Schnug, E., & Haneklaus, S. (1994). Sulphur deficiency in *Brassica napus* - biochemistry, symptomatology, morphogenesis. *Landbauforschung Völkenrode* (special issue 144).
Schnug, E., & Haneklaus, S. (1998). Diagnosis of sulphur nutrition. In E. Schnug, & H. Beringer (Eds.), *Sulphur in agro-ecosystems. Vol. 2 of the series mineral nutrition in ecosystems* (pp. 1–38). Dordrecht: Kluwer Academic Publ.
Schnug, E., & Haneklaus, S. (2005). Sulphur deficiency symptoms in oilseed rape (*Brassica napus* L.) – the aesthetics of starvation. *Phyton, 45*(3), 79–95.
Schnug, E., & Haneklaus, S. (2016). Emulation and backtracking of HPLC chromatographic profiles for glucosinolates from total sulphur concentrations in oilseed rape seeds. *Journal for Cultivated Plants, 68*, 63–71.
Schnug, E., Haneklaus, S., Schimmel, H., Linsinger, T., Mabon, N., Marlier, M., & Wathelet, J.-P. (2000). Standard reference materials for sulfur and glucosinolates. In C. Brunold, H. Rennenberg, L. J. De Kok, I. Stulen, & J. C. Davidian (Eds.), *Sulfur Metabolismus in higher plants – Molecular, ecophysiological and nutritional aspects* (pp. 251–253). Leiden, The Netherlands: Backhuys Publishers.
Schnug, E., & Kallweit, P. (1987). Ergebnisse eines Ringversuches zur röntgenfluoreszenzanalytischen Bestimmung des Gesamtglucosinolatgehaltes von Rapssamen. *Fett-Wissenschaft Technologie, 89*, 377–381.
Schnug, E., Murray, F., & Haneklaus, S. (1993). Preparation techniques of small sample sizes for sulphur and indirect total glucosinolate analysis in *Brassica* seeds by X-ray fluorescence spectroscopy. *Fat Science Technology, 95*, 334–336.
Sebanek, J., Kutacek, M., Luxova, M., Erdelska, O., Zladky, Z., & Prochazka, S. (2012). *Experimental morphogenesis and integration of plants.* Elsevier.
Stauss, R., Bleiholder, H., van der Boom, T., Buhr, L., Hack, H., Hess, M., … Weber, E. (1994). *Einheitliche Codierung der phänologischen Entwicklungsstadien mono- und dikotyler Pflanzen.* Geigy AG, Basel: Ciba.
Traka, M., & Mithen, R. (2009). Glucosinolates, isothiocyanates and human health. *Phytochemistry Reviews, 8*, 269–282.
Wagstaffe, J. P., Boenke, J., Schnug, E., & Lindsey, A. S. (1992). Certification of the sulphur content of three rapeseed reference materials. *Fresenius' Zeitschrift für Analytische Chemie, 344*, 1–7.
Wathelet, J.-P., Marlier, M., & Biston, R. (1987). *Preparation de materiaux pour une intercomparison preliminaire entre les methods de dosage des glucosinolates dans les colzas et evaluation des resultants.* Rapport préliminaire. Contrat 2806/1/5/307/87/7-BCR-B(10).
Wathelet, J. P., Marlier, M., & Severin, M. (1989). Préparation d'un colza á teneur certifiée en glucosinolates totaux (CRM 190). *Revue Francaise des CORPS GRAS, 36*, 309–312.
Zukalova, H., Vasak, J., Nerad, D., & Stranc, P. (2002). The role of glucosinolates of *Brassica* genus in the crop system. *Rostlinna Vyroba, 48*, 181–189.

SUBJECT INDEX

'*Note*: Page numbers followed by "f" indicate figures, "t" indicate tables.'

B

C

D

E

AUTHOR INDEX

'*Note*: Page numbers followed by "f" indicate figures and "t" indicate tables.'

A

B

C

D

E

G

H

I

J

K

L

M

N

O

P

Q

S

U

V

W

Kolibri 360 GmbH
Ferdinand-Jühlke-Straße 7
99095 Erfurt, Deutschland
produktsicherheit@kolibri360.de